Society and Genetic Information

Codes and Laws in the Genetic Era

CPS BOOKS

Andrea Krizsán (ed.) (2001)
Ethnic Monitoring and Data Protection. The European Context.
Published jointly with the
Human Rights Information and Documentation Center (INDOK)

Miklós Sükösd and Péter Bajomi-Lázár (eds.) (2003)
Reinventing Media. Media Policy Reform in East-Central Europe

Andrea Krizsán and Violetta Zentai (eds.) (2003)
Reshaping Globalization. Multilateral Dialogues and New Polocy Initiatives

forthcoming

Alina Mungiu-Pippidi and Ivan Krastev (eds.)
Nationalism After Communism. Lessons Learned

Daniel Smilov (ed.)
Party Funding, Campaign Finance and Corruption in Eastern Europe

Society and Genetic Information

Codes and Laws in the Genetic Era

Edited by

JUDIT SÁNDOR

CEU PRESS

CENTRAL EUROPEAN UNIVERSITY PRESS

BUDAPEST – NEW YORK

The publication of this volume was made possible by a grant from the Open Society Institute.

Copyeditor: *Arton Vidal*
Design and typesetting: *Judit Kovács/Createch*
Production: *Miklós Vörös*

Printed in Hungary by
Createch Ltd.

English edition published in 2003 by
Central European University Press
An imprint of the Central European University Share Company
Nádor utca 11, H–1051 Budapest, Hungary
Telephone: (+36-1) 327-3138 or 327-3000
Fax: (+36-1) 327-3183
E-mail: ceupress@ceu.hu
Website: www.ceupress.com

400 West 59th Street, New York NY 10019, USA
Telephone: (+1-212) 547-6932
Fax: (+1-212) 548-4607
E-mail: mgreenwald@sorosny.org

Library of Congress Cataloging-in-Publication Data

Sándor, Judit, 1962–
Society and genetic information: codes and laws in the genetic era
/ edited by Judit Sándor. p.416 ; 16.5 x 23.6 cm.
Includes bibliographical references and index.
ISBN 963 9241 75 X (pbk.) (CPS Books, ISSN 1587 6942)
1. Human genetics—Social aspects. 2. Human genetics—Moral
and ethical aspects. 3. Human genetics—Databases—Law and
legislation. 4. Human genetics—Databases—Access control.
5. Privacy, Right of.
[DNLM: 1. Databases, Genetic--ethics. 2. Databases, Genetic—legislation & jurisprudence.
3. Genetic Privacy—ethics. 4. Genetic
Privacy—legislation & jurisprudence. QZ 50 S2185s 2004]
I. Title. QH438.7.S26 2004 303.48—dc22
2003023799

Foreword

If it is true, as is often asserted, that society lags behind science, and law behind society, then the area of genetic research is likely to be a prime case. The Human Genome Project has opened up a range of issues from cloning to stem cell research, from genetic screening to sex selection, that pose important questions for structures, ethical reflection and legal order, while in the fast-moving field of genetic research, others are constantly being added. Some questions concern the research itself: should it progress without hindrance regardless of where it leads or the consequences of what follows from it? If limits and constraints are to be imposed, what form should they take and how could they be enforced?

Other questions concern the consequences for society of the sudden availability of genetic information that reaches far into a person's past and enables predictions about the future; information that can clearly be put to many social and economic uses, some to the good, others not. One does not have search far to find cases of abuse, when science is misused for dubious social ends. Since science has to be understood and interpreted, it is not surprising that it can be misunderstood and misapplied even with the best motives, while with less honorable motives the scope for misuse is truly frightening. Clearly there is a need for reflection on ethical issues which should form the basis for new legal principles and regulatory structures.

The most cursory consideration of these matters shows just how pressing they are; it also reveals their complexity. This book is an important and timely addition to a growing, but still inadequate, literature on the subject. The editor, Dr. Judit Sándor, in an exemplary introductory chapter, reviews the issues and provides a perceptive and up-to-date analysis of them. After referring to the unique problems "in handling, protecting, and using information" posed by genetic data, Dr. Sándor points tellingly to the lessons to be learnt from ill-fated attempts at eugenics and the dire social consequences that can result. She then sketches the background to genetic research and draws the important distinction between its use for medical purposes, on the one hand, and its wider, non-medical uses, on the other hand. While the former is not without problems, it is the latter that creates new and unique concerns.

The discussion then turns to ethical and legal issues, which include matters such as discrimination and privacy, disclosure and commercial exploitation. In each case, whether it is the implications of using genetic information in employment, insurance, or intellectual property, the issues are accurately and concisely portrayed, providing an excellent introduction for the reader with little prior knowledge, while stimulating deeper reflection in the more experienced.

There then follows a collection of essays, each of which examines in depth a more specific issue arising from genetic research and the uses to which it can be put. These range from an assessment of the impact of genetic research, and the Human Genome Project in particular, to the ethical and cultural consequences. The final set of essays, written by highly-regarded lawyers and scholars, addresses in depth some of the legal issues and dilemmas concerning genetic databanks and biobanks, and how they have been dealt with in a selection of countries. The most complex and important issues for legal provision and regulation are analyzed with insight and wisdom. The book as a whole is a fine example of how research from different disciplines can be brought together to inform and enlighten on a matter that must be regarded as one of immense significance for our times.

Denis J. Galligan
Chairman of the International Advisory Board of the
Center for Policy Studies, Central European University,
and Professor of Socio-Legal Studies and Director of the
Centre for Socio-Legal Studies, University of Oxford

Contents

About the Authors

Ragnar Aðalsteinsson

Supreme Court Lawyer, Chairman of the Board of the *Icelandic Human Rights Center*; formerly Chairman of the *Icelandic Bar Association*

Roberto Andorno

Visiting Professor of International Bioethics at the Department of Medical Ethics of the *University of Göttingen*, Germany; Member of the UNESCO International Bioethics Committee; and Consultant to the WHO Ethics Council

Alan Doyle

Scientific Program Manager, Biomedical Resources at the *Wellcome Trust*, London, United Kingdom

Peter Greenaway

Chief Scientific Officer and Assistant Director of Research and Development at the *Department of Health*, Government of the United Kingdom

Elmārs Grēns

Professor of Molecular Biology at the Faculty of Biology of the *University of Latvia*, Rīga, Latvia; Director of the International Center of Biomedicine and Biotechnology and Scientific Director of the *Biomedical Research and Study Center at the University of Latvia*; and Member of the Latvian Science Council, Rīga, Latvia

Krista Kruuv

Chairman of the Management Board, *Estonian Genome Project Foundation*, Tartu, Estonia

Sheila A. M. McLean

Professor and Chair of Law and Ethics in Medicine at the School of Law of *Glasgow University*, United Kingdom; Director of the Institute of Law and Ethics in Medicine at *Glasgow University*; Chair, *Scottish Criminal Cases Review Commission*; and Consultant to the World Health Organization and the Council of Europe

Ants Nõmper

Attorney, *Raidla and Partners Law Office*, Tallinn, Estonia and Lecturer at the *University of Tartu*, Estonia

Valdis Pīrāgs

Medical Director of the *Latvian Genome Project*, Associate Professor at the Faculty of Medicine, *University of Latvia*, and Head of the Clinic of Internal Medicine, *Paul Stradins University Hospital*, Rīga, Latvia

Aivita Putniņa

Associate Professor of Social Anthropology at the Department of Sociology, Faculty of Social Sciences, *University of Latvia*, Rīga, Latvia; Vice President of the *Latvian Association of Anthropologists*; and Scientific Director of the Oral History Project at the Institute of Philosophy and Sociology, *University of Latvia*, Rīga, Latvia

István Raskó

Director of the Institute of Genetics and Vice General Director of Teaching and Education at the *Biological Research Center* in Szeged, Hungary, and Visiting Professor at *University of Ulster*, Coleraine, United Kingdom

Frances Rawle

Research Program Manager of the UK Biobank Project, Strategy and Liaison Group, Medical Research Council, London, United Kingdom

Judit Sándor

Associate Professor at the Faculty of Political Science, Legal Studies and Gender Studies of the *Central European University*, Budapest; Member of the Hungarian Science and Research Ethics Council; Member of the High-Level Expert Group on Health of the European Commission; and Consultant to the UNESCO

Jürgen W. Simon

Professor of Economic Law, Private Environmental Law, and Biotechnological Law; Executive Chair of the Department of Law at the *University of Lüneburg*, Germany; and President of the Research Center on Biotechnology and Law at the *European Academy of Environment and Economy*, Lüneburg, Germany

Béla Somfai

Professor Emeritus of Christian Ethics at Regis College, *Toronto Graduate School of Theologogical Studies*, Toronto, Canada; Lecturer at the *Theological College of Szeged*, Hungary and at the Faculty of Medicine of the *University of Szeged*, Szeged, Hungary; Roman Catholic Priest and Member of the Society of Jesus

Imre Szebik

Research Associate at the Department of Bioethics of Institute of Behavioral Sciences, *Semmelweis University*, Budapest, Hungary; and Member of the Clinical Trials Research Group of the Biomedical Ethics Unit at *McGill University*, Montreal, Quebec, Canada

David M. R. Townend

Lecturer at the Department of Law and Sub-Dean of Postgraduate Studies at the *University of Sheffield*, Sheffield, United Kingdom

Pál Venetianer

Research Professor at the *Biological Research Center*, Hungarian Academy of Sciences, Szeged, Hungary; Vice Chairman of the Council of the European Molecular Biology Organization; and Chairman, National Committee of Gene Technology, Budapest, Hungary

Gregor Wolbring

Biochemist at the Faculty of Medicine, *University of Calgary*; Adjunct Assistant Professor of Bioethics at the Department of Community Rehabilitation and Disability Studies of the *University of Calgary*, Alberta, Canada; Executive Director of the *International Center for Bioethics, Culture, and Disability*; and Member of the Ethics Working Group of the Canadian Commission to UNESCO

List of Tables and Figures

List of Abbreviations

ACLU	American Civil Liberties Union (United States)
ACGT	Advisory Committee on Genetic Testing (United Kingdom)
ADA	Americans with Disabilities Act (United States)
ASHG	American Society of Human Genetics
BMA	British Medical Association
CDBI	Comité Directeur pour la Bioéthique (Council of Europe's Steering Committee on Bioethics)
CEDAW	Convention on Elimination of All Forms of Discrimination Against Women
CGS	Clinical Genetics Society (United Kingdom)
CIOMS	Council for International Organizations of Medical Sciences
CoE	Council of Europe
COM	European Commission Document
CRG	Council for Responsible Genetics
DNA	Deoxyribonucleic Acid
EC	European Community
ECHR	European Court of Human Rights
ECHR	European Convention on Human Rights
EEC	European Economic Community
EEOC	Equal Employment Opportunity Commission (United States)
EGE	European Group on Ethics in Science and New Technologies (European Commission)
EPO	European Patent Office
ESC	European Social Charter
ESHG	European Society of Human Genetics
ETS	European Treaty Series
EU	European Union
FDA	Food and Drug Administration (United States)
GAEIB	Group of Advisers on the Ethical Implications of Biotechnology (European Commission)
GAIC	Genetics and Insurance Committee (United Kingdom)
GIG	Genetic Interest Group (United Kingdom)
GTAC	Gene Therapy Advisory Committee (United Kingdom)

HFEA	Human Fertilisation and Embryology Authority (United Kingdom)
HGAC	Human Genetics Advisory Commission (United Kingdom)
HGC	Human Genetics Commission (United Kingdom)
HGP	Human Genome Project
HGSA	Human Genetics Society of Australasia
HIPAA	Health Insurance Portability and Accountability Act (United States)
HLA	Human Leukocyte Antigen
HUGO	The Human Genome Organization
IBC	International Bioethics Committee (UNESCO)
ICCPR	International Covenant on Civil and Political Rights
ICESCR	International Covenant on Economic Social and Cultural Rights
ICJ	International Court of Justice
IFHGS	International Federation of Human Genetics Societies
IGBC	Intergovernmental Bioethics Committee (UNESCO)
ILO	International Labour Organization (United Nations)
KGF	Keratinocyte Growth Factor
MPIF	Myeloid Protein Inhibitory Factor
MRC	Medical Research Council (Hungary)
MRC	Medical Research Council (United Kingdom)
NHS	National Health Service (United Kingdom)
NBIC	Nanotechnology, Biotechnology, Information Technology, and Cognitive Sciences
OECD	Organisation for Economic Cooperation and Development
PGD	Pre-Implantation Genetic Diagnosis
PKU	Phenylketonuria
PPPC	Public and Professional Policy Committee (at ESHG)
RELAGH	Red Latinoamericana de Genética Humana (Latin American Human Genetics Network)
SEC	European Commission Working Document
SNP	Single Nucleotide Polymorphism
TRIPs	Trade-Related Aspects of Intellectual Property Rights
UK	United Kingdom
UN	United Nations
UNEP	United Nations Environment Programme
UNESCO	United Nations Educational, Scientific, and Cultural Organization
US	United States
VEGF	Vascular Endothelial Growth Factor
WHO	World Health Organization (United Nations)
WTO	World Trade Organization

Preface

Advances in biomedical sciences often produce changes in our values because they affect our understanding of the human body, of health and disease, of personal identity, of heritage, or even of life itself. Human genetics is an excellent example of how such an advance creates a very lively ethical debate on many issues including reproduction, medical treatment, health enhancement, personal identity, and, more specifically, stem cell research, cloning, genetic testing and screening, sex selection—only to mention some of them.

Human genetics is at a research stage. It is very common in medicine that quite a long time elapses between the emergence of scientific knowledge, the formulation of diagnosis, and the development of treatment. There is no doubt that today genetics is the cornerstone for advances in the life sciences; but the benefits of human genetic research are not yet clearly perceived. As the gap between the publication of research results and the development of therapeutic solutions widens, the fears and expectations of society towards human genetics increases. Uncertainty fuels ethical debate and genomics is not an exception.

A good example of this situation is the Sixth Framework Programme of Research of the European Union. Life sciences research projects in this program usually include a component of genomics. In the evaluation process of this research, ethical aspects are often heavily debated. A very lively discussion is taking place now on the conditions to be fulfilled in order to fund research on human embryonic stem cells. Reproductive cloning, germ-line research, and the creation of embryos for research, for example, are excluded from funding under this program.

Ethical issues become even more conspicuous when genetics do not have a health enhancement goal. This is the case when genetic information is used in the workplace or for insurance purposes. These two areas share the possibility of discrimination against the sicker and thus weaker. It is difficult to legislate in this uncertain environment, especially when contradictions exist between the various ethical principles prevailing in the Member States of the European Union. The power to decide on ethics issues lies with the individual Member States, but it is complicated to operate, especially in research, without a certain degree of harmonization.

The medicine of the future is based on human genetics. This field of scientific research, however, raises a number of ethical and legal issues that are at the center of many debates on bioethics today. The translation of these ethical debates into law is even more problematic. The only rational way to deal with human genetics is through a multidisciplinary approach which includes different value systems. We have seen in the past that many of the problems stemming from scientific research exist because people did not understand the scientific issues at stake. Scientists, ethicists, philosophers, and lawyers need to engage in discussions with each other in order to explore the possible solutions. This book is an excellent example of this endeavor and paves the way for a dialogue not just among the experts but also between them and the wider society.

Dr. Octavi Quintana-Trias
Director of Directorate F – Health
Directorate-General for Research
European Commission of the European Union

Introduction

Genetic Information: Science, Society, and Legal Norms

Judit Sándor

1. Introduction: The World of Genetic Information

The results and interpretation of the Human Genome Project, the various forms of cloning, stem cell research, genetic screening, pre-implantational tests, sex selection, and other advances in medical genetics, all challenge the existing social and cultural meanings of personal identity, reproduction, health, therapy, and family bonds. So far, the treatment of these fundamental issues has been separated into two different arenas of discussion. While scientific discourses focus on the novelty of genetic information and the medical-technological aspects of the new genetic era, other experts in the fields of the humanities and social sciences, such as philosophers, lawyers, and anthropologists, express their concerns about the social and cultural impacts of this epoch.

This book endeavors to address the issues of genetic information from a multidisciplinary perspective. With prominent biologists, medical doctors, lawyers, anthropologists, philosophers, sociologists, and theologians from different European countries and Canada among the contributors, a unique opportunity opens up for exchanging views on the complex biological and social impacts of the future proliferation of genetic information. The authors of this book explore the various uses and applications of genetic information in the biomedical sciences, in insurance, and in family law.

The beginning of the twenty-first century will be marked in human history for collecting, classifying, and interpreting genetic data, including the human genome. Most of these tasks are still ahead of us, but it is not too early to think about the ways we accumulate, store, and handle genetic data. A number of factors warrant the consideration of special protection. One is the fact that some genetic information relates to the future, thus the individuals concerned might be in a less favorable situation later than they are now. The law has arrived at a number of solutions for eliminating past and current injuries of discrimination, but the legal techniques of sanctioning prejudice against disadvantages arising *in the future* have not been developed yet.

Genetic data is precious at the individual level because, in addition to providing an exact diagnosis of an existing illness, it may also give us a glimpse of the future by showing our predisposition to certain other conditions. Although the real capabilities of science currently are far behind what expectations might have led us to believe, it is still possible that the genetic background of several diseases, but also of variations and differences among human beings and particularly between genders, will be illuminated. However, it is important to stress that we do not know much about the relationship between a person's genome and the *expression* of that person's genetic make-up. Most of the conditions that are labeled as genetic involve some environmental and even cultural factors.

Genetic data poses unique problems in handling, protecting, and using information.[1] Even the term 'genetic data' in itself is difficult to define. Certain family medical records can also be qualified as such, but so can predispositions to diseases or monogenic disorders (that is, disorders caused by the lack of, or error in, a single gene). The lack of precision in defining genetic data also makes it difficult to establish clear rules for handling of these data. Genetic data is also unique in that while other medical data is related only to one person's health status, the knowledge of this type of data potentially involves more people. David Heyd[2] pointed out in his classic study that unlike a traditional medical test, genetic testing and screening may affect other persons besides the tested patients. Learning of an inherited disease has a serious effect on the lives and decisions of also those family members who otherwise may not have wanted to submit themselves to testing. Once someone decides to inquire about his or her genetic make-up, it will ultimately influence other family members in their life style, reproductive and family planning, or health insurance status and decisions.

Health status based on genetic characteristics has already become an element in the public perception of health. In order to forge a better chance to have a healthy child, pre-implantation diagnosis is often suggested by doctors. Genetic diseases have special characteristics. They raise ethical considerations different from those raised by non-genetic diseases. First of all, these diseases are regarded as irreversible and more or less objective conditions. The threat of developing symptoms affects the individual's life, especially in case of late-onset diseases. Genetic anomalies may be inherited by the offspring, and consequently may effect a number of decisions concerning reproduction and health care.

1. William F. Mulholland and Amy S. Jaeger (1999) "Genetic Privacy and Discrimination: A Survey of State Legislation," *Jurimetrics Journal*, 39(3): 317–326.

2. David Heyd (1992) *Genethics: Moral Issues in the Creation of People* (Berkeley: University of California Press).

1.1 Lessons of Eugenism

Since the commencement of the Human Genome Project, the never-before-seen novelty of the Genetic Era has been emphasized by scientists and politicians, in public debates and in the media. However, to determine what is really new is always a confusing question. One may refer to different episodes of eugenism[3], which since the times of antiquity have provided numerous examples for the use of genetic enhancement, even if under the current state of science we would not necessarily call these practices genetic-based enhancements.

These historical episodes demonstrate that inheritable characteristics, even in the total absence of scientific evidence, often served as a basis for various forms of discrimination,[4] as well as for eugenic sterilization, selective killing, and other injustices that later were regarded as crimes. In 1927 Justice Holmes delivered the infamous verdict that "it is better for all the world, if instead of waiting to execute degenerate offspring for crime, or let them starve for their imbecility, society can prevent those who are manifestly unfit from continuing their kind. The principle that sustains compulsory vaccination is broad enough to cover cutting the fallopian tubes. Three generations of imbeciles are enough."[5] No wonder that fears of eugenism still play an important role in the public understanding of genetics.

Among the natural sciences, the history of biomedical research is especially burdened by examples of risky and hazardous interventions. It is equally important to note, however, that the different eugenic episodes in biomedical thought and practice were often supported by sympathetic branches within the social sciences, as both were influenced by the same general discourse. The mentally ill were sterilized; newborn babies with disabilities simply did not receive medical treatment. Still, despite the regrets that followed the tragedies, despite the acknowledgement of misconceptions and mistakes, nothing could hinder the continuation of scientific discoveries and nothing could tame human curiosity.

3. Much before Nazism, a different form of eugenism appeared frequently in the social sciences. The Frenchman Arthur de Gobineau published his work on the inequality of races in 1853. He considered the relevance of blood as of utmost importance not only on a family level but also on the level of the entire nation. He thought that *métissage* or *mélange des sangs* could lead to degeneration of the whole nation. In 1902 the Englishman Archibald Garrod pointed out that certain diseases appear soon after birth and therefore can be regarded as inherited. For more details, see Catherine Bachelard-Jobard (2001) *L'eugénisme, la science et le droit* (Paris: Presses Universitaires de France).

4. During its early history, United States immigration policy excluded certain populations from entering the country, for instance the Chinese between 1882 and 1902. Then in 1924 the American Immigration Act was clearly based on the principle of preserving racial purity.

5. *Buck v. Bell* (1927) 274 U.S. 200.

If one examines the chronicle of these human tragedies it can be observed that a specific form of reductionism, namely 'biologism' played a crucial role in the process. Biologism, or the naturalization of social and cultural differences, became a central part of social ideologies as well, further reinforcing the taken-for-granted nature of these scientific practices. Of course, by looking at the roots of this research one can easily see that *unethical* scientific research was also *bad* scientific research. The other conclusion, which can be illustrated by numerous episodes of the eugenic movement, is that bad science was reinforced by bad social science and consequently legitimized legal norms that were later regarded as unconstitutional. The abuses that were investigated by lawyers in the Nuremberg trials later were interpreted by scientists as examples of bad science and were considered as severe violations of human rights as well.

Even the more recent history of medical research reveals a disturbing pattern of discrimination against racial and ethnic minorities, and against women. And this indicates that the frontiers between science and law are not self-evident. The formulation of research and control groups are regarded as part of the scientific method, although these categories may be culturally, morally, or legally problematic. This is why it is crucial that there be communication between various disciplines on theoretical and methodological issues. In one infamous example, a research project was both discriminatory and scientifically unsound. In the *Tuskegee Study*, which ran from 1950s until the early 1970s, researchers studied the effects of untreated syphilis in a group of African-American men. The researchers never disclosed to the research subjects that they continued to suffer from a treatable but serious illness. The researchers were able to study the long-term effects of infection with syphilis by withholding treatment from the research subjects.[6]

Examples can also be found of biological prejudices when otherwise relevant scientific information leads to severe mistakes in social policy simply because of adopting a scientific paradigm in a broader area without testing the verity of the extension. In 1965 researchers found a high prevalence of the *karyotype XYY* among prison inmates in Britain. Many people, scientists and laymen alike, quickly took this finding to justify the link between an extra Y chromosome and a tendency to hyper-aggressivity and violence. This assumption was rejected by genetic researchers, but still resulted in an enduring popular belief. Some other authors believe that the increased risk of arrest or conviction may stem from increased likelihood of getting caught.[7] This story

6. David J. Rothman (1982) "Research Ethics and Social Deprivation: Were Tuskegee and Willowbrook 'Studies in Nature'?" *Hastings Center Report*, 12(2): 5–7.

7. David Wasserman and Robert Wachbroit (2001) "Introduction: Methods, Meanings, and Morals," in David Wasserman and Robert Wachbroit (eds.) *Genetics and Criminal Behavior* (Cambridge: Cambridge University Press), p.9.

is an eloquent example of the interrelation of bad science and bad social (criminal) policy.

If one looks at the cases of scandals and abuses in science, one can notice that not only ethical problems arose but also scientific usefulness was questioned. Consequently, if basic ethical values are compromised we can not talk about good science. In other words, science is and must be a multidisciplinary enterprise today.

Multidisciplinarity also involves a stronger emphasis on ethical and legal norms. After the Second World War, as a result of the Nuremberg trials, German doctors and scientists were held liable for their abuse of human subjects in biomedical experiments. The Nuremberg Code declared the necessity of *voluntary consent* and disclosure of the general nature of the experiment. The principles of conducting research on human subjects were further developed in the Helsinki Declaration in 1964.

The significance of bioethics, as well as the combination of the freedom of research principle and the necessary legal restrictions, are best illustrated today by the Council of Europe's Convention on Human Rights and Biomedicine.[8] By adopting this Convention in 1997, the Council of Europe established the most important ethical norms for biomedicine. The Convention has been supplemented by a number of additional protocols since then, dealing with issues of cloning and transplantation. The Charter of Fundamental Rights can also be regarded as a safeguard in the field of life sciences and biotechnology. The European Union and the various European states have, over the course of the past decade, adopted directives and passed laws that prove an enduring commitment to protect fundamental values in the field of biomedicine and biomedical research.[9]

1.2 The Advent of the Genomic Era

When on the twenty-sixth of June in 2000, the leaders of the public Human Genome Project and the private company Celera announced completion of the 'working draft' of the human genome sequence, the dominant attitude was of scientific pride and enthusiasm. The results of this historically relevant project and the analysis of the human genome were published by the public project on February 15, 2001 in the magazine *Nature* and by Celera Genomics on February 16, 2001 in the magazine *Science*.

8. Convention for the Protection of Human Rights and Dignity of the Human Being with regard to the Application of Biology and Medicine: Convention on Human Rights and Biomedicine (ETS No. 164). Signed in Oviedo, April 4, 1997. See the Appendix of this volume, and for the official text: http: //conventions.coe.int/treaty/en/treaties/html/164.htm.

9. For more details, see the Appendix.

Since the beginnings of the genome projects, the key application of human genome research has been to find *disease genes*. The genome sequence has also helped to reveal the mechanism that may lead to some chromosomal deletion syndromes. Consequently more accurate and timely diagnoses can be made and hopefully better therapies can be developed. The other important domain is the *pharmaceutical use of genetics*: based on our newly acquired genetic knowledge, the pharmaceutical industry is now heavily focused on developing a limited set of new drugs. The knowledge of the complete set of human genes and proteins will greatly expand the search for suitable new drug targets.

The *medical use* of the genetic information, inasmuch as it only serves healing, is rather easily morally justified. Medical intervention aimed at curing ill children or cancer patients (for example), even if this demands gene therapy, is easier for the public to accept than scientific research that does not offer the immediate hope for medical application, or the use of genetic information in other spheres of life. The public is much more wary of the use of genetic information for *non-medical, non-research* purposes, and the law is also likely to set much stricter standards for the use of genetic knowledge here. It is precisely because of these unique concerns that this volume will focus primarily on the use of genetic information *outside* healthcare, and on unusual applications beyond traditional medical treatment.[10]

We should note here that one of the major dilemmas in contemporary bioethics is whether one should distinguish between therapy and enhancement. Francis Fukuyama, who is a member of President Bush's Council on Bioethics, tries to defend the separation between the two in his book titled *Our Posthuman Future.*[11] Drawing moral distinctions between treating patients and enhancing their characteristics is not only an ethical and philosophical issue, it is also preeminently relevant for legal policy makers. This is why we turn to a discussion of the ethical and legal questions involved in genetic research and in the various uses of genetic information.

2. Ethical and Legal Questions

In the legal regulation of scientific activities and developments, law usually reacts to the actual social potential of science. However, in addition, it also takes an overview of the basic ethical norms affecting future risks, basic rights and social values,

10. Jean-René Binet (2002) *Droit et progrés scientifique. Science du droit, valeurs et biomédecine* (Paris: Presses Universitaires de France).

11. Francis Fukuyama (2002) *Our Posthuman Future: Consequences of the Biotechnology Revolution* (New York: Farrar, Straus and Giroux).

health policy, and scientific research, and then attempts to develop legal norms based upon these. But legal changes have been much more limited in scope than is required by the pressure for innovation in science. If a new technical or scientific development arises, the reaction of legal thought is likely to tend towards an incorporation into the already existing legal principles and analysis, rather than to the development of new legal institutions. Reluctance of legal innovation is related to the need for coherence within the legal system and that is often one of the major guiding forces of new legal policies. Furthermore, internal coherence of the legal system is one of the most important guarantees for the principle of justice.

Moving to the field of genetic information, the development of new, genetics-based medical treatments, the monitoring of genetic data, genetic testing and screening, the establishment of genetic databanks, all have impact on fundamental legal questions, such as the equal treatment of individuals, privacy, and access to health care.[12] The new genetic knowledge we face now affects essential human conditions: our family relations, our reproductive decisions, but also more specific domains, such as insurance, labor relations, and intellectual property rights.

As more and more genetic information is produced and its uses become increasingly widespread, complex legal questions of data protection emerge. This is partly because the human DNA has a symbolic significance. Furthermore, as already mentioned above, the term 'genetic data' may not be so easily defined.[13] Genetic data may be obtained from DNA and chromosome analyses and certain clinical tests, as well as from traditional sources of genetic information, such as family history.[14] One of the most problematic features of genetic characteristics is that they may be inherited by the offspring, and it may affect the individual's health status and his or her spouse's chances of having a healthy child. In terms of causality, one may make a distinction between the type of genetic information that is produced by genetic tests for monogenic recessive or dominant conditions where degrees of risk are high, and the

12. Bertrand Mathieu (2000) *Génome humain et droits fondamentaux* (Aix-en-Provence: Presses Universitaires d'Aix-Marseille).

13. The British Human Genetics Commission defines the notion of "personal genetic information" as "information about the genetic make-up" of an identifiable person, whether derived directly from DNA (or other biochemical) testing methods or indirectly from any other source. See Human Genetics Commission (2002) *Inside Information: Balancing Interests in the Use of Personal Genetic Data* (London: HGC), p.25. And the website http://www.hgc.gov.uk/insideinformation/iichapter1.pdf.

14. *Genotype information* that involves the analysis of the DNA provides very detailed information about the individual and about inherited characteristics. The *phenotype information* does not require the direct examination of the DNA, sometimes it is observable through physical examination (e.g. eye color). The third type of genetic information is the *family medical history*, which can show certain patterns of inheritance. See Human Genetics Commission, *op. cit.*, p.26.

type that can be derived from testing as susceptibility data. In this latter case, genetic information is not only an indicator of one individual's current state of health; it is also an indicator of that individual's likely future health.[15]

Specific legal definitions and interpretations of genetic data have already appeared in numerous international legal documents.[16] It is acknowledged in them that genetic information may be gained secretly (without the acknowledgment of the individual) and from a very small amount of material. Due to the stability of the DNA it can be recovered from stored material or even from archaeological deposits. Genetic information reveals not only the actual health status of the individual, but it also has a predictive value. Genetic information may tell a lot about the biological origin of the individual, including paternity information.[17]

The particular legal treatment of genetic data goes beyond the conceptual problems of genetic information. Legal philosophy and policy-making now focuses primarily on the various *uses* of genetic data: in making a diagnosis, in the health care of family members, in biomedical and other research, in setting up biobanks, in public health care, in forensic applications, in insurance and employment.

Before analyzing the handling of genetic information in specific biomedical and non-medical domains, two legal principles need to be analyzed: the principles of *privacy* and of *non-discrimination*.

3. The Principles of Privacy and Non-Discrimination

3.1 Privacy and Genetics

Mark Rothstein applies the metaphor of the 'secret' in explaining the relevance of the genetic revolution after the completion of the Human Genome Project. "Before the last half of the twentieth century, human genetic secrets belonged exclusively to

15. For medical conditions controlled by a single gene (such as Huntington's disease), the indication can be precise. For medical conditions that involve a more complex combination of genes or environmental factors (such as heart disease), the indication might involve only an increased probability that a future medical problem may arise.

16. UNESCO Universal Declaration on the Human Genome and Human Rights (1997); Council of Europe, Recommendation No. R (92) 3 on Genetic testing and Screening for Health Care Purposes (February 1992); WHO, Proposed International Guidelines on Ethical Issues in Medical Genetics and Genetic Services (1997).

17. More about the relevance of genetics in a paternity case can be seen in the *Mikulić v. Croatia* case (ECHR Case No. 53176/99, Judgment No. 074a(2002), dated February 7, 2002). See the press release at http://press.coe.int/cp/2002/074a(2002).htm or the full judgment at http://hudoc.echr.coe.int/Hudoc2doc2/HEJUD/200210/mikulic - 53176jv.chb1 07022002e.doc.

nature, they were locked securely within the genes . . . Today the concern has shifted to a different type of genetic secrets. The secrets no longer belong solely to nature. Some belong to humans as well."[18]

Right to privacy usually refers to the control over someone's *own* personal information. Genetics has opened a vast field of research on individual genetic characteristics, which has expanded access to personal genetic information, destabilizing our traditional notions of privacy. Genetic screening and testing,[19] in particular, undermine the individual-based concept of privacy in at least two respects. First, genetic screening of one individual may affect not only close family members, but also relatives who do not even have regular contact, who may have different religions and different levels of education, and might live in distant places, even in different societies. Second, genetic testing and screening does not really constitute a right to decide. When the results of testing or screening are disclosed to adults, the treatment of genetic disorders may often be impossible or the solution offered in those cases are not regarded as therapy at all. In case of prenatal testing, the only 'treatment' available is a genetically induced abortion. In case of pre-marriage testing, the 'treatment' is not to get married or not to have children if, for instance, both parents turn out to be carriers of a disease gene. It follows that an individual's control over his or her personal information should be guaranteed during the entire course of her medical consultation and examination. This control includes the right not to know and the right to avoid those tests where no therapeutic advantage can be offered.

Genetic testing may often have life-long relevance as genetic data can effect a person's lifestyle, future plans, partner choice, plans for children, career choice, and even educational ambitions. The question from a legal standpoint is thus how genetic data should be protected and used.[20]

18. Mark A. Rothstein (1997) "Genetic Secrets: A Policy Framework," in Mark A. Rothstein (ed.) *Genetic Secrets: Protecting Privacy and Confidentiality in the Genetic Era* (New Haven and London: Yale University Press), pp.451–495.

19. The American Task Force on Genetic Testing, a joint working group of the Department of Energy and the National Institutes of Health, has offered a working definition of a genetic test. The definition restricts genetic testing to processes which are carried out for the direct analysis of human DNA and other compounds such as RNA, chromosomes, proteins and certain metabolites, with a view to achieving a number of clearly identified end points; namely, the prediction of inherited disease, the detection of carrier status or the diagnosis of actual inherited disease. This can encompass not only the testing of individuals but also the screening of at-risk populations, and will include prenatal and antenatal screening and the testing of families with recognized histories of genetic disease. For more, see the website http: //www.hopkinsmedicine.org/tfgtelsi/fedregister/index.html.

20. Wolfram Henn (1999) "Genetic Screening with the DNA Chip: A New Pandora's Box?" *Journal of Medical Ethics*, 25(2): 200–203.

I believe that a unique legal status must be afforded to genetic data[21], for it is *not only health data*, but also can be used to *identify the individual*. This complex characteristic of genetic data requires a specifically designed confidentiality regulation. Moreover, potentially unlawful access should be eliminated and the individuals' own access rights should be reaffirmed. The health care institution or the research staff can easily process genetic data. Therefore the individual should have a right to know that a test is going to be performed on him or her, but nevertheless the affected individual should be allowed to refuse to know the test results. In other words, the individual should have 'a right not to know'.

Another significant difference between genetic data and traditional medical data is that besides providing insight into the health of the individual examined, genetic data may also indicate the medical conditions of the family members of that individual, among them even unborn babies. Thus we could say that certain affected individuals might not even know of the existence of health data related to them. The genetic tests for diseases like Huntington chorea or cystic fibrosis involve a mutation in a single gene. It means that if someone inherits the mutation, he or she will certainly get the disease. Other available tests may reveal diseases, such as Alzheimer's disease, that are caused by mutations in more than one gene. These pose different problems since the test result may reveal only that the individual is susceptible to the disease, but it is not certain that he or she will get it.

There is a general belief that we consider all medical information to be private, and genetic information is not an exception in this respect. This notion starts to be problematic if we regard the fact the genetic information is obtained not necessarily through an individual DNA testing but often together with the analysis of the family medical history. However, closer observation of public attitudes towards genetic information may challenge this general view. While most people would insist on the confidentiality of traditional medical information, especially the information about some physical or mental disability, it may not be the case with genetic information.

Although there is as yet no internationally accepted comprehensive legal solution to the unique data protection problems raised by genetic information handling, the beginnings of one can be recognized. The suggestion made by a Canadian commis-

21. Specific status was guaranteed to genetic data in the International Declaration on Human Genetic Data, adopted by the UNESCO in October 2003. (For the official text of this document, see the Appendix.) For different analyses of the special relationship between right to privacy and the protection of genetic data, see Marie-Isabelle Malauzat (2000) *Le droit face aux pouvoirs de données génétiques* (Aix-en-Provence: Presses Universitaires d'Aix-Marseille); and José Antonio Seoane Rodríguez (2002) "De la intimdad genética al derecho a la protección de datos genéticos. La protección iusfundamental de los datos genéticos en el derecho espanol (A propósito de las SSTC 290/2000, de 30 de noviembre)," *Revista de derecho y genoma Humano* (Bilbao), 16: 71–105.

sioner concerned with the right to privacy in 1995 was particularly significant. He offered that the goal of genetic data collection should be communicated to the person affected even if the Privacy Act does not so prescribe.[22]

The Human Genetics Commission (HGC), the UK's genetics watchdog, has published a report titled *Inside Information: Balancing Interests in the Use of Personal Genetic Data*.[23] It contains proposals as to in what circumstances a person's personal genetic information should be open to use and by whom, and in what circumstances this information should be protected from outside use. The HGC report recommends that obtaining, testing or storing someone's DNA without their knowledge or consent should become a crime, except as allowed by the law for forensic purposes. It also says that employers should not insist that employees take genetic tests as a condition of employment or promotion; insurers should not use the results of genetic tests for their own benefit; and that no one should be unfairly discriminated against on the basis of his or her genetic characteristics. According to the HGC, medical research should be allowed to use genetic information, while ensuring that those individuals are protected whose genetic information is stored in medical databases, such as in the UK Biobank. Databases established for medical or research purposes should not be used for any other purpose, including access by the police. All such databases should be overseen by an independent body, such as an ethics committee.

Genetic information and privacy pose so many new legal dilemmas that *genetic privacy* has become a *terminus technicus* to refer to the specificity of legal problems related to the question of how to maintain control over the genetic information of the individual.[24]

3.2 The Prohibition of Genetic Discrimination

The ethicist Thomas Murray described human genetics as a "science of inequality: a study of human particularity and difference."[25] Indeed, the danger of genetic discrimination exists in several spheres of life, but the recognition of such discrimination is not as easy as it is, for example, in the case of racial discrimination. The same person who, because of a definite personal attribute, is a member of a genetic minority,

22. See The Privacy Commissioner of Canada (1995) *Genetic Testing and Privacy* (Ottawa: Ministry of Supply and Services), at the website http://www.privcom.gc.ca/information/02_05_11_e.pdf.
23. See footnote 14, and the report at http://www.hgc.gov.uk/insideinformation/index.htm.
24. One of the most comprehensive treatments of this complex field has been published by Graeme Laurie. See Graeme T. Laurie (2002) *Genetic Privacy: A Challenge to Medico-Legal Norms* (Cambridge: Cambridge University Press).
25. Thomas Murray (1992) "Genetics and the Moral Mission of Health Insurance," *Hastings Center Report*, 22(6): 12–17.

might from another point of view belong to the genetic majority. At the same time, belonging to a minority might give rise to positive social discrimination, as in the case of a rare but beneficial genetic attribute.

Throughout history people have been segregated on basis of race, sex, religion, and ethnicity. With the recent emergence of genetic screening and genetic engineering, a new form of potential discrimination has appeared: discrimination based on genotype. This type of discrimination may occur in health care services, and in the fields of life insurance policy, labor law, reproductive rights, family law, and so on. Genetic discrimination is especially problematic, as not only diseases but also certain human characteristics and abilities have a genetic component. The danger with this development is that genetic discrimination potentially involves overall discrimination.

Discrimination on the basis of genetic attributes, however, differs from previously known types of discrimination in that although these attributes are insurmountable, they are for the most part *invisible*. If these conditions of an individual become public, uncertain and unpredictable prejudice may arise against the affected individual, and this unpredictability allows for abuses. Another possible effect of the abundance of genetic knowledge can be that if some genetic traits are significantly more common in a certain ethnic group, then the exclusion of this trait from an insurance contract, for instance, may result in a new form of indirect, ethnic discrimination.[26]

Moreover, difficulties might emerge concerning the *time dimension* of genetic information: while in comparison with a traditional disease, which may prevent someone from fulfilling a job in the present, a genetic predictive test may reveal a probability to develop the disease in the future, and thus a looming loss of capacity to work. As genetic discrimination may differentiate between people based on predictions of a future handicap, it can be regarded as a *new form* of discrimination since the classical grounds for discrimination refers to present or past disadvantages.

Genetic data, as a form of prognostic information, has a *financial value*. Although commercialization does not necessarily lead to discrimination, the processes by which economic interests may generate discrimination should be carefully scrutinized. Thus, we need to consider which diseases are regarded as priorities in medical, pharmaceutical research, who should be offered, and what sort of genetic testing.

There is another form of discrimination that may occur in the use of genetic data. Most people know what to make of the results of, say, a blood sugar test. However, data coming from a genetic test is much more complex; its reading and interpreta-

26. In 2000 the European Parliament accepted a resolution to propose a directive that would include the prohibition on the use of personal medical data such as genetic data to enable insurers to act in a discriminatory way.

tion requires special expertise. Therefore it can be foreseen that there will be groups of individuals who are less likely to be able to get access to genetic counseling and even if they have access they may lack the necessary education to understand the information. As a consequence, they will suffer an additional form of discrimination that is the *discrimination in the accessibility* of sensitive and complex medical data.

The European Union was surprisingly quick to include genetic discrimination among the traditional forms of discrimination prohibited under the Charter of Fundamental Rights.[27] However, if we examine the notion and the existing jurisprudence of discrimination we may not find the concept of genetic-based discrimination so evident. Perhaps the only stable element of this form of discrimination that is supported by consensus is that racial, ethnic, and national minorities should not suffer further discrimination by genetic tests.

Fear of discrimination, of course, is not a sufficient reason to reject the scientific knowledge gained from biomedical and genetic research, such as the Human Genome Project. What is at stake is, rather, how one can avoid the discriminatory effects of the project and how our genetic interventionism can be tamed and used for good purposes.

4. Using Genetic Data for Research Purposes

While general data protection norms specify the *medical use* of health care data, *research purposes* are regarded as distinct from the preventive and therapeutic uses. In Article 2 of the Convention on Human Rights and Biomedicine,[28] it is stated that "The interests and welfare of the human being shall prevail over the sole interest of society or science." It also follows from the European Union data protection directive[29] that data collected and processed for preventive treatment, diagnosis, or specific research purposes may only be used in the manner that is specified.

27. Article 21 on Non-Discrimination in the Charter of Fundamental Rights of the European Union states that "Any discrimination based on any ground such as sex, race, colour, ethnic or social origin, genetic features, language, religion or belief, political or any other opinion, membership of a national minority, property, birth, disability, age or sexual orientation shall be prohibited." For the official text, see *Official Journal of the European Communities, C 364, 18.12.2000*, p.9, at http://www.europarl.eu.int/charter/pdf/text_en.pdf.

28. See footnote 9.

29. Directive 95/46/EC of the European Parliament and of the Council of 24 October 1995 on the Protection of Individuals with Regard to the Processing of Personal Data and on the Free Movement of Such Data. For the official text, see *Official Journal of the European Communities, L 281, 23.11.1995*, pp.31–50, at http://europa.eu.int/comm/internal_market/privacy/docs/95-46-ce/dir1995-46_part1_en.pdf.

When personal data has to be used it is not satisfactory to claim that the data will be used 'for research purposes'. The aims of the research should also be explained. Another problem may arise when further or different uses of already archived samples and tissues are demanded. Obtaining valid and informed consent is often problematic in this case. Sometimes the mere disclosure of the names of the sources might be a violation of their privacy rights.

In clinical practice it is often difficult to separate the therapeutic and research use of tissues and the related health data information. Therefore, in some countries, there is an explicit prohibition on using data that had been collected for the purpose of diagnosis and therapy for research purposes. Concerns for data protection and the separation of various uses of sensitive personal information are important legal concerns, and not only in Europe. For instance, according to Article 22 of Quebec's Civil Code,[30] no part of the human body (including, organs, tissue or substances) taken from a person during the course of therapy can be used for research if the concerned individual did not consent to it.

National bioethics commissions in many countries have tried to grapple with the dilemmas of how to obtain valid consent for multiple research projects and what are the acceptable guarantees for data protection in the field of research. In 2000, the Bioethics Committee of the Japanese Council for Science and Technology adopted the "Fundamental Principles of Research on the Human Genome." In the document, Principle 12 deals with the "leakage of personal information."[31] The United Kingdom Human Genetics Commission proposed in its above quoted report that a *general consent* is acceptable if the genetic data used in research has already been anonymized.[32] The Estonian Genome Project introduced the concept of *open consent*. Open consent refers to the fact that under the Estonian Human Genes Research Act individuals consent to become general gene donors. As many of the specific research applications are not yet known and cannot be specified in the consent form, therefore, it covers genetic research in a broad sense. In addition to this, in the Estonian model, gene donors cannot provide partial or conditional consent.

30. Adopted in December 1991.

31. "Research institutions and researchers should take necessary measures in order to prevent the leakage of personal information. In the event of the leakage of personal information, firm disciplinary measures, including demotion, should be taken against the person(s) who leaked the information, the researchers undertaking the said research, the custodians of the personal information, the director of the research institution and other personnel associated with the leaked information. Any individual whose personal information is leaked or who sustained damage from the said leakage is legally entitled to receive compensation or indemnity." (Official translation from Japanese.)

32. Human Genetics Commission, *op. cit.*, p.95.

5. From Assisted Reproduction to Cloning

Perhaps there is no other field in which the development of the interaction between science, ethics, and law has been as socially significant and politically charged as in the case of reproduction. Assisted procreation provided a unique opportunity for biomedical research, but at the same time created new ethical dilemmas concerning such issues as the use of pre-implantation genetic tests. Starting with the technical possibility of medically assisted reproduction, we have, practically without noticing it, achieved a demand for sperm and egg donors. Then through the 'micro manipulation' of sperm (intracytoplasmic sperm injection, ICSI) we have come to the first artificial reproduction. In the end this process has brought us to the reproductive technique now considered scandalous, in which just one single reproductive cell, an egg, plays a role. Observing this development, the question arises: if the new techniques that separated procreation from sexuality and have been so easy for the public to accept, then why is cloning so different?

It is notable how easily public opinion sailed through all these enormous changes, and without particular hesitation accepted the fact that now not every child is born through natural reproduction, and that pregnancy is not always induced in 'a natural way'. Most regulations on reproduction in Europe have effected a belief that the procedures are a sort of *quasi* medical treatment to assist infertile partners who could not have children naturally, and for whom infertility treatment has proven to be ineffective.[33] The role of assisted procreation and embryo research in the genetic era was also recognized by Jürgen Habermas in his recent book titled *The Future of Human Nature*[34]. He envisages the development of liberal eugenics as a result of autonomous choices made by individuals for selection rather than just a request for therapy in the field of biomedicine. This trespassing of the frontiers between therapy and selection[35] will reinforce the cultivation of the increasingly medicalized choices that will create a new and extensive domain of biomedical interventions. As a result of these choices, liberal eugenics will affect not only the individual who makes the choice but also interpersonal relationships.[36] "The irreversible choice a person makes for the desired makeup of the genome of another person initiates a type of relationship between

33. Philosophical issues related to infertility and rights are elaborated in a recently published book: Mary Warnock (2002) *Making Babies: Is There a Right to Have Children?* (Oxford: Oxford University Press).
34. Jürgen Habermas (2003) *The Future of Human Nature* (Cambridge: Polity Press). For the original German version, see Jürgen Habermas (2001) *Die Zukunft der menschlichen Natur. Auf dem Weg zu einer liberalen Eugenik?* (Frankfurt am Main: Suhrkamp).
35. For instance sex selection of the future child and pre-implantational selection of certain characteristics.
36. Habermas, *op. cit.*, p.63.

these two which jeopardizes a precondition for the moral self-understanding of autonomous actors."[37] The procreator becomes a creator.

'Creation' can be exercised now in numerous forms in the *in vitro* clinics. One of the most recent and ethically contested technique was the *in vitro* 'treatment' of using two eggs from two different 'mothers'. In March 2002, the monthly journal *Human Reproduction* announced a report on a research project,[38] which made it clear that some reproductive medical scientists do apply gene therapy, even though it had been strongly condemned on ethical grounds. The researchers stated that genetic modification had been used in the 'test-tube baby program' to eliminate those individual characteristics of sterile women's eggs that were disadvantageous from the standpoint of conception. Therefore they introduced certain mitochondria to the eggs that originated from another mother.

The ethical concerns focused on the participation of two mothers in a single reproductive procedure. The participation of more than two people in reproduction, however, is an inherent part of reproductive medicine. Extra-body, *in vitro* fertilization makes it possible to implant the embryo into a body different from that of its biological mother,[39] to use sperm from an anonymous donor, and not from the father, or to fertilize the egg of an anonymous donor, and not one from the mother.

Since the birth of the first test-tube baby in 1978[40] reproductive medicine has developed several multiple-parent models for the 'treatment of infertility'. Thus, the family palette has become even more colorful, and now in addition to natural parents, adoptive parents, blood parents and foster parents, the term 'genetic parents' has

37. *Ibid.*

38. Jason A. Barritt and his colleagues reported the technique of using two eggs and that has led to the birth of 30 babies. See Jason A. Barritt, Steen Willadsen, Carol Brenner, and Jacques Cohen (2001) "Epigenetic and Experimental Modifications in Early Mammalian Development, Part II: Cytoplasmic Transfer in Assisted Reproduction," *Human Reproduction Update*, 7(4): 428–435, at http://www3.oup.co.uk/humupd/hdb/Volume_07/Issue_04/pdf/070428.pdf.

39. In a settlement of a surrogate pregnancy case in California, *Johnson v. Calvert*, the egg-donating genetic mother demanded that the motherhood of the baby be determined by the court, as did her daughter who bore the baby. To complicate the matter both demanded a genetic determination of motherhood. Both parties supported their claim with medical opinion. Mark and Crispina Calvert established a surrogate motherhood contract with Anna Johnson to bear their baby. After birth, however, both mothers were determined to have the baby on the basis of blood. Each mother supported her case with expert medical opinion, and 'proved' their 'natural' motherhood based on differing points of view. See *Johnson v. Calvert* (Ca.1993) 851 P.2d 776, 790 and also http://philosophy.wisc.edu/streiffer/BioandLawF99Folder/Readings/Johnson_v_Calvert.pdf.

40. Although the development of various techniques of *in vitro fertilization* (IVF) and embryo transfer (ET) started after the second world war, nevertheless the first baby who was born as a result of this technology was Louise Brown. After her birth in 1978, reproductive techniques revolutionized the 'treatment' of infertility.

appeared, and we now have egg donors, sperm donors, and 'just' surrogate mothers. If we add embryo donation to the list, we have four new cases of parenthood, three of which are genetic.

With this acceleration in the technological development of assisted procreation, the possibility of human reproduction by cloning has become more than a theoretical one. Some thirty years ago James Watson, one of the describers of the double helix of DNA, already suggested the prohibition of cloning human beings. At that time, nobody took his warning seriously. The situation changed overnight when the little sheep created through cloning, Dolly, became a press sensation all over the world. President Clinton, with surprising speed, banned the use of federal funds on human cloning.

George Annas, the American bioethicist, made a noteworthy remark about Dolly. He felt that the lamb generated a surprising amount of public outrage, even among scientists, because in contrast to the mass of genetically modified animals she was named Dolly, and thus gained a personality. After all, in the original scientific announcement she was still not given a name—the cloned lamb was only called 'Number 6LL3'. In addition to being evocative of a toy 'doll', a sort of admission of willingness to take responsibility for her appeared, for in contrast to the monster created by Dr. Frankenstein, she was given a name.[41]

A unique aspect of the cloning debate is that artistic imagination was enthralled by the technique long before it was possible to pursue. Perhaps this is precisely why the majority of fears and threats are not related to what is truly possible to achieve now. We have not been able to clone a human yet, even if from time to time a scientist eager for fame makes the attempt. Not long ago yet another form of embryonic stem cell research failed. If in the future the so-called 'therapeutic cloning' succeeds, it is good to know that this does not mean that we will necessarily be able to, or even want to use the technique to implant any particular ethical or emotional characteristics into the embryo—that is, to produce cruel or angelically good people. Besides this, cloning for reproductive purposes is not permitted in North America, Japan, or Europe.

Since 2001 the public debate over reproductive cloning has strongly tended toward debate on stem cell research[42], and other therapeutic cloning techniques. The first breakthrough was signaled by the debate on embryo cloning in Britain, which

41. For more about the cloning debate, see Leon R. Kass and James Q Wilson (1998) *The Ethics of Human Cloning* (Washington, DC: The AEI Press).
42. For one of the most comprehensive record of this discussion, see Suzanne Holland, Karen Lebacqz, and Laurie Zoloth (eds.) (2002) *The Human Embryonic Stem Cell Debate: Science, Ethics, and Public Policy* (Cambridge, MA: MIT Press).

has been further stimulated by the Human Fertilisation and Embryology Act passed in 1990.[43] For the first time a question was raised whether the research on human embryos can be extended for harvesting of surplus embryos for stem cell research. British researchers hope that the permission granted for this research will enable them to find treatments for Parkinson's and Alzheimer's disease, and for certain types of diabetes. The law is only valid for stored embryos that would not otherwise be used, and experiments are strictly to be pursued only within 14 days of conception. Therapeutic cloning, which I will not deal with here in detail, is being pursued to produce organs and tissue and not to make a copy of a human being. Since through this technique it appears that not only embryonic cells, but also adult cells can be used, the major legal concern is not the protection of embryos, but rather the need to eliminate abuses in the enormous trade in stem cells.

Most frequently the protection of human dignity and human individuality is brought up as an opposition to human cloning. The uniqueness of each individual as an individual is not, however, a universal value. Moreover, people are not only biological and genetic beings, but are also connected to society through unbreakable bonds that strongly determine their characters: factors such as their upbringing, their environment, the age in which they are born. Among other things, theoretical issues such as these had to be struggled with at all international forums that decided to stop human cloning.

Not by chance, the first Additional Protocol to the Oviedo Convention was on the Prohibition of Cloning Human Beings.[44] The Protocol is categorical, but the reasoning evoking the ban is not made clear by its text. Article 1 declares that "[a]ny intervention seeking to create a human being genetically identical to another human being, whether living or dead, is prohibited."[45] The Protocol determines 'genetically identical' to mean that two human beings share with each other the same nuclear gene set: the nuclei of the cells contain identical genetic material. Thus cell and tissue cloning is not banned by the Protocol. Article 11 of UNESCO's Universal Declaration on the Human Genome and Human Rights also bans cloning for *reproductive* purposes, as it is "contrary to human dignity."[46]

43. For the official text, see http://www.hmso.gov.uk/acts/acts1990/Ukpga_19900037_en_1.htm.

44. Additional Protocol to the Convention for the Protection of Human Rights and Dignity of the Human Being with regard to the Application of Biology and Medicine, on the Prohibition of Cloning Human Beings (ETS No. 168). Signed in Paris, January 12, 1998. See the Appendix of this volume, and for the official text: http://conventions.coe.int/treaty/en/treaties/html/168.htm.

45. *Ibid.*, Article 1.

46. Universal Declaration on the Human Genome and Human Rights, adopted unanimously and by acclamation on November 11, 1997 by the 29th session of the General Conference of the UNESCO. See the Appendix for the official text.

As long as the development of the embryo was hidden deep in the womb until the moment of birth, philosophical[47] and theological approaches dominated the formation of normative concepts regarding reproduction. Since we became capable of observing the embryo with ultrasound, and by learning its sex were even able to give it a name, our normative ideas about reproduction have started to be formulated in biological terms.[48] But this is not always successful, and I believe it is not even desirable in every case. We may consider the embryo brought about by cloning to be simply a human being who is born as a result of a 'single-gamete, sex-free reproductive procedure'. But we should not forget that while this latter, euphemistic qualification is biologically correct, it does not necessarily supply us with effective legal terminology.

This is not a unique case of divergence between biological and normative terminology. According to Hungarian law, for instance, the embryo comes into existence through fertilization (literally: "every human embryo is an embryo from the moment of fertilization to the twelfth week of pregnancy").[49] As it follows, the clone-embryo is not a human embryo in the legal sense, because it is not developed from a fertilized egg. But once the embryo is implanted into the mother's womb, the law considers it to be a fetus, and thus a developing human being from the twelfth week of pregnancy.

Furthermore, the recent advances in biomedical and genetic research on the human embryo as well as the widening application of pre-implantation genetic testing raised an issue that is destined to become one of the major ethical concerns in the future. And this is the question of whether we should be allowed to use genetic knowledge for human enhancement. Pre-implantation genetic testing may reveal not only a possible lethal disease in the embryo but also some less severe conditions—some of them not even regarded as diseases but only features seen as disadvantageous for the future human being. Consequently, it is difficult to draw the line between healing and enhancement. This conceptual problem of distinguishing between prevention, therapy, and eugenic uses of genetics will soon become a matter of law and health policy.

6. Family and Genetic Information

The discoveries of genetics have made it imaginable and possible for a growing number of people, both parents and children, to find new evidence for biological

47. Aristotle believed, for instance, that the personality developed along with the embryo as it gained human form. He felt that this process lasted forty days in the case of men, and eighty days for women. This concept was adopted by early canon law.

48. For a comprehensive overview of the debate on the legal status of the fetus, see Jean Reith Schroedel (2000) *Is the Fetus a Person? A Comparison of Policies Across the Fifty States* (Ithaca: Cornell University Press).

49. Fertilization means the joining of two gametes. Accordingly, a cloned human being is not covered by this law.

relations in their genetic roots through establishing genetic identity. Biological evidences to support the legitimacy of social bonds between parents and children have always been called upon, but the recently emerging possibility of taking DNA tests may finally reveal the truth for those who want to know it. And so far DNA paternity tests are regarded as the most reliable tests.[50]

In the future, genetic data might be also used to identify not only fathers, but also mothers.[51] It will be an important change especially in those countries where mothers may deliver a child by using a pseudo name. Biological links, of course, had been significant even before using genetic evidence for establishing family links became possible. Family law recognizes this relevance by providing an exclusive right to develop a relationship with the child. It follows from the *de jure* elimination of the discrimination between the children born within and out of the wedlock that legitimization of paternity *post nuptias,* and legitimization based on recognition, is universally accepted in Europe. Moreover, legal systems generally provide equal rights for children born within and outside wedlock.[52]

Lawyers often ask now whether Article 8 of the European Convention on Human Rights[53] (right to respect for private and family life) is applicable to the offspring of donors, saying that it, now brought into UK law by the Human Rights Act of 1998, guarantees the right to form a personal identity. They are also asking whether Article 14 (prohibition of discrimination) can be used to argue that the donor's children should have the same rights as adopted children to trace their genetic parents.

Natural science's ability to provide proof of origin has developed enormously over the past decades as a result of the possibilities provided by genetics.[54] The worldwide demand for the establishment of paternity through genetic testing is expected to

50. The possibility that two unrelated people possess the same DNA band pattern has been calculated to be, on average, 30 billion to one.

51. The practice to keep the mother's identity in secret has a long history, the idea behind this solution might be to avoid infanticide and the shame that a single mother might face.

52. In 1979, the European Human Rights Court recognized both the violation of privacy and discrimination on the ground of birth status in the *Marckx v. Belgium* case (ECHR Case No. 6833/74, Judgement dated June 13, 1979). Published in Series A, vol. 31 (1979–80, 2 EHCR 330), available at http://www.pravnadatoteka.hr/pdf/aktualno/eng/20021104/Marckx_v_Belgium.pdf.

53. Convention for the Protection of Human Rights and Fundamental Freedoms (ETS No. 5). Signed in Rome, November 4, 1950. For the official text, as amended in 1994 by Protocol No. 11 (ETS No. 155), see http://conventions.coe.int/treaty/en/treaties/html/005.htm.

54. While in the fifties in Hungary, blood tests could at most determine who was *not* the father of a given child in 10–13 percent of the cases, today through blood serum testing and DNA exams paternity can be established with a fair degree of certainty.

grow worldwide, even in cases where no family dispute has arisen.[55] The possibility of finding proof of ancestry, even posthumous proof, and the chance to discover family relations have recently increased significantly. This is in itself perfectly reasonable, but if genetic family relations are overemphasized, this could potentially lead to a dangerous devaluation of non-biological familial ties. The demand for genetic testing of origin is now so irresistible, that at times it takes precedence over respect for the memory of the deceased. Genetic determination of paternity can take place well after death, even after burial.[56]

The social bonds of love, care, and acceptance, as well as the father's declaration of paternity or adoption, are traditionally important aspects of family membership. This is why up until very recently biological proof of family relations has been employed only in exceptional cases. Several years ago, however, biological ancestry became so important that many adopted children also now want to know their 'real' origins. The United Nations' Convention on the Rights of the Child[57] recognizes the child's right to know his or her identity. This right, however, can conflict with other rights of the individual, such as the right to privacy, and with family integrity. So far, in only a very few cases have people attempted to find who their genetic relatives were. DNA identification, however, provides such solid proof that the temptation to uncover family secrets has become great. It appears that in contrast to other approaches, this method of testing has gained validity. The problem increasingly arises in the case of embryos created using donor cells, where the demand to become acquainted with biological identity can arise in adulthood.

In the jurisprudence of the American courts, there are already numerous examples that demonstrate the wide array of legal problems of how genetically determined medical information can be used by family members for medical-preventive purposes. Disclosing the result of the individual genetic test, for example, may spare relatives the harm that may result from continued risk and uncertainty. The court in these cases usually faces the problem of the duty to disclose information to the relatives of a patient who may have some degree of genetic risk. In the *Safer v. Pack*[58] case, the appellants, a patient and husband, contended that *multiple polyposis*, a con-

55. The Swiss Federal Constitution provides everyone the right to have access to information about their ancestors.

56. One well-known case was the post-mortem genetic examination of Yves Montand, ordered by a French court in order to come to a decision in a paternity case.

57. Convention on the Rights of the Child. Adopted and Opened for Signature, Ratification, and Accession by UN General Assembly Resolution 44/25 of 20 November 1989. For the official text, see http://www.unicef.org/crc/fulltext.htm.

58. *Safer v. Pack*, Superior Court of New Jersey, Appellate Division (1996) 291 NJ Super 619.

dition which the appellant patient suffered from, was a hereditary condition that, if undiscovered and untreated, invariably led to metastatic colorectal cancer. The appellant patient's father had also suffered from this condition. The appellant contended that the physician who was treating the appellant patient's father knew the hereditary nature of the disease at the time. She also contended that the physician was required, by medical standards then prevailing, to warn those at risk so that they might have the benefits of early examination, monitoring, detection, and treatment, that would provide opportunity to avoid the most baneful consequences of the condition. The court agreed with the appellant. The court held that a physician had a duty to warn those known to be at risk of avoidable harm from a genetically transmissible condition. The court held that there was no essential difference between the type of genetic threat at issue here and the menace of infection, contagion, or a threat of physical harm. Accordingly, the court reversed the order of the trial court, which had previously dismissed the complaint.

In the course of adoption, the adoptive parents might wish to receive proof through genetic 'testing' that they are adopting a healthy child. I believe this is objectionable, and is in contradiction with the policy goals of adoption. The goal of adoption, according to the law, is to establish a familial relationship between the adopters and the adopted. The primary objective of this legal institution is to provide children with a family environment, especially children whose parents are either no longer living, or who are incapable of raising them.[59]

In the first major, already cited report of the UK Human Genetics Commission, it has been estimated that one in ten paternity tests are carried out on children covertly, that is without the consent or knowledge of one of the parents (usually the mother). While the standard practice is to carry out tests on a blood sample or mouth swab, results can be obtained from a single strand of hair or even by analyzing saliva on a used coffee cup. The HGC called for measures to prevent employers and insurers using genetic information from discriminating against people. It proposed that independent bodies should oversee medical research databases, and called for new legislation to prevent police access to such DNA collections. With these recommendations, the HGC is hoping to balance the need for genetic privacy with the need for continued research into the effect of our genes on our health. If scientists are to identify the genes involved in common illnesses such as heart disease, then they will need to examine DNA samples from large numbers of people. The safeguards suggested by the HGC should ensure that such knowledge benefits everyone's health without compromising the privacy of those involved in such studies.

59. Family Law, Section 46.

7. Commercial Uses of Genetic Information

7.1 Insurance[60]

The biomedical field in the genomic era is predominantly governed by 'human rights spirited' health law. One need not be a rights-skeptic to notice that the norms in the medical sector will not automatically influence the norms in the non-medical sector. On the contrary, parallel to the development of important international and domestic legal sources, a gap has gradually opened between the human rights approach and the commercial laws. As a consequence of these trends, a question has to be raised: to what extent principles of medical law apply to related non-medical areas, such as insurance?

The problems of private insurance cannot be explained easily with traditional human rights concepts. In much of Europe, health services rely on general health insurance based on citizenship rights. This is why the measurement of genetic risk is of little significance at present, for it will not lead to higher insurance rates or exclusion from insurance. In the case of life, accident, and disability insurance the situation is markedly different. Here the broad legal principle of *uberrimae fides* rules. This principle assumes the greatest degree of trust between the contracting parties. This is why the individual to be insured may not keep secret any information that may be of significance in the determination of risk.[61] Therefore, if an individual is in possession of a genetic 'finding' that effects the risk incurred by the provider of life insurance, then in theory the individual may not keep the information secret from the insurer.[62]

Cases may also arise in which the genetic test would indicate more favorable insurance payments than those traditionally calculated. If, for instance, the insurer's questions regarding illnesses that have occurred in the family provide a negative picture, but through genetic testing it can be shown that the person to be insured does not carry the gene responsible for the development of the illness, then he or she might avoid paying high premiums set because of the high risk that would otherwise have been determined.

According to Article 12 of the Convention on Human Rights and Biomedicine, a diagnostic genetic test is to be carried out based on proper genetic consultation, and

60. Interviews with Hungarian insurers helped me better understand their point of view. I would like to give special thanks here to Ildikó Takács, who assisted in preparations for these interviews.

61. Tony McGleenan, Urban Wiesing, and Francois Ewald (1999) *Genetics and Insurance* (Oxford: Bios).

62. Neil A. Doherty and Lisa L. Posey (1998) "On the Value of a Checkup: Adverse Selection, Moral Hazard, and the Value of Information," *Journal of Risk and Insurance*, 65(2): 189–211.

only for medical or scientific research purposes. Hence, the insurer may not compel the insured to undergo genetic testing. It is true that this has not entirely solved the problem because those few who are already in possession of unfortunate results from genetic tests are in a disadvantageous position in establishing the insurance contract. If we add the norms of data protection to this Article, then we find that from the restriction of data provision to the goal of research, existing data can only be used for other purposes with the express permission of the affected individual.

Another problem arises from the fact that insurers ask their clients a fairly broad spectrum of health-related questions. Although the majority of health data requested is traditional medical information, more and more such information in the future will also be genetic in nature.[63] That is, as we learn more about the genetic underpinnings of certain diseases, the genetic component of such data will also increase.

While the applicable medical information forms differ from insurer to insurer, some insurance companies do expressly ask about congenital diseases. Obviously, the insurer is interested in such data because a judgment of the future state of health is very much part of the insurer's job. Certain insurance companies will only consider such data as a domestic effect, while others will use it as indirect proof of higher health risk. If, for instance, there have been a number of cases of cancer or diabetes in the family, it can affect insurance premiums.

7.2 Employers and Genetic Information

Usually the employer does not have the right to see an employee's genetic data. One of the few exceptions could be that it is required for health reasons. In practice, however, it is not easy to separate genetic data collected for various reasons, including the checking of health conditions. This is not even a simple task in the case of traditional medical information. If, for instance, the employer provides life insurance for a number of employees, the employer can fairly easily come to a conclusion about the health of an individual employee through the behavior of the insurer without gaining access to the employee's health documentation. If, for instance, an insurer does not wish to insure three out of forty employees, or is only willing to do so at a higher premium rate, although not warranted by the employees' age, it would be rather easy to come to the conclusion that these employees have a considerable health

63. We may note here that Hungarian insurers have long been interested in diseases in the family. They ask about how long parents and siblings lived, and about the causes of death. This is, in point of fact, a form of genetic data, for in addition to collecting information on their client, they examine the family's health conditions as well.

risk. This fact alone can indirectly affect the employer's decision-making regarding an employee's future career. According to Claude Michele Poissonnet,[64] the genetic test performed in relation with employment jeopardizes the basic rights of the employees, especially the respect for private life.[65]

In the case of mental illness, because of the added threat of stigmatization, individual research projects should examine whether an exposure of genetic factors might lead to further discrimination. Genetic reductionism is especially dangerous in the genetic examination of mental illness. Thus, if genetic research is made on mental illness, its findings always have to be compared with the perspectives of alternative approaches, in order to make room for alternative explanations of the mental condition. One unique characteristic of research on mental illness is that it is difficult to connect the physical, biochemical, or genetic factors to the more complex socially and culturally specific aspects of the determination of psychological illness.

Without an exploration of the connection between genetic factors and environmental effects, genetic information can appear to provide the final word on a condition. This could lead to the conclusion that research results should not be disclosed to relatives. The issue, however, is not so simple. There are cases when a relative may have a legally supportable claim to information that also says something about him or her.[66]

7.3 Intellectual Property Rights

After a long and difficult gestation, the European Directive on the legal protection of biotechnological inventions was delivered (Directive 98/44/EC[67]). It is far from easy to analyze its underlying principle. The problem here primarily arises from the fact that in the first paragraph of Article 5,[68] the document finds the mapping of the human body and its constituent parts unworthy of intellectual property right

64. Claude Michele Poissonnet (2003) *Le dépistage génétique en milieu de travail* (Paris: L'Harmattan).

65. In France in 2002, a Parliamentary Act was passed in order to ban such type of genetic discrimination. See Article 4 of *Loi No. 2002-303 du 4 mars 2002 relative aux droits des malades*.

66. According to Hungarian law, for example, relatives cannot be banned from receiving health information also effecting them.

67. Directive 98/44/EC of the European Parliament and the Council of 6 July 1998 on the Legal Protection of Biotechnological Inventions, *Official Journal of the European Communities, L 213, 30.07.1998*, p.13–21. See the Appendix for the official text.

68. Article 5(1): "The human body, at various stages of its formation and development, and the simple discovery of one of its elements, including the sequence or partial sequence of a gene, cannot constitute patentable inventions."

protection. The second paragraph[69] states, however, that the technology that makes it possible to extract these constituent parts, such as genes or gene lines, even to the degree of sequencing, is worthy of protection. The morality of patenting human gene sequences was already examined in 1995 in the *Relaxin* case.[70]

The debate surrounding the Directive is well illustrated by the fact that not long ago the Netherlands attempted, before the European Court in Luxembourg, to eliminate the Directive on biotechnological inventions. The Netherlands argued that plants, animals, or human genetic matter should not be patentable. Italy and Norway supported the Dutch initiative. The Dutch party argued that the right to the integrity of the human body is violated if inventions created using human biological material can enjoy intellectual property right protection, and further that the Directive is also in violation of the principle of subsidiarity. The Dutch here argued that the Directive is not clear, and contradicts itself. These arguments did not impress the court in Luxembourg, and the Directive remains in force.[71] The Court found that the expressions the Netherlands found fault with, such as *ordre public* (public order), to be satisfactorily clear in the Directive.

Article One of the UNESCO Declaration on the Human Genome and Human Rights states that the human genome symbolically represents the "common heritage of humanity." However, the UNESCO Declaration itself contains a small loophole through which research can escape. According to Article 15, "States should take appropriate steps to provide the framework for the free exercise of research on the human genome with due regard for the principles set out in this Declaration, in order to safeguard respect for human rights, fundamental freedoms and human dignity and to protect public health. They should seek to ensure that research results are not used for non-peaceful purposes."

It appeared that a UNESCO meeting held in Paris in September, 2001 might result in a new turn in the patenting of the human genome and human genes. The International Bioethics Committee of UNESCO then considered taking up the matter with the World Trade Organization. It intends to insure that Article 27 of the so-called TRIPs (Trade Related Aspects of Intellectual Property Rights) agreement would clearly state that: "the human genome is not patentable on the basis of the

69. Article 5(2): "An element isolated from the human body or otherwise produced by means of technical process, including the sequence of a gene, may constitute a patentable invention, even if the structure of that element is identical to that of a natural element."

70. *Relaxin* case (1995) *Official Journal of the European Patent Office*, 6: 388.

71. *The Netherlands v. Parliament and Council*, Case No. C-377/98 at the European Court of Justice. See http://www.curia.eu.int/en/actu/communiques/cp01/aff/cp0148en.htm for the press release on the judgment of October 9, 2001.

public interest considerations set out therein, in particular, public order, morality and the protection of human life and health."[72]

8. On the Composition of the Book

8.1 Genetic Information and Biomedical Science

The first part of the book includes an analysis of the impact of Human Genome Project by two prominent scientists from Hungary. *Pál Venetianer* looks at the major steps in describing and understanding the human genome and demonstrates its relevance in the future of health care. He regards the present development in the field of genetics as a scientific revolution, however, based on different criteria than the one Kuhn applies. The author argues that DNA sequencing, cloning from somatic cells, stem cell technology, and DNA-array (as well as RNA- and protein-array) technology is based on an evolution of technological and methodological innovation and not on a revolutionary change of scientific paradigm.

István Raskó examines the use and application of genetic knowledge in various fields of life, predominantly in the field of medicine. However, the author's analysis goes far beyond the idealized vision of genetics, as he openly discloses fallacies, misbeliefs, and even uncertainties that exist in biomedical science and in the application of genetic knowledge. He admits that it is not at all evident how to interpret the pool of genetic data. For instance, the spectrum of gene mutations show a population-dependent distribution, that is, patients, say, in the United Kingdom may have gene mutations different from that in Hungary. Therefore tests useful in one population could be useless in another. Besides the biomedical domain, István Raskó also explores other fields where genetic information might have important uses, such as forensic medicine, criminology, and genetic archeology. Finally, he explores the problem of how genetic knowledge is communicated to the public and the scientists' role in this communication.

8.2 Genetic Information: Ethical and Cultural Aspects

The general public is usually excited about new scientific results, but often becomes critical about the ways genetic information is used and utilized. Scientific discoveries in genetics, and their applications in particular, thus have many policy-

72. Proceedings of the International Bioethics Committee of UNESCO (IBC), 8th Session. (September 14, 2001).

related implications. In order to develop a policy or a protocol on the use of genetic data, at first scrupulous research has to be conducted on the social and ethical impacts of collecting and analyzing genetic information on the individual and the society. The lay public may understand the notions of embryo and stem cell and the individual's own relation to his or her genetic make-up differently from the way scientists do. It is often the case that scientific discoveries are confronted only with the position of the Church, while ethical, sociological, anthropological, legal, and even theological aspects are not heard.

Today two important therapeutic uses of biotechnology, the stem cell-based therapies and the gene transfer techniques, stir the most lively ethical debates. In the *Sixth Framework Program* of the European Commission, preventive and therapeutic tools are regarded as priorities. Nevertheless, because of the ethical concerns of using human embryonic stem cells, funding has been suspended in these contested areas. Stem cells are especially useful for research and therapeutic purposes because they are non-differentiated, they can divide and multiply for a long period and, under certain physiological or experimental conditions, they can also give rise to more specialized differentiated cells.[73] Special ethical problems appear in the autologous collection and transplantation of haematopoietic stem cells.[74]

In the second part of the book, *Imre Szebik* examines the moral dilemmas related to various gene transfer techniques and compares the somatic and germ-line gene transfer techniques. Germ-cell gene transfer techniques are unique in the sense that their effect is not restricted to the patient treated but may affect many generations. This is why many ethicists argue that the use of germ-line gene transfer techniques should be prohibited as future generations may not able to consent to this intervention. Further ethical problems might arise from the use of the so-called *mitochondrial DNA* (mtDNA) procedures in medically assisted procreation. As a result of the intervention, the offspring will harbor genetic substance from three persons: from her mother, father, and the women who donated her enucleated egg cell.

In public debates, references to various religious thoughts often play an important role. However, the interpretation of our newly acquired genetic knowledge within the conceptual framework of the existing religious traditions is often very difficult. *Béla*

73. European Commission (2003) *Commission Staff Working Paper: Report on Human Embryonic Stem Cell Research*, SEC(2003)441, 3 April 2003 (Brussels: Commission of the European Communities). Available at http://europa.eu.int/comm/research/conferences/2003/bioethics/pdf/sec2003-441report_en.pdf.

74. Avis No. 74. of the French Consultative National Ethics Committe (Comité consultatif national d' éthique pour les sciences de la vie et de la santé) Les banques de sang de cordon ombilival en vue d'une utilisation autologue ou en recherche.

Somfai's chapter provides the reader with a unique analysis of religious thought on stem cell research. He explores the differences and similarities between the Buddhist and Hindu, Islamic and Jewish, Roman Catholic, Protestant, and Eastern Orthodox perspectives. The author also explains the Japanese practice of *mizuko kuyo* (practice that gives respect to the aborted fetus), which does not have a direct theological foundation, but nevertheless evokes various religious ideas. Although it is not easy to find a common position among the different religious traditions, it might be stated that most religions presume the idea of universal humanity: regardless of genetic differences, all members of humanity have common characteristics, and these common elements define our human identity.

8.3 Law in the Genetic Era

The advent of the genetic era has given rise to significant legal dilemmas. Among them we can mention such questions as who may own genetic data and information; in what circumstances can a genetic test be performed on children; how can genetic-based discrimination be avoided; and to what extent and in what ways can we protect genetic data. In the third part of the book, these legal questions are discussed in the context of both international legal standards and some national laws.

In the first contribution, *Roberto Andorno* looks at the UNESCO Declaration on the Human Genome and Human Rights, and first describes the circumstances in which the Declaration was formed. After providing the reader with this important historical background, the author then presents and analyzes the basic concepts and main propositions of the UNESCO Declaration: the human genome as a common heritage of humanity; human dignity; informed consent and the right not to know; confidentiality; prohibition of reproductive cloning; germ-line interventions; solidarity; and role of the Member States.

In the next chapter, *David Townend* explores some possible answers to the legal theoretical question: who owns genetic information? Genetic information is used by the individual, by the doctor, by the insurance company, and undoubtedly it has some commercial value. Given this wide range of affected individuals, groups of individuals, and institutions, it is not evident what legal concept best protects this information. The traditional route for protection of genetic information is privacy. This involves the control of information and the respect afforded to this information. Genetic information, however, is made subject also to property rights through patents where control over information is regulated differently than in case of personal genetic data.

In the third essay of this section, *Sheila McLean* analyzes the legal problems of conducting genetic testing on children. The author emphasizes that genetic tests are special among medical tests in that the link between diagnosis and therapy here is uncertain and therefore when the child is unable to consent to testing, it is very unlikely that testing serves the interest of the individual child directly. Genetic conditions have life-long consequences and therefore decisions made on behalf of the child may influence many important aspects of the child's future. Thus, the main questions that legal scholars and policy-makers have to face is: who and in what circumstances has the authority to make decisions on the child's genetic testing?

In the next chapter, *Gregor Wolbring* proposes a disability rights approach to the formulation of laws prohibiting genetic discrimination. The main thrust in the author's argument is that the prevailing discourse of disability treats disabled people as a medical problem, which can be solved only by some form of medical or technological intervention. This view in itself is discriminatory because it distinguishes between various biomedical and genetic conditions, implicitly selecting between 'good' and 'bad' traits. The author recommends the replacement of this medical model of disability with a social justice model of disability, which puts the issue of disability discrimination in a human rights perspective. The current focus on 'fixing' disability up to a certain norm gets an ironic twist in the third, 'transhumanist' model of disability, which emphasizes that the emerging genetic and non-genetic enhancement techniques might make what is presently seen as 'subnormal' actually 'supranormal' in the future.

In the last contribution to this section of the volume, *Jürgen Simon* provides an important contribution to the contemporary legal discussions of how much and what kind of genetic information insurance companies may use. In Germany, the Insurance Contract Act stipulates that the insurance applicant is obliged to give all information he has about the state of his health that is relevant for the insurance contract and the insurer. Obligation means that its defiance could exclude the insurance benefits. According to the Insurance Contract Act, all circumstances are relevant that the insurer has expressly and in a written form asked for. Apart from the analysis of the German law, Simon also offers a comparison between the different national legal models of how the insurance business treats genetic test results.

8.4 Genetic Databanks

The fourth part of the book deals with different models of national genome projects and biobanks. Scientists, pharmaceutical companies, public health experts, insurance companies, as well as the society in general, are all interested in the collection and classification of genetic data in large databases and in the biomedical and genetic research that is based on them. Tissues, blood samples, and other biological material obtained in medical and therapeutic institutions may be rich sources of genetic information. Biotechnological companies that aim to uncover the relationships between specific genetic sequences and particular diseases, might be particularly interested in the commercialization of genetic information. The donors of biological material, thus of genetic information, however, should be able to give a second consent if these data are used for research purposes beyond the therapeutic sector.

The establishment and operation of national genetic databanks and other institutionalized genome projects, whether they are used solely to store genetic data, to process such data, or to operate as a tissue bank, require special regulation. Further and more problematic legal issues may arise if a genetic databank is used to conduct research on mental illness or to explore the genetic background of human behavior.

The first national law in this specific field is the Act No. 139 of 1998 on a Health Sector Database of Iceland.[75] *Ragnar Aðalsteinsson* provides a history of the Icelandic genome project and analyzes the legal and social concerns of the first population-based gene collection enterprise. It is well known that relatively homogeneous communities are especially interesting resources for genetic research. This was recognized in 1996 by Dr. Kari Stefansson who established the company deCode Genetics, Inc. in order to finance genetic research in Iceland. The company asked for and received an exclusive license to build a database from the medical records of the Icelandic population. According to opinion surveys, the population of Iceland originally supported the idea to provide a 12-year exclusive license to deCode. The Health Sector Database Act, however, introduced an opting-out model (which presumed the consent of the donors) instead of an opting-in model (which would have required the explicit, informed consent of the donors). The support for the large-scale data collection was taken for granted and any individual who did not want to participate had to express refusal. In 2001 another Act was passed which allowed the use of the collected biological samples for the purposes of clinical tests or treatment.

The Estonian Genome Project is presented by *Krista Kruuv*, the executive director of the project and *Ants Nõmper*, a lawyer who actively participated in working out the legal issues of the project. Besides providing a useful comparison of the Estonian project with the Icelandic database, the contribution also analyzes the Estonian Human Genes Research Act.[76] Under this Act, every gene donor has the right to remain anonymous after coding; to permit disclosure of his or her identity; to disclose the fact of being or not being a gene donor and the circumstances thereof, unless otherwise prescribed by law; not to know their genetic data; to access their data for free except genealogies stored in the Gene Bank; and to receive genetic counseling upon accessing their data.

75. For the official text of the Health Sector Database Act of Iceland, see Appendix to this volume.

76. For the official text of the Human Genes Research Act of Estonia, see Appendix to this volume.

In the Latvian Human Genome Research Act[77], gene donors' rights are based on the informed consent procedure. This is an agreement to provide a tissue sample for the Genome Database, to obtain health description and genealogy, and to use the tissue, health description, and genealogy for genetic research, public health research, and statistical purposes. In this model, all personal data is replaced with a code, which enables the reverse identification of the gene donor, including the name, personal code, and residence. The code shall be indicated on the written informed consent of the gene donor. Under the Latvian law, gene donors have the right to access their data stored in the Genome Database and the right to genetic counseling.

National biobanks and genome projects are usually analyzed in a scientific or legal context, but the social and cultural dimensions of these projects are very rarely explored. However, it is particularly important to discuss these issues, as population-based genetic research constructs group identities. *Aivita Putniņa* explores a specific example of how national biobanks might take part in constructing cultural identities. The author distinguishes three basic discourses that justify the Latvian Genome Project. The first one is related to the particular socio-historic context of Latvia (political concerns about independence, positioning Latvia on the global map, and nationalism are reflected in the incentives of the genome project). The search for a 'Baltic gene', for example, has a strong political and cultural meaning. The second discourse concerns the strategic placement of biomedical science in the public and political fields. And finally, the third dimension focuses on public health policy aspects.

In the last contribution to this section, *Alan Doyle, Frances Rawle*, and *Peter Greenaway* provide an insight into the operation of the UK Biobank. This includes collecting samples and analyzing multifactorial diseases of adult life in 500,000 volunteers aged 45–69 selected at random from the UK population. From this study, important personal information, such as lifestyle information, may be derived. The UK data collection includes follow-up by tracking through healthcare records over an extended period including the use of existing disease registers. In the UK Biobank, both the database and the biological samples will be made accessible to the academic and commercial research communities under a carefully planned ethical and legal framework on an anonymized basis. As volunteers will be recruited on an entirely opt-in basis (the consent has to be explicit), there has been and will continue to be considerable open debate and consultation on all of the issues surrounding the project.

77. The Human Genome Research Act came into force on January 1, 2003, and was amended in June, 2003. With the amendments, the Act will finally enter into force on January 1, 2004. For the official text of the Act as well as its amendment, see the Appendix to this volume.

Those countries where the questions of genetic discrimination and data protection have not yet been discussed in front of the broad public, could very easily become 'genetic resource countries' for other countries that conduct advanced genetic research. Therefore it is of utmost importance that legal and ethical issues of genetics are discussed in a comparative context.

9. Closing Thoughts

In the future genetic information will become increasingly important to society, for it can enhance our understanding of the appearance and development of disease and can increase the effectiveness of treatment. In addition to researchers of the biomedical field, social scientists will also eagerly use this information for their research. Of course, serious economic interests also underlie curiosity on genetics. Employers and insurers have an interest in using the genetic data of their employees, clients, and future partners in order to reduce risk, or optimize the use of labor.

We should be aware that as our knowledge of genetics accumulates, we ourselves also become more 'transparent'. Regulations providing for the right to control the body are not up to the task of assisting us in developing rules on the use of human gene samples. It is important to recognize that medical and genetic information may be detached from the human body: even though the body dies, the gene set lives on, thus even a deceased person could be a source of valuable information. Uncovering all the secrets of our genes will still take some time, but it appears that the twenty-first century will present us with an ever more precise picture of our genes.

Our knowledge of human genetics will doubtless provide us with a great many advantages, but if we deprive our individuality of every non-genetic attribute and separate it from its human and cultural connections, science could fall into the trap of genetic reductionism. As in the case of every new scientific paradigm, if scientists and laypeople alike take the genetic 'code' out of the social context they can give rise to unforeseen and unwelcome consequences.

The genomic era provides us with a vast range of biomedical and socio-cultural benefits. But in order to enjoy these benefits, it is necessary to spread education across the disciplines, and to allow for negotiation and mutual interpretation between them. I believe that this multidisciplinary book on society and genetic information will be a valuable contribution to various scientific and political discussions, and it will be of interest to lawyers, policy makers, philosophers, and social scientists as well.

Acknowledgements

I would like to express my special thanks to Szilveszter E. Vizi and to Yehuda Elkana for their kind professional support to the research project and workshop on which this book is based. I am grateful to Zsuzsanna Gábor and Lilla Jéri from the CEU Center for Policy Studies for their unselfish contribution to organizing the workshop. I would also like to thank Ildikó Takács and Péter Flanek for making the workshop even more memorable with their indispensable technical assistance. I owe a great deal to Miklós Vörös for his thoughtful comments to improve the original manuscript, and his careful reading of the texts. This book would not have been materialized without the generous support of the CEU Center for Policy Studies IPF Faculty Fellowship established by the Open Society Institute. The IPF Faculty Fellowship was originally the idea of Katalin Koncz, to whom I am particularly grateful.

I.

Genetic Information and Biomedical Science

Genetic Revolution?
Consequences for Science and Society

Pál Venetianer

1. Introduction

According to one of the most influential historians of science, Thomas Kuhn, scientific revolutions happen when the conceptual framework of a scientific discipline—in other words: the paradigm—completely changes, the old paradigm fails to explain the new facts, and an entirely new theory is required. Although there can be little doubt that our generation is witnessing a true revolution in biology, which profoundly changes the ways and means of biological research and will undoubtedly deeply influence the future of the entire human race, this revolution is the consequence of mere technological and methodological innovations.

The five most important methods and technologies contributing to this biological revolution are: first, *the techniques of in vitro DNA recombination* (genetic engineering); second, *DNA sequencing*; third, *cloning from somatic cells*; fourth, *stem cell technology*; and fifth, *DNA (and RNA, and protein)-array technology* (DNA-chips). The last three of these techniques are fairly new, and their practical consequences have not developed yet to their complete fruition, therefore I am not going to deal with them in this essay. The first two, however, are about a quarter of a century old, and the dynamic development of these techniques led to the recently announced breakthrough: the compilation of the draft DNA sequence of the entire human genome by two teams, in February 2001.[1]

2. Sequencing the Human Genome

Let me first give a brief outline of the history of these technological innovations. Figure 1 summarizes the landmarks in the history of scientific discoveries and in the development of new methods.

1. International Human Genome Sequencing Consortium (2001) "Initial Sequencing and Analysis of the Human Genome," *Nature*, 409 (February 15, 2001): 860–921; and J. Craig Venter *et al.* (2001) "The Sequence of the Human Genome," Science, 291(5507, February 16, 2001): 1304–1351.

Figure 1. Historic landmarks

1871	Miescher discovers the DNA
1944	Avery identifies DNA as the genetic material
1953	Watson and Crick construct the double-helix model of DNA structure
1975–77	Sanger and Gilbert develop the techniques of DNA sequencing
1977	Sanger determines the first full DNA sequence (φx174 bacteriophage)
1984	First proposal to sequence the human genome
1987	First human genetic map with 403 markers
1988	First automatic DNA sequencers on the market
1990	October 1: Official start of the Human Genome Project (planned completion: 2004)
1992	Cohen's team reports the first complete physical map of the human genome
1995	Venter's team reports the first complete bacterial DNA sequence (Haemophilus influenzae)
1996	The first complete eukaryotic DNA sequence (yeast)
1997	The DNA sequence of the first multicellular animal (worm)
1998	Venter announces that he will finish the sequencing of the human genome in his private for-profit company (Celera)
1999	September 8: Celera starts sequencing the human genome
2000	June 26: Venter (Celera) and Collins (HGP) jointly announce the (essentially) successful completion of the human genome sequencing program
2001	February 15–16: Publication of the results of the two teams

The revolutionary innovation, it should be pointed out, that made the deciphering of the genetic information possible was the discovery of fast and efficient methods for sequencing DNA by Sanger and Gilbert in 1975–77; their work was awarded the Nobel Prize a few years later. While Gilbert's method has disappeared in the dustbin of history, Sanger's has survived essentially unchanged in principle but incredibly improved in efficiency. When Sanger's team finished the sequencing of the first complete genome (that of a small bacteriophage), their output was about 500 nucleotides in a human year. At the beginning of the Human Genome Project in 1990, the best teams could sequence 20,000 nucleotides in a year. At present, Celera produces more than 1000 nucleotides in a second.

This powerful method has been used to sequence more and more complex DNA molecules and genomes over the past 25 years. The progress is illustrated in Figure 2.

Figure 2. The progress of DNA sequencing

1977	ϕx174 bacteriophage	5,386	base pairs (bp)
1981	human mitochondrion	16,569	bp
1982	λ bacteriophage	48,509	bp
1986	tobacco chloroplast	155,844	bp
1990	cytomegalovirus	230	kilo base pairs (kbp)
1992	yeast, 3rd chromosome	315	kbp
1995	*Haemophilus influenzae* (bacterium)	1,830	kbp
1996	*Saccharomyces cerevisiae* (yeast)	13,478	kpb
1997	*Escherichia coli* (bacterium)	4,639	kpb
1998	*Caenorhabditis elegans* (worm)	97	Mega base pairs (Mbp)
1999	*Drosophila melanogaster* (fly)	180	Mbp
2000	*Arabidopsis thaliana* (plant)	115	Mbp
2001	*Homo sapiens*	3,200	Mbp

The difficulties of the task of sequencing a DNA molecule increase not linearly but exponentially with the complexity and size of the DNA, and the more than 3 gigabases of the human genome represent an unbelievably large number. Just to illustrate this enormity: with normal print, this sequence would fill a library of about 2000 average-size volumes, and the reading of this text with a speed of one nucleotide per second, would last for 100 years.

Solving the structure of the human genome is a very complicated process in comparison with the smaller genomes analyzed thus far, not only because of its length, but also because of the presence of large amounts of repetitive sequences. These latter makes the final arrangement of the individual sequence elements into continuous stretches of final sequences (called contigs and scaffolds in the jargon of the sequenators) extremely difficult. The achievement of this goal required several new strategic innovations, tremendous progress in the speed of sequencing, and enormous computing power.

As it is well known, the essentially complete determination of the draft sequence of the human genome was announced on June 26, 2000, by two independent teams. One was the publicly funded international Human Genome Project (HGP), with the participation of 20 large research groups from six countries (USA, Great Britain, Japan, Germany, France, China) led by Francis Collins; the other was a private company, Celera, formed and directed by Craig J. Venter (since then the company and Venter have parted ways, and Celera became a drug-development company).

The results of these competing two teams were published—after protracted negotiations—simultaneously, in February 2001, in the journals *Nature* and *Science*, respectively.[2] By the time of this publication, the public databases had already contained the DNA sequences listed in Figure 3. The total amount of DNA sequences in these databases exceeded 10^{10} nucleotides by mid-2000, and the doubling time of the size of these databases is now less than one year.

Figure 3. Fully sequenced DNAs (until the year 2000)

599	viruses and viroids
205	plasmids
185	organelles (plastids, mitochondria)
33	eubacteria
7	archaebacteria
1	fungus (Saccharomyces)
1	plant (Arabidopsis)
2	animals (Caenorhabditis, Drosophila)

Figure 4 shows that compared to the fully sequenced DNAs listed above, the sequence of homo sapiens is still far from complete.

Figure 4. What was reported in February 2001?

	Celera	HGP
Draft sequences	2,654 Mb	2,693 Mb
Coverage	83 percent	84 percent
Number of genes	39,100	31,800
Raw sequences	27,272 Mb	23,147 Mb
Finished sequences		842 Mb
Gaps	116,000	150,000

2. See footnote 1.

It is not easy to compare the achievements of the two teams, because the methods and the nomenclatures used by the teams are very different. It is obvious, however, that both published sequences represent less than 90 percent of the total, and the level of reliability is lower than expected. Both sequences contain a very large number of gaps, to be filled by further work. In fact, several members of the scientific community expressed serious doubts about the appropriateness of the publication at this stage, especially about the claims of the HGP of successful completion. Nevertheless, the publication has proven to be helpful for a large number of useful applications, and it certainly demonstrates that the giant task was solvable, and will be solved within a reasonable time, therefore it was justified. Of course the hype and the fanfare surrounding the announcement and the publication should have been treated with more than a grain of salt, and with some reservations about the validity of some of the claims. Since the publication we have already witnessed some critical reevaluations, criticisms, and mutual reproaches, but this is only the businesslike operation of the scientific community.

Here we should concentrate on two important questions: first, what should be done next, and second, what are the possible applications of this information for human medicine?

3. Future Tasks and Possible Applications

As to the upcoming tasks: obviously the human sequence should be finished. This will probably take at least two more years, and it is possible that some parts will still not be reliably sequenced. Then the sequence must be 'annotated'. This means that its functional units should be delineated, the beginning and end of each gene determined, the true number of the functional genes firmly established. This task will be quite difficult and may take several more years. The annotation, however, still does not say anything about the functions of the individual genes. In other words, in reading a text, the knowledge of the proper sentence-structure does not mean that it is understood. As according to the present estimates, there are 30–40,000 (or maybe more) genes, solving this problem will probably keep molecular biologists busy for the next few decades—firmly establishing the discipline of 'functional genomics'.

So far, we have discussed the human genome *in abstracto*, that is, the DNA sequence of 'Everyman'. We know, however, that each individual differs from everybody else, and for every possible practical application, these differences (albeit small, not exceeding 0.15 percent) are the most important. The collection of such differences (in the jargon of the molecular biologist, 'snips', that is, Single Nucleotide Polymorphisms or SNPs) is a crucial, formidable, and potentially profitable task;

therefore some of the big pharmaceutical companies are investing tremendous amounts of money in it. And finally, the large sequencing teams are ready to determine the DNA sequence of other species. The medically significant mouse and rat draft sequences have already been finished up to the perfection of the human one. The next tasks in the pipeline: the chimpanzee, the chicken, the honey-bee.

The second main question: the possible applications for human medicine. In dealing with this very complex problem, at first it must be clearly stated that DNA sequences will not cure sick people. This is, of course, a truism. Nevertheless, the general public expects that mapping the human genome sequence will be immediately followed by miracle cures, and unfortunately some experts and scientific journalists only nourish these false hopes. Figure 5 is a summary of the real possibilities.

Figure 5. Perspectives for human medicine

1. Model organisms
   - Homologous disease genes
   - Animal models of disease
   - Pathogenic microorganisms
2. Molecular mechanisms of hereditary diseases
   - Monogenic *(strictu senso)* genetic diseases (Huntington chorea, cystic fibrosis, etc.)
   - Hereditary components of polyfactorial diseases (cancer, diabetes, cardiovascular diseases, etc.)
3. Drug discovery and development
   - Identification of genes coding for new hormones, cytokines, etc.
   - Identification of new drug targets
   - Pharmacogenomics
4. Gene therapy
   - Somatic
   - Germ-line

For the general public, it is not obvious at all how tremendously important are the DNA sequences of various model organisms. A surprisingly large number of human genes—among them genes the mutations of which are responsible for diseases—have homologous counterparts in organisms as distant as yeast, the worm, or the *Arabidopsis* plant. This fact might help scientists finding cures or drugs for these diseases. Let me cite only one example: a new drug under development against Alzheimer's disease was found with the help of the genome analysis of the worm, *Caenorhabditis elegans.*

Another important help in drug development is the sequencing of pathogenic microorganisms. Again only one example: the complete sequence of the bacterium

Neisseria meningitidis, responsible for the dreaded meningitis, led to the development of a new vaccine against this disease. In fact such vaccines can now be routinely developed against any pathogenic bacterium with a known DNA sequence. The impact of identifying disease-causing mutations, the application of this information in diagnostics, and the understanding of the molecular basis of these monogenic and especially of multifactorial diseases are discussed in detail by István Raskó's article in this volume.[3] Nevertheless, it might be worthwhile to say a few words about the ways the human genome sequence can be used in drug development.

First, genes for hormones and cytokins can be found in the sequence, and these genes can be cloned and used as a direct source of medications. Figure 6 shows some examples of such drugs.

Figure 6. Examples of medically useful proteins found in genome search

Name	Abbreviation	Application
Keratinocyte growth factor	KGF	Healing wounds
Myeloid protein inhibitory factor	MPIF	Inhibiting proliferation during chemotherapy
Vascular endothelial growth factor	VEGF	Treatment of ischemic limbs

Second, new drug targets can be found using the traditional methods of drug development, but relying on an immense amount of new information. Perhaps it is not generally known that the targets of all the drugs available now on the market represent less than 500 proteins, whereas the total number of human proteins probably exceeds 100,000.

Third, and probably the most important potential use of the new wealth of information is represented by a new discipline called *pharmacogenomics*. This term means the exploitation of individual genetic differences, some of which code for phenotypically not observable traits, but might cause different responses to the effects of medical drugs. Figure 7 lists some of these effects.

Figure 7. What problems might be solved by pharmacogenomics?

• Depending on the medication, 9 to 80 percent of patients are refractory to accepted drugs
• Two million prescriptions per year in the US are actually harmful for the patients
• 100,000 deaths per year in the US are direct consequences of accepted medications
• This costs an estimated 100 billion dollars per year in the US

3. See pages 67–77.

4. Implications for the Scientific Community

In this volume, several contributions discuss the possible legal, social, and ethical implications of our newly obtained knowledge, therefore it is not the task of the present author to deal with most of these issues. In the following discussion, I shall try to restrict myself to commenting on some of the implications for the scientific community itself. In other words, to what extent are we—molecular biologists—now obliged to change our ways of working, thinking, publishing, and communicating.

4.1 Working in Giant Teams

This has long been a characteristic of nuclear physics, but it is a new phenomenon in biology. Sanger's team for the sequencing of the genome of the smallest bacteriophage consisted of nine people, and some of the early sequencing tasks were accomplished by single individuals. In contrast, the sequencing of the yeast genome involved over 600 scientists.

4.2 Transfer of Front-Line Research from Academia to the Corporate World

This phenomenon has also characterized other disciplines, such as materials science and informatics in the last decades, but it is entirely new in biology.

4.3 Involvement of Informatics: In Silico Biology

In the international job market of the research and development field, the hottest items, the most highly valued skills are currently those of the bioinformaticists, that is, scientists equally skilled in informatics and biology. The best biological journals are replete with papers, reporting results without a single 'wet' experiment, relying only on data obtained from databases (the new catchword is 'data-mining') and using only the computer as experimental tool.

4.4 Results in Databases, Not in Publications

The journals are unable to cope with the tremendous amounts of data; they have long ago ceased to publish sequences. These are available only in databases.

4.5 Difficulties in Handling and Presenting the Data

This difficulty concerns not only the DNA sequence data. Most experimental results with DNA chips have not found yet the proper ways of presenting themselves in an intelligible and usable form. A few years ago Edward Ueberbacher, from the Oak Ridge National Laboratory in the United States, told a reporter: "There's a big data wave coming, and anyone with any sanity should get out of its way." Unfortunately we cannot do this.

4.6 Retreat of Hypothesis-Driven Research

Experimental scientists, especially molecular biologists, traditionally treated with deep contempt the stamp-collector-type attitude of those colleagues whose job was simply to amass a large amount of data and describe observed natural phenomena with the attitude that 'well, some useful generalization will sooner or later emerge'. This attitude is nowadays widespread, highly reputed, and fashionable.

4.7 Communication with the Lay Public

Finally, a few words have to be said about an important issue: the communication of the scientist with the general public. The situation has never been as difficult, as heavily loaded with serious problems, as it is today. The main problem is that genetics is fashionable; the public and therefore the media are eager consumers of the exciting headline news offered by this science and its representatives. The temptation to serve this appetite is all the more alluring because good public relations and frequent appearance in the media are now important factors in the career, reputation, and grant-winning potential of a scientist, therefore very few are willing to resist this temptation. This situation nearly always leads to irresponsibly bold and exaggerated statements, claims, false and illusory promises. But even if the scientist's pronouncements and explanations are entirely correct, cautiously worded, and sufficiently modest, even if the mediating journalists do not distort them, even in this—very rarely occurring—best of possible cases, there is a very serious danger of misunderstanding with harmful effects.

The danger lies in the fact that the understanding of probabilities is very difficult, or perhaps even impossible, for the general lay public. Therefore, there is a dangerous tendency towards believing in a sort of unjustified and one-sided genetic determinism. If the man in the street reads or hears about the role of genetic factors in high blood pressure, drug abuse, or homosexuality, then the result might easily be increased prejudice, fear, or defeatism. To read in a scientific paper about a significant positive correlation between long digits of the hand, the ability to play football, membership in symphonic orchestras, left-handedness, autism, and homosexuality, might be interesting for the expert, amusing for the rare well-informed layman, but it might create very dangerous misconceptions in the minds of most people. In fact, the results of some available public opinion surveys already indicate that certain genetic test results (about the predisposition to high blood pressure and arteriosclerosis) might adversely affect the life-styles of many people. They might think: "Why should I exercise, or refrain from eating what I like, if I am doomed anyway?"

It is certainly not to suggest that genetic testing or the publication and popularization of interesting new results is dangerous *per se*. It is only to emphasize that we are moving on dangerous territory and therefore we should proceed with utmost care, and with the cautious calculation of all possible consequences.

What Can We Do with Our Genetic Knowledge?

István Raskó

1. Introduction

The last few years witnessed tremendous advances in genetics. The expansion of our genetic knowledge was predominantly a result of the information boom that followed the completion of the genome projects. New technological breakthroughs facilitated in-depth analyses of different genetic pathways, which in turn led to a better understanding of the uses of genetic information. It is also imminent that the accumulated genetic knowledge will be utilized in the wider society, beyond the walls of scientific laboratories. This chapter provides a far from complete overview of the major areas where society will undoubtedly benefit from the new discoveries.

It is generally accepted that the most valuable use of genetics will be in the medical field in order to improve the health services. There are great expectations now that the first draft of the human genome has been published. Sometimes these expectations are exaggerated, mainly because of the media hype surrounding the scientific discoveries. Nonetheless, even in this early stage, there are clearly predictable areas where new useful developments could be expected in the coming few years.

2. Advances in Genetic Testing

In medical genetics, understanding the molecular pathology of different diseases is an important task. At present we know these in most autosomal recessive, dominant, and sex-linked monogenic disorders. On the basis of this knowledge, new tests have been developed and successfully used in pre-implantation, post-implantation, and prenatal genetic diagnosis.

Cystic fibrosis could be considered as the exemplary test case for human molecular genetics. Using this disease as an example, all the pros and cons of genetic tests could be presented. The gene, which is responsible for this deadly disease, is located on human chromosome seven. In normal circumstances it is responsible for the maintenance of ion balance in epithelial cells, by controlling a chloride channel in the cellular membranes. The gene spans a 6129 bp DNA region and has 27 exons. There is a particular founder mutation present in the gene, which shows a North/

South-East distribution. The rate of heterozygotes is quite high: out of 23–25 people there is one heterozygote. These individuals are symptom-free. It is suggested that the high number of heterozygotes originates from a founder effect, a selective advantage. By having a marginally reduced ion flux across intestinal membranes, heterozygotes may be more resistant to typhoid infections. The problems arise when both copies of the gene harbor a mutation. The clinical picture is greatly variable: from a mild bronchitis, through male infertility, to a repeated, severe and later antibiotic-resistant pneumonia with a deadly outcome.

It has been predicted that if single gene disorders became amenable to study at the molecular level, the reasons for their phenotypic variability would become apparent. Unfortunately, the progress in this important aspect of medical genetics has been slow, as it is in the case of cystic fibrosis as well. There is still very little information about the inherited and acquired factors that modify the expression of these conditions. It would be useful to have some knowledge of the way in which the clinical phenotypes of more simple monogenic diseases can be modified. We need to understand to what extent acquired and environmental factors are implicated.

Apart from the most prevalent mutation (which is an amino acid loss in the protein) there are unevenly distributed mutations throughout the gene. The disease is frequent in European (Caucasoid) populations and it is unknown or very rare in Asia. The spectrum of mutations in other places of the gene shows a population-dependent distribution. This means that the mutation spectrum of patients in the United Kingdom, for example, is different from that in Hungary. Therefore tests useful in one population could be useless in other. The other problem is that in prenatal diagnosis, if one can exclude the presence of the frequent mutation, the fetus can still be defective originating from mutations in the other part of the gene. In order to exclude the existence of all the disease-causing mutations a complete gene-scan would be necessary. This however would be extremely costly. The situation, however, is not so disappointing.

There is a technological breakthrough that holds promise for the genetic diagnosis of certain genetic diseases. DNA chip, or microarray technology is based on the principles of chip technology in the electronic industry. During the application of this method, short DNA sequences are synthesized on, or attached to, the surface of a solid support. These DNA sequences could be artificially synthesized as DNA oligonucleotides spanning the entire genomic length of the particular gene, or could represent a complete gene-expression library. Isolated and labeled DNA sequences of the patients could be hybridized to the oligonucleotides and then analyzed by the means of a laser-guided scanner and appropriate computer software. The molecular diagnosis could be provided in no more than 15 minutes.

Although medical science is already able to apply prenatal or pre-implantation genetic tests in the case of most monogenic diseases, at present we are very far from having the means to identify the myriad of susceptibility genes or alleles that participate in the development of the most common complex diseases: such as cancer, cardiovascular diseases, asthma, diabetes, neurodegenerative and psychiatric disorders. According to the presently accepted theory, complex diseases are developed on the basis of the presence of several contributory genes and their interactions with environmental factors. In this case environment means all *in utero* effects, the nutritional and lifestyle environment, plus the natural environment in a broader sense. There is some skepticism as to whether all the genetic interactions taking place in the development of a particular complex disease could ever be achieved. Some of the problems we may encounter during the exploration of complex polygenic systems can be predicted by considering them in their evolutionary context. There is a very good chance that—as in the case of monogenic diseases, such as the above-mentioned cystic fibrosis, which reached their current high frequency through the selective advantages of heterozygotes for better survival in certain environments—susceptibility genes of complex diseases could have provided similar advantages. Considering the likelihood that the susceptibility genes of most common complex disorders may reflect ancient and diverse polymorphisms which evolved during the early evolution of mankind, it would be difficult to identify the key or master gene or genes (if there is any) which is involved in disease susceptibility or resistance. It is not known which environmental factors contribute and how to these diseases and it is also not known which metabolic pathways of the body are converting the environmental effects into harmful metabolites in complex diseases. By the means of DNA chip technology, the characterization of genetic susceptibility alleles is imminent.

One of the greatest challenges of modern medicine is *molecular genetic diagnosis* and appropriate therapy of malignant diseases. It is known that tumors, which could also be considered as polygenic complex disorders, arise because of the mutations and malfunctions in members of three major groups of genes: the gain of function mutations of cellular oncogenes (new functions of tumor causing genes); the loss of function mutations of tumor suppressor genes; and the loss of function mutations of DNA repair genes, which are the caretakers of the genome. So far around 100 oncogenes and 30 tumor suppressors and around 130 DNA repair genes have been identified. After the publication of the first draft of the complete human genome by sequence analysis, no new genes participating in carcinogenesis as oncogenes or tumor suppressors have been detected. It is possible, however, that by the means of the new concept of functional genomics, genes could be discovered that do not go through mutation but their expression during the tumorogenic process will be changed. The complex information about those genes could facilitate the establishment of a new oncogenic taxonomy and it could be foreseen that most effective antitumor therapies will be based on the 'genetic fingerprints' of particular tumors. By the means of DNA microarray technology, already several tumors have been characterized and reclassified according to their gene expression patterns.

Apart from the three major families of genes, there are other contributory genes in carcinogenesis, such as genes of cell-cycle-checkpoints, apoptosis, angiogenesis, and cell signaling, just to mention a few. It is worth mentioning that as the human genome sequencing project is coming to its end, a cancer genome project has been launched, of which the first useful result (identification of a gene, participating in the signaling of malignant melanomas) has already been published.

The main difficulties are in interpreting the results of genetic analysis, and in knowing what use to make of it. Sometimes the outcome is clear-cut and the potential alternatives can be clearly distinguished: patients are either carriers for a particular disease allele or not; they will certainly develop the disease or not; there is a risk to their offspring or not. But often, it will be a matter of probabilities; especially in the very common autoimmune diseases, where particular HLA haplotypes predispose to, but are neither necessary nor sufficient for, the development of a disease. In such cases, or if a patient has a dominant disease allele which is expressed with low penetrance, what advice can be given? And to whom should it be given?

One obvious standpoint is that genetic information is a *most personal aspect* of the life of an individual, and individuals should be entitled to keep it to themselves. Employers may see things rather differently. Nobody disputes that genetic diseases, when undoubtedly present, may legitimately affect job prospects or insurance premiums. A color-blind man is unlikely to be employed as a pilot or coach driver, and in the overriding interests of public safety, he should not be. What exactly are the ethics of employing someone with a cancer-prone genetic defect in a nuclear power plant? Likewise, insurance companies have their own perspective on the probabilities of disaster. People who are ill pose greater insurance risks, whether they suffer from genetic or acquired ailments. Thus, insurance companies might contemplate the question: why is it acceptable for an insurance premium to reflect the objective probability of your succumbing to a disease you already have, but not the objective probability of your developing a disease that your genes dispose you to? There is no simple answer to this question. Nevertheless, as genetic discrimination is not tolerated in many countries now, it seems that a company can reasonably refuse employment or insurance to somebody who has a disease, but not to somebody who is unusually likely to get it.

There are other problems, as well. *We are all mutants*, or at least carriers of rare alleles. Work it out for yourself; suppose for simplicity's sake that there are 10,000 functional genes in the human genome, that all genes are polymorphic, and that 99 percent of the population are homozygous for the most common allele at each locus. Then the chance that any single individual has 'normal' alleles for every gene is $0.99^{10,000}$, which is about 1 in 44,000,000,000,000,000,000,000,000,000,000,000,000,000,000.

As genetic testing spreads, more and more people will be found to carry small mutations in some gene or other, which are not themselves *known* to cause diseases, but are not normal. The only way to distinguish a rare neutral polymorphism from a rare disease-causing mutation is to wait and see, but for how long? And how serious is the disease likely to be? We have met this problem in a less attenuated form, as it affects the carriers of rare, relatively ineffectual mutations in the cystic fibrosis gene; there are probably no easy answers. Nor are there easy answers to the question of how to treat somebody who, although not being aware of it, has a high probability or even certainty, of developing a distressing ailment in several decades. This problem has become peculiarly acute in the counseling of Huntington's disease patients. With the recent discovery that the disease (untreatable, dominant, but late-developing) correlates with a triplet expansion in the IT15 gene, genetic testing of relatives who are as yet without symptoms has become very easy. A simple PCR test will establish the diagnosis. But it is obvious that an asymptomatic relative should never be tested without his or her consent, and only after informed consideration; and it is also generally agreed that the children of Huntington's disease patients should not be tested.

Nevertheless, it is worth identifying families that carry disease genes and, bearing in mind their full right to respect for individual autonomy, privacy, and confidentiality, giving them also their right to make informed choices. Avoidable suffering is not a desirable goal. Mankind has suffered from genetic diseases for long enough; now the time is approaching when we can begin to be rid of them.

3. Developments in Gene Therapy

A new approach in *drug therapy* could be predicted. This approach is based on the new molecular targets provided by the functional genomic use of the human genomic sequence data. The pharmacogenomic approach has become a focus of the attention of medical therapists. It has been long known that there is an individual genetic-based difference towards the sensitivity or the resistance in drug effects. This trait could also be considered as a complex polygenic characteristic. Much of the genetic variability among human beings comes from single-nucleotide polymorphisms, known as SNPs. These SNPs are natural changes in the arrangement of the four bases from which DNA is made. A single change in DNA base sequence can alter the function of a protein. This can lead to a disease, an altered presentation of a disease, susceptibility to a disease, or to altered response to medications. By comparing the prevalence of certain SNPs in treatment-resistant and responsive patients, the relative risk of a specific SNP for a certain trait, like treatment resistance, could be determined. Then the functional relevance of a certain SNP can be traced to a specific protein. Drug therapy could then, by considering a medication's actions, be tailored for a specific need. The most important genetic variations in drug metabolism are caused by the allele variations of those genes that code for enzymes, such as cytochrom P450 and N-acetiltransferase. The pharmaceutical factories are trying to identify population-dependent characteristics regarding drug response. Eventually this will lead to the development of personalized medicine, and from this not only the health care system but also the manufacturers could benefit.

By establishing what the major contributory genetic factors for certain diseases are, a new type of therapy is appearing on the horizon. Some of the reservations that could be raised against genetic techniques in medicine have also been applied to the prospect of using molecular techniques to correct human genetic diseases. *Gene therapy*, which has become gradually possible, is now a reality. Nevertheless, scientists and clinicians aiming to develop and apply various techniques of gene therapy have had to overcome a formidable array of regulatory obstacles. This is entirely understandable: mention genetics and people think of eugenics, and then of racial policies carried out in the name of genetics, in confident ignorance or uncontrolled lunacy. Also, the fundamental medical principle that one should avoid doing harm whenever possible makes people very cautious about making any deliberate change in the human genome. How long the period of extreme caution will last is uncertain; with experience, some current fears may prove to be less firmly based than seems possible at the moment.

There may be a parallel here with the intense public concern in the 1970s, when genetic engineering of bacteria became possible—a concern so great as to lead to a temporary moratorium, initiated by the scientists working in this field, and very stringent restrictions, which have slowly been moderated. There are already several thousand protocols that regulate the pre-clinical or clinical phase in the new field of gene therapy.

The aim of gene therapy is to replace the malfunctioning disease-causing gene with the 'healthier variant'. Two possible forms of gene therapy could be distinguished. In case of *somatic cell therapy* the cells of the body are modified, while in *germ cell therapy* (which is rejected by many countries on basis of ethical concerns) the embryo producing germ cells could be genetically altered.

The least controversial form of gene transfer into human somatic cells, and the first to be carried out, is the insertion of marker genes, in order to allow the fate of the marked cells to be determined after their return to the patient. This has no direct therapeutic value; but it is an essential first step towards establishing therapeutic protocols. For example, in one case marker genes are inserted into human hepatocytes, grown in culture, to test the efficiency with which they can be reincorporated into the liver.

Insertion of functional genes, to correct a genetic deficiency, requires that a careful series of choices be made once the appropriate gene has been identified. First, is the disease one for which the existing means of treatment are inadequate? Currently, many diseases affecting one step of a metabolic pathway can be treated either by removal of a potentially toxic precursor (as in phenylketonuria, where patients are simply kept on a phenylalanine-restricted diet), or by supplying a missing end-product (as in the treatment of adrenal hyperplasia with the appropriate steroids). Sometimes, it is possible to supply the missing protein directly, as in the cases of hemophilia or insulin-dependent diabetes. In such cases, gene therapy may not yet be cost-effective even if it is possible.

Second, are there suitable somatic cells to act as recipients? For some purposes, they would ideally be stem cells capable of being grown in culture and returned to the patient; currently, this can be done easily with bone marrow (the precursor of so many blood cells) where there is much transplant experience, or fibroblasts, which are astonishingly tough. Keratinocytes, which can easily be removed and returned through well-established procedures of skin grafting, can be modified to secrete proteins that are needed in the circulation and do not have to come from any particular place; gene therapy for hemophilia via keratinocyte transfection seems quite possible. Other important cells, such as hepatocytes or muscle stem cells, are more problematic. Circulating lymphocytes is easy, but it is not the case with stem cells. Somatic cells in their normal position in the patient may do, if a vector can reach them; for the lung cells in cystic fibrosis or emphysema, there is little alternative. But if the disease depends upon the expression of a gene at a critical time in development, therapy may be too late.

Third, is the disease caused by the simple absence of a gene, or by the expression of a mutant gene? In the first case, it will simply be necessary to insert a functional gene somewhere in the appropriate cells. But in the second case, it would be necessary to replace the mutated gene by targeted recombination. If this has to be done in all the mutant cells, prospects are obviously rather bleak, unless (as with blood cells) all the stem cells can be temporarily removed.

Fourth, but by no means least, is there a suitable and safe vector available for the gene to be transferred? Nowadays genetically modified viruses, artificial chromosomes, liposome mediated gene transfer techniques, or naked DNA injection methodologies are available. All the initial experiments have involved defective retroviral vectors, in which the viral genes were replaced by others; these viruses are obviously incapable of replicating. But retroviruses must be grown in cells, and cannot be purified entirely. If helper viruses were used to help assemble the defective retrovirus, obviously they might be hazardous. Nevertheless, even defective retroviruses might inactivate a functional, important gene by inserting into it; they might even activate cellular proto-oncogenes. Fortunately, tumor development is a multi-step process. Experiments with gene therapy of animal models of human diseases have been fairly reassuring on this point. It is also possible to design vectors that incorporate inducible genes that are lethal to cells expressing them; thus, any malfunctioning cells produced by gene transfer can be eliminated, if that becomes necessary.

Quite rapidly, more and more genetic diseases are being identified as suitable for gene therapy. Individually, they are all fairly uncommon; though in the affluent West, genetic diseases are now the cause of 10 percent of all pediatric hospital admissions, and they represent a major cause of death in childhood, since infectious diseases have been suppressed. However, the combined burden of genetic disease is of course trivial compared to the suffering from traditional illnesses and simple malnutrition in less fortunate parts of the world. But that does not mean that all those working on genetic diseases should abandon their posts, and seek to remove world poverty. The causes of the latter are essentially social and political, that is to say, they cannot be changed by genetic expertise, and indeed only slowly and unpredictably likely to respond to rational considerations. Still less does it mean that money spent on relieving genetic diseases should be spent on aid to the world's poorer countries; for money spent on the former has a good chance of achieving its aims, and is unlikely to be diverted to Swiss bank accounts.

Gene transfer to correct rare but relatively simple genetic diseases is one thing. Improving the therapy of common cancers is another. Current methods of surgery, radiotherapy, and chemotherapy aim simply to remove all the affected area, or to kill anything that multiplies. Sometimes they work. Nevertheless, nobody would be willing to endure the side-effects of these forms of cancer therapy, if more subtle and specific treatments were available. It is not surprising, therefore, that most of the existing gene therapy protocols were developed and are being used for the treatment of different malignancies.

According the latest surveys both in the United Kingdom and the United States, the general public widely and easily accepts the use of genetic knowledge in medicine. However, in spite of this general acceptance, several prominent people accuse medical geneticists of 'playing God'. To defuse this argument, it is the best to cite James Watson, one of the discoverers of the DNA structure:

> We are increasingly going to be accused of unwisely playing God when we use genetics to improve the quality of either current or future human life. Partly these accusations reflect the objections of the individuals who don't think we

have the right to do 'God's' work. But I also sense that sometimes the uneasiness comes from the fear that we might someday use genetic procedures in Hitler-like ways, using our scientific powers to further discriminate against unpopular political and racial groups. But diabolical as Hitler was, and I do not want to minimize the evil he perpetuated using false genetic arguments, we should be not held in hostage to his awful past. For the genetic dice will continue to inflict cruel fates to all too many individuals and their families who do not deserve this damnation. Decency demands that someone must rescue them from genetic hells. If we do not play God, who will?[1]

4. Non-Medical Uses of Genetic Knowledge

Genetic knowledge could be applied in *criminology.* Everybody can remember the O.J. Simpson case in the US, where DNA fingerprinting provided the crucial evidence. DNA fingerprinting should in principle establish the indisputable identity of a criminal. With PCR, it is possible to generate micro-satellite patterns from tiny samples found at the scene of the crime—a drop of blood, semen or saliva, a hair root—and the odds against a suspect accidentally having the same genetic fingerprint are very remote indeed. A genetic fingerprint should be as clearly incriminating as a real fingerprint, and a great deal harder to avoid leaving.

In practice, there have been two problems. DNA fingerprinting needs a really high standard of competence, which has not always been reached. And second, one must never underestimate the ingenuity of a defense lawyer with a client to lose . . . The prosecution may claim that the likelihood that a random member of the population has the same genetic fingerprint as the defendant, is one in 400,000,000. But what if the prosecution lawyers base their calculations on DNA samples taken from people with different ethnic background than the defendant has? Would not someone more closely related to the defendant be more likely to have the same fingerprint? And having established a reasonable doubt, must the court not acquit the defendant? Sometimes, it does (again, just remember the O. J. Simpson case).

The importance of genetics in *forensic medicine* became evident after 9/11. DNA fingerprinting identified the victims of this disaster. In another case in which a big restaurant chain mixed kangaroo meat with beef meat in hamburgers, DNA-based technology proved to be crucial in establishing evidence.

1. James Watson as quoted in Donald A. Chambers, Kenneth B.M. Reid, and Rhonna L. Cohen (1995) "The Biomedical Revolution at Forty Years," in Donald A. Chambers (ed.) *DNA, The Double Helix: Perspective and Prospective at Forty Years* (New York: New York Academy of Sciences), pp.441–472, see p.449.

Genetic knowledge has important uses not only for human beings but also for non-human species. Hunger is one of the greatest threats in the developing world, which opens an opportunity to seek to enhance agricultural crops and economically important animals by genetic technologies. By the use of *transgenesis*, drought resistant crops and farm animals with healthier meat could be produced and these are already on the market. The medical significance of genetically modified plants is immense. Just consider the necessity of gluten-free food in the diet of a person with celiac disease, or the possibility to immunize children against diarrhea simply by providing genetically modified bananas harboring the immunogenic epitope of enteritis causing bacteria. The question is whether the public of those countries in which most of the population is still starving will accept the existence of genetically modified organisms and GM food.

Those who agree with Bertrand Russell that the world might, for all we know, have been created only in the last few moments, complete with fossils, historical records, and our fallacious memories of the past, may regard ancient human history as irrelevant to the present problems. But many find great interest in contemplating the human evolutionary trajectory. To these, the insights from molecular genetics have been fascinating, if disputable. The most publicized discovery has been 'African Eve', a hypothetical woman deduced from *mitochondrial genetics*. Mitochondria have small, distinct genomes that do not recombine and are inherited almost exclusively in the female line. Therefore, mitochondrial genetics provides a simple system for establishing temporal relationships between human beings. Analyzing patterns of mitochondrial RFLP polymorphisms in different human races by large computers, researchers reached two surprising conclusions. First, all human populations outside sub-Saharan Africa (except for 'recent' African-American migrants and their descendants) are more closely related to each other than they are to any African population. Second, all mitochondrial evolutionary lineages converge on one point that, calculating with plausible rates of evolutionary mutation, can be dated about 200,000 years ago. From this point various African mitochondrial types developed and from one of these types eventually the non-African lineages separated.

These discoveries led to the emergence of a new science, *genetic archaeology*, which amalgamates archaeology with molecular genetics. This science uses genetic knowledge to understand the history of mankind and its different populations. Classical genetic markers, such as blood groups and enzyme polymorphisms, already indicated the possible origins of the gene pools of today's populations. The inclusion of DNA-based polymorphisms, particularly in the Y chromosome and in the mitochondrial genome, greatly improved the quality of evidence, and together with the archaeological findings, it finally was possible to establish the 'out of Africa' hypothesis for the origin of the diverse populations living on the Earth. It is now known that around ten thousand individuals migrating out from Africa founded the European populations.

The general public seems to understand and accommodate very slowly the great advances that come from the genomic revolution. The most important issue is how to interpret the consequences of genetic discoveries for society. Scientists must have a leading role in this task. The author of this chapter fully agrees with Sir Walter Bodmer: "All scientists should at least take some responsibility for explaining their science intelligibly to the general public. In this way, we might hope that the extraordinary advances in dealing with human disease, improving agriculture and conceptually in understanding our own evolution and evolution in general, will be taken advantage of to the full for the greater benefit of mankind."[2]

2. Sir Walter Bodmer (1995) "Where Will Genome Analysis Lead Us Forty Years on?" in Donald A. Chambers (ed.) *DNA, The Double Helix: Perspective and Prospective at Forty Years* (New York: New York Academy of Sciences), pp.414–426, see p.425.

II.

Genetic Information: Ethical and Cultural Aspects

Religious Traditions and Stem Cell Research

Béla Somfai

1. Introduction: Religion and Genetics

All religious beliefs are dealing with questions about the ultimate realities of our world. They seek an explanation of the relationship between men and God (or the deities) and provide some information about the ultimate power that creates, sustains and governs the world. They are asking questions about our place in the universe, the meaning and purpose of human life and death, the possibility of existence beyond death, about the origin and destiny of the universe. Their answers are usually sustained through faith rooted in some form of manifestation or revelation of divine knowledge and will, and point beyond the horizons of time and space.

The genetic revolution raises similar questions, but strictly within the parameters of time and space. We are rapidly deciphering the language of evolution and sooner or later will come to understand much better its history, correct or change its course. In possession of this tool even now we have new insights into the secrets of the gradual 'spiritual' transformation of the world. In the language of contemporary theology this transformation manifests itself in evolution. The idea originates in the ancient philosophical and Christian thought that looked on the hierarchy of vegetative, sensitive and intelligent life as gradually higher manifestations of the spiritual component inherent in the world. According to this insight the lowest form of life, vegetation, is endowed with 'vegetative soul'. This enables the plants to maintain their life, grow and multiply. The animal world is provided with 'sensitive soul'. In addition to the powers of the vegetative soul, this component provides the animal kingdom with the capability of sensitive perception and various reactions to external stimulation and dangers. The 'intelligent soul' is considered the source of our intelligence, consciousness and inner freedom. According to this notion, the spiritual soul carries the personal imprint of the Creator for each and every person. The idea of spiritual transformation remains central to Christian thought even today, though with significant refinements. It plays a significant role in the theological interpretation of evolution as the gradual unfolding of creation; it also provides a foundation for the due respect and protection of human life. This insight, basically a gift of Aristotelian philosophy, did exercise influence on some other religious traditions as well.

Although the insights of modern genetics remain within the parameters of time and space, genetic science does provide religious world-views with better understanding of our relationship with the biosphere. It makes it easier to locate us more accurately within the world of the living and in the whole universe. It will also lead us to a fuller comprehension of the meaning and purpose of life in general. Once we come to a more reliable possession of the language of evolution, we will better understand what it means to have been created out of 'the dust of the earth'; and perhaps we will have more defined parameters for the divine command to have 'dominion' over the created world. The prophets of the new scientific imagination even predict that in the not too distant future we will have the powers to re-create ourselves and overcome many of the burdens of sickness and suffering.

Even if the questions raised by genetics remain within the horizons of our world, the picture that emerges through this new knowledge compels religious believers to reinterpret their notion of creation, of original sin and the mandate to continue the work of the Creator. It was therefore not by accident that since the beginning of this revolution Judeo-Christian religious ethics, soon followed by representatives of other religious traditions, paid increasing attention to its achievements. In the early seventies, when the simian virus SV40 was successfully spliced into the DNA molecule of a laboratory strain of *escherichia coli*, the predicted possibilities of this new technological achievement were already called the "eighth day of creation."[1] Since then of course we can hear again and again the accusation that the powers unleashed by the genetic revolution will finally make it possible for us to put forth our hand "and take also of the tree of life, and eat, and live for ever" (Gen. 3,22.). For the fundamentalist religious mindset, manipulation with the genetic inheritance appears to be nothing less than the very arrogant and dangerous game of 'playing God'.

On the other hand, for the more informed representatives of religious traditions the challenge is placed into an entirely different context. Ever since the discovery of the double-helix structure of the deoxyribonucleic acid, we are faced with a host of unprecedented scientific, ethical, legal, political, and even economic choices. The power to alter the biological basis of life has created new responsibilities, said John Habgood, the Anglican Archbishop of York, by earlier training a biologist and pharmacologist. In 1992 at an ecumenical conference on religion and genetics in Houston, he reminded his audience of the fact that knowledge is not an abstract pattern in the mind or in the computer. Rather,

> [I]t is an understanding of what to do with such patterns, how to use them. It is a claim to be able to make responsible decisions about some more or less

1. Daniel J. Sullivan (1977) "Gene Splicing: The Eighth Day of Creation," *America*, December 17, 1977, pp.440–443.

reliable bit of the natural world. And, where that knowledge relates to human beings, it is important that its connections with other aspects of human life are not broken or ignored.[2]

Will the society have the wisdom, the powers of discernment, and the appropriate commitments to apply its new knowledge and capabilities for ethical ends? Habgood implied an affirmative answer and stressed that for this reason scientific research is always connected with ethical responsibilities.

2. Religious Views on Stem Cell Research

Before we turn our attention to the religious views on stem cell research we have to make a few introductory remarks. Ancient religious traditions do not know of embryos, of stem cells or, among other similar things, of the problems connected with creating human life outside the human uterus. Their stance on ethical issues related to this area has to be construed on the basis of insights obtained from the authoritative sources of their traditional religious and moral value-systems. They construct their stance on genetics and on stem cell research from their answers to the ultimate questions, and from the moral value perceptions that evolve from these answers. In other words, the manner in which they perceive and express their relationship to God, the meaning they see in life and death, in marital relationships and human procreation, in family and social relationships, and in their place in the universe will have significant influence on how they respond to the moral challenges of contemporary genetics.

The *centrally organized religious communities* may also receive information, inspiration or even explicit ruling on such issues from their governing authorities. They can be formulated either by conciliar procedures, similar to modern parliamentary methods of decision making, or by invoking some form of authority vested in those who carry the task of governance and teaching in a community. Since it is a rather complex process of interaction between their religious and ethical value systems and the relevant scientific information that formulates these statements, their moral stance is subject to reinterpretation and change. This is also the case with moral judgments on stem cell research. This too will depend to a significant extent on the interpretation of scientific data. This situation is not different, for example, from the question of the obligatory protection of human life. During the long history of the moral debate on due protection of developing human life, its degree and manner was always determined to a great extent by the scientific information available in a given time period.

2. Quoted in James R. Nelson (1994) *On the New Frontiers of Genetics and Religion* (Grand Rapids: William Eerdmans Co.), p.28.

In making these decisions the Protestant and Orthodox Christian communities follow conciliar procedures. In the Roman Catholic Church guidelines, directives, or doctrinal decrees can be formulated either by the Roman Pontiff or by a conciliar process similar to that of Orthodox Christian communities. The Buddhist, Hindu, Islamic, and Jewish communities do not have central governing bodies or teaching authorities. Therefore their position on questions related to stem cell research has a certain degree of incertitude. I begin with the position of communities without central decision-making authority.

2.1 Buddhist and Hindu Perspectives

The Buddhist way of life insists on the following of the 'noble eightfold path' of perfection. This leads to the elimination of suffering and sorrow, but only by the use of spiritual means. Among its moral commandments it includes the prohibition of killing, and of causing any harm to living things. The followers of its Jainist branch carry this prohibition to an extreme. They sweep the path in front of their steps to prevent the accidental killing of insects, and wear a mask over their mouth to prevent even their unintended swallowing. It is out of this stance that they formulate their position on ethical issues in genetics, and severely criticize the destruction of embryos for the purpose of scientific research.

For the first time in its 46 years of existence, the German Buddhist Union took a public stance on such matters at a membership conference in the spring of 2001. The 52 member organizations were unanimous in articulating their concerns with certain developments in genetic research and biotechnology. They justified their stance by Buddhism's deep sense of responsibility towards the protection of all life. They oppose the legal approval of research that involves the destruction of embryos. They do not accept pre-implantation diagnosis, cloning for purposes of therapy or reproduction, germ-line therapy, and condemn the acquisition of patent rights on the know-how pertinent to life and living beings. They wish to maintain the legislation that protects embryos and prohibits the import of embryonic stem-cell lines.

Buddhism rejects the killing or wounding of any living being and maintains that we must set limits to human needs and accept boundaries for the acquisition of human knowledge. Without opposing the work of science generally, they warn against the harmful social consequences of genetic intervention for eugenic purposes. These will result "not in the elimination of sickness, but of sick people," and in the destruction of human solidarity. Scientific efforts to produce a superior human race, free of suffering and faults, are necessarily destined to failure, because human perfection and freedom from suffering can only be achieved by spiritual means. Such efforts would turn science into substitute-religion. Therefore they call on the ethical responsibilities of all members of the scientific, economic, and political community and exhort them "not to sacrifice earthly life to short-sighted economic interests or to the dogma of helping and healing at all costs." Our obligation, they emphasize, is to secure conditions for life that equally correspond to the dignity of human and animal life[3].

Hinduism is a derivative of many religious traditions on the Subcontinent. One of the common elements is the belief in the transmigration of souls. This belief provides the foundation for a peculiar respect for animals, but to my knowledge an ethical consensus in the area of genetic research has not evolved up until now.

2.2 Islamic Perspectives

Islam does not have an official position on these questions. Two major schools of thought—Sunni and Shi'i—dominate Islamic thought rooted in the Koran and in the teachings of its earlier commentators. Their position on these matters is derived from the rulings of *Shari'a*, the Islamic law. Both schools of thought are recognized as authentic and regard early human life as being in a process of development to full personhood. The fetus reaches this point around the fourth month, when ensoulment occurs and the mother experiences its first movements. Even prior to this not clearly defined point in fetal development, life must be protected, though the experts of these schools are not in complete agreement on the extent of due protection. They agree however on the prohibition of abortion after the fourth month in pregnancy. There is also a considerable degree of disagreement among them on the possibility of destroying embryos not used for *in vitro* fertilization for purpose of pure scientific research. Most authors however allow interference with the life of a not implanted embryo for the purpose of improving human health. This effort is considered a participation in God's action of creating and healing. Stem cells may be collected even from the remains of illegally aborted fetuses. Most modern Muslim opinions would also allow collection of embryonic stem cells from not implanted blastocysts, and would consider research with them at this early stage of life ". . . as an act of faith in the ultimate will of God as the Giver of all life, as long as such an intervention is undertaken with the purpose of improving human health".[4] In most Islamic states legislation laid down by the political authority has greater or lesser degree of religious

3. "Stellungnahme der DBU zur Entwicklung der gegenwärtigen Geforschung und Bioethik," Die Mitgliederversammlung der Deutschen Buddhist Union (DBU), Steyerberg, April 29, 2001. See www.buddhismus-deutschland.de/main.html.

4. Abdulasis Sachedina (2000) "Islamic Perspectives on Research with Human Embryonic Stem Cells," in *Ethical Issues in Human Stem Cell Research, vol. III. Religious Perspectives* (Rockville, MD: National Bioethics Advisory Commission), p.G3. See http://www.georgetown.edu/research/nrcbl/nbac/stemcell3.pdf

binding force, because their legal system is rooted in the Koran. Therefore, legislative differences pertinent to these matters would also create variations on the interpretation of the religious binding force of a particular prescription in a given political community.

2.3 Jewish Perspectives

According to *Jewish perspectives*, four relevant criteria must be considered from religious law in the ethical analysis of research. First, the God-given obligation to protect our bodies: they belong to God and we are only stewards of them. Second, the belief that we are partners with God in healing and have the duty to use any natural or artificial means for this purpose. Third, all human beings are equally valuable because they carry the image of God. Finally, since we do not know all the consequences of our actions, before anything else we must avoid causing harm. From these criteria four potential moral impediments arise to stem cell research: the moral status of the fetus and of the act of abortion; possible complicity with evil; the commandment to respect the dead; and the moral status of the embryo.

According to the *Conservative Jewish* interpretation of the Law, there is no objection to collection of stem cells from the inner cell mass of the blastocyst outside or inside the human body, because it does not grant moral status to the embryo prior to 40 days of gestation. The moral status of these cells during this early stage is the same as that of male or female gametes. According to *Orthodox Judaism*, on the other hand, the Law considers the fetus a person with full moral rights after the time of 'quickening', and it may not be aborted except to protect the pregnant woman's health. The life-saving use of stem cells procured from illegitimately aborted fetuses, however, is permitted. The commandment to protect and preserve life overrides concerns with abortion or obligatory respect towards the dead. Therefore, also the orthodox interpretation of Jewish Law will allow organ and tissue procurement, or the collection of stem cells from corpses or from aborted fetuses in order that the living may benefit. Although both schools of thought recognize that obtaining the necessary consent for such procedure may be problematic, Conservative and Orthodox Jewish scholars do agree that extra-corporeal embryos have no human status. They are similar to the gametes; therefore they may be discarded, frozen, or used as life-saving sources for procurement of stem cells. There is no clear legal prohibition against creating extra-corporeal embryos for research purposes either. Stem cell research, therefore, is not considered intrinsically immoral by Jewish religious law. Problems arise, however, when the benefits of such research are not accessible to all people who would need them. There is also general consensus that this research on genetic engineering should not be used to enhance human characteristics. The question of whether it may be used to improve health, or must be restricted to life saving purposes is subject to some debate among Jewish scholars.[5]

2.4 Christian Perspectives

Protestant positions have a wide range starting from a highly restrictive stance, which draws the line close to the official position of the Roman Catholic Church, all the way to a non-restrictive one, similar to liberal Jewish views. *The neo-orthodox* school argues against stem cell research because it threatens the life of embryos, the weakest members of our community: "no community, whether family, village or state, is really strong if it will not protect its weak and even its weakest members." The person's existence, in fact the existence of all of us, begins not with capacities, but by being someone with a history. No benefit to others can justify their destruction. Similarly, the use of aborted fetuses for research would imply that we use human life for the benefit for others twice; first, to benefit the women who aborted them, and then to benefit society through embryonic germ cell research.[6] For the *liberal Protestant positions* the moral status of the embryo is not clearly defined. They deserve some respect and protection, but in view of clearly defined benefits they can be sacrificed, though not without some remorse. Similarly safeguards can be erected for the use of aborted fetuses in order to prevent the increase of abortions, or threat of some other kind to nascent life. Such decisions must be preceded by public debate in order to exclude considerations of individual or market interests. Global access to the benefits of such research has to be secured. In its substance the liberal Protestant position on these issues is identical with that of the liberal Jewish view.[7]

The *Eastern Orthodox* and the *Roman Catholic* positions are basically the same. All human beings are created in the image of God, and their life begins with the coming into existence of the zygote. In the uterus or outside of it, a zygote is committed to a developmental course, which, with the grace of God and human help, will lead to the coming into existence of a person. Unborn human life is entitled to the same

5. Rabbi Elliot N. Dorff (2000) "Stem Cell Research," in *Ethical Issues in Human Stem Cell Research, op. cit.*, vol. III. pp.C1–C5; Rabbi Moshe Dovid Tendler (2000) "Stem Cell Research and Therapy: A Judeo-Biblical Perspective," *op. cit.*, vol. III., pp.H1–H5. Laurie Zoloth (2000) "The Ethics of the Eighth Day: Jewish Bioethics and Genetic Medicine, A Jewish Contribution to the Discourse," *op. cit.*, vol. III. pp.J1–J26, provides a thorough analysis of Jewish bioethical literature. Since it is not a document of religious interpretation in a strict sense, I have left it out of consideration.

6. Gilbert C. Meilaender, quoting Karl Barth, in Meilaender (2000) "Testimony," in *Ethical Issues in Human Stem Cell Research, op. cit.*, vol. III., pp.E1–E6, see p.E3.

7. Ronald Cole-Turner (2000) "Testimony," in *Ethical Issues in Human Stem Cell Research, op. cit.*, vol. III, pp.A1–A4; Meilaender, *op. cit.*

protection as are those already born. Therefore, the derivation of embryonic stem cell lines is immoral, because it sacrifices the life of human embryos. The *in vitro* fertilization process that produces 'surplus' embryos for this purpose is immoral for the same reason. Stem cells may be procured only from spontaneously aborted fetuses, from adults and placental tissue and blood. The remains of induced abortions may not be used for this purpose, because that would imply complicity with an immoral decision. Established embryonic stem cell lines that already exist, however, may be used for the benefit of mankind according to orthodox views. It would be wasteful to destroy or discard them. It is on this point that the orthodox view differs from Roman Catholic teaching.[8]

According to the *official Roman Catholic* position, obtaining stem cells from embryos that remain after *in vitro* fertilization involves the intentional destruction of a genetically unique living member of the human species that deserves full protection from the beginning of its existence. No amount of benefit to others can justify this deed. The creation of human embryos for the sole purpose of experimentation is a heinous crime against humanity according to Pope John-Paul II. Collection of gonadal tissue from the remains of aborted fetuses involves complicity with abortion for two reasons. First, the possibility of obtaining this valuable material at the occasion of a morally unjustifiable abortion might render the abortion-decision more palatable, establishing thereby a motivational connection between the two procedures. Moreover, the necessary coordination of the method and of the timing of abortion will create a symbolic connection between them, *even* when the interest of harvesting the tissue did not influence the decision.

This judgment of the Church, however, is not directed against stem cell research as such, but is concerned only with the use of certain kind of sources for obtaining stem cells, and with the methods of collecting them. Embryos and aborted fetuses may not be used; stem cells, however, may be obtained from adults, from miscarriages, or from placental tissue or umbilical cord blood. Compared with the embryonic stem cells, these are usually on a slightly advanced level of differentiation. The medical advisory board of the Vatican recently issued a statement urging that research be directed to these pluripotential adult stem cells, because in their opinion this work might offer more immediate therapeutic benefits without requiring the destruction of embryos.[9] The Roman Catholic and the Orthodox churches insist on the equitable

8. Demetrios Demopulos (2000) "An Eastern Orthodox View of Embryonic Stem Cell Research," in *Ethical Issues in Human Stem Cell Research, op. cit.*, vol. III., pp.B1–B4.

9. Pontifical Academy for Life (2000) *Declaration of the Production and the Scientific and Therapeutic Use of Human Embryonic Stem Cells* (Vatican City: Pontifical Academy for Life), p.10, with a brief summary of pertinent 'official' Church teaching. See at http://www.vatican.va/roman_curia/pontifical_academies/acdlife/documents/rc_pa_acdlife_doc_20000824_cellule-staminali_en.html.

access to all benefits of such research. Besides the cloning of embryos, the Roman Catholic Church proscribes interspecies cell nuclear transfer, somatic cell nuclear transfer for reproduction, and the eugenic use of the results of genetic research. It has no objection, however, to somatic nuclear transfer for research leading not to the cloning of a person, but for example to the production of transferable organs.[10]

On the use of embryos that remain after *in vitro* fertilization, or on the question of collecting stem cells or gonadal tissue from aborted fetuses, not all Catholic scholars agree with the official position of their Church. The possibility of using not implanted embryos (blastocysts) for research is rooted in Church tradition, they claim. For many centuries it was held by Christian theology that the required degree of protection to ensouled and unensouled fetal life was different. It is only when a certain degree of developmental complexity is reached that the fetus becomes capable of receiving its 'intelligent soul' from the Creator. At the earlier stages of its development the fetus was endowed only with 'vegetative' or 'sensitive soul'. In modern terms, it is at this last stage that it becomes a person. According to ancient and medieval biological knowledge, this was presumed to happen roughly 40 to 60 days into pregnancy. (It is this tradition that influences also the Jewish and Muslim stance on our question.) For this reason, while abortion after ensoulment was allowed only to save the mother's life, prior to that event less stringent reasons were also acceptable for the procedure. The measure of due protection therefore was determined in accordance with the degree of animation, and the criteria for the differentiation were determined with the help of available scientific knowledge.

Following the same wisdom today, some Catholic scholars claim, we should take into account the information contemporary life sciences provide on the earliest stages of human life, and presume that some degree of developmental complexity is needed before a zygote can claim the degree of protection and respect that is owed to a person. The early human embryo is not yet "an individualized human entity with the settled inherent potential to become a human person."[11] Using the terminology of traditional Scottish philosophy, blastocysts belong to the human species; they share our *common human nature*, but do not yet possess those specific characteristics that differentiate them from other members of the human community.[12] At this stage of

10. Edmund D. Pellegrino (2000) "Testimony," in *Ethical Issues in Human Stem Cell Research, op. cit.*, vol. III., pp.F1–F5; Kevin William Wildes (2000) "Testimony," in *Ethical Issues in Human Stem Cell Research, op. cit.*, vol. III., pp.I1–I4.

11. Margaret Farley (2000) "Roman Catholic Views on Research Involving Human Embryonic Stem Cells," in *Ethical Issues in Human Stem Cell Research, op. cit.*, vol. III., pp.D1–D5.

12. Thomas A. Shannon (1998) "Remaking Ourselves? The Ethics of Stem Cell Research," *Commonweal*, December 4, 1998, pp.9–10; Thomas A. Shannon (1993) "Method in Ethics, A Scottish Contribution," *Journal of Theological Studies*, 44(1): 272–293.

their development twinning remains a possibility. Therefore, zygotes brought into existence *in vitro*, and left without the possibility of reinsertion into the uterus could be used for the interest of the species under certain conditions.

Without the unique properties that render this being an *individual*, that is, being able to become one and only one member of the human family, one cannot speak of a person. Although all the necessary prerequisites for personal existence cannot be identified, individual identity is certainly one of them. Therefore, until the capacity of twinning is present in an embryo, personal existence and a corresponding degree of protection cannot be claimed for it. Such a situation in itself, however, does not justify the destruction of embryonic life. Human life must be protected with utmost care from conception, affirmed the second Vatican Council. Some interpreters of this statement identify 'conception' with the third cell division, others with the event of implantation.

The Council statement preceded the birth of Luisa Brown by more than a decade. Therefore, it is safe to say that the Council did not consider all the ethically relevant factors of artificially produced extra-uterine human life, especially not the situation of embryos that remain 'surplus' after *in vitro* fertilization. It was considering the protection of human life only in the womb, where all the necessary requisites for normal development are provided. As one of the byproducts of the *in vitro* technology, nowadays we can say that being in the uterus is a necessary condition for the birth of a human person. Not having a chance for implantation, the life history of an embryo is necessarily on a dead-end course. Embryos in this situation have no possibility of becoming members of the human community. In this situation they cannot establish connection with the maternal environment and secure the necessary biological, psychological, and social requisites of hominization; they are deprived of the primal biological, emotional, and intellectual formative relationship with the person of the mother. Excluded from the evolutionary process that leads to hominization, they are destined to die within three weeks or later, if kept frozen. It is the absence of uniqueness (a prerequisite for individuality) together with the situation of not being related to a mother (who secures the necessary conditions of birth) that provides the argument for the decrease of obligatory respect and protection of 'surplus' embryos. It is in this situation—and only in this one—that the moral possibility of using them for research arises according to these Catholic scholars.

This approach presupposes that personal status, or the possibility to achieve it, are the only reasons why intentional destruction of human life is prohibited. In absence of such a possibility, life-destroying experimentation becomes possible on embryos, at least when saving peoples' lives cannot be secured by other means. This assumption however, is open to objections advanced by conservative Catholic and neo-orthodox Protestant thinkers. Gilbert Meilander, a Lutheran scholar, points out in agreement with others that respect for human life and intentional destruction of an embryo, even when it is inevitably destined to die, are incompatible attitudes. This incompatibility in his opinion is poorly masked by reference to the expected benefits of life-destroying experimentation.

It is possible to preserve respect for human life when its destruction is caused by a mistake or by an accident, when it is the only means to avoid greater damage (my own or other people's death), or when it is imposed as an obligation in case of a 'just war', or as a result of capital punishment—but it becomes impossible when it is done simply to satisfy some perceived future need. When benefits of an action result at the cost of eliminating the person who would benefit from it we are faced with the same contradiction that lies behind each case of 'active euthanasia'. Benefits, slight or significant, always presuppose the existence of the one who enjoys them. When incipient life is destroyed for hoped-for benefits that might materialize from the research, the present existence of a human being is sacrificed for the benefit of future ones. To make this decision acceptable we have to claim that this 'being' does not have 'full moral status' in the human community—to diminish, even if not completely deny its membership in the human community, the basic reason for which respect is owed to it.[13]

Meilander's opponents, on the other hand, point to the problems that arise from the claim that an equal measure of respect is owed to persons and to human beings that are at the earliest stage of their existence even when they have been left without the possibility of implantation. Human experience proves that it is possible to destroy a sacred object precisely for the purpose of paying respect to it, in order to prevent its disrespectful use by unworthy hands, for example. Respect can be paid even in the process of executing someone: for example, not by hanging but by firing squad, or by performing the rituals given to all deceased members of a community after the execution. Hunting cultures used to manifest a respectful relationship to animals they had to kill for food. Cree and Micmac hunters addressed their prey by familial names, made conciliatory speeches to, or prayers for them, and in general followed various ritual methods of killing. North American Indians in general were convinced that the supernatural owners of these animals protected them and would punish unnecessary killing, improper ways of destruction or wasting the food thus prepared.

The Japanese *mizuko kuyo*, the practice of giving respect to the aborted fetus displays a similar tension-filled attitude. It is a memorial service to aborted fetuses,

13. More on this debate can be found in Gilbert Meilaender's (2001) article "The Point of a Ban: Or How to Think about Stem Cell Research," *The Hastings Center Report*, 31(1): 9–16; and in the reply by Michael J. Meyer and Lawrence J. Nelson (2001) "Respecting What We Destroy: Reflections on Human Embryo Research," *The Hastings Center Report*, 31(1): 16–23.

initiated mostly by women who have had abortions. The practice includes prayers, floral offerings, burning incense, lighting candles, erecting wooden plaques that carry prayers or conciliatory messages to the deceased fetus, such as "please sleep peacefully" or "please forgive me." The ritual also includes "reverential bows to the *mizuko jizo*, a statuary image that represents the soul of the deceased fetus and the deity that cares for departed children."[14] Although the ritual varies widely they all aim to 'comfort and honor' the spirits of aborted fetuses. The public performance of the ritual also intends to demonstrate that they did not act unfeelingly or callously but with a sense of sorrow and responsibility. This practice is a reminder also for the radical pro-choice movement, in the sense that "abortion rights should be placed within a moral framework that urges all discussions about 'destroying a manifestation of life' be characterized by 'grief and respect'."[15]

3. Conclusion

Without offering any evaluation of this debate, I proceed with some concluding remarks. These religious opinions offer a wide range of judgment on the destruction of human life. The Judeo-Christian tradition (also reflected in the Muslim view) anchors its position in the belief that the person carries the image and likeness of its Creator. Consequently, unconditional respect and protection is owed to his or her life whenever and under whatever conditions this image comes into existence. However, as we have seen, there is no consensus in this tradition on the criteria that establish the appearance of this image. The most inclusive position associates this with the fertilization process, the most restrictive one with the time of 'ensoulment' or even with birth. Buddhism tends to emphasize not the personal status, but our inherent obligation to protect all forms of life. Zen Buddhism, on the other hand, considers the duty of respect and protection of human life even in a context of conflicting social values.

In this brief survey I examined the tension between the interests of genetic research and the due respect and protection of human life at its earliest stages in light of the ethical wisdom of religious traditions. I focused my inquiry on stem cell research. In some senses most religions presume that there are certain characteristics common to

14. Meyer and Nelson, *op. cit.*, p.20, citing Helen Hardacre (1997) *Marketing the Menacing Fetus in Japan* (Berkeley: University of California Press), p.2.

15. Meyer and Nelson, *op. cit.*, p.20, citing Naomi Wolf (1998) "Our Bodies, Our Souls," in Louis P. Pojman and Francis J. Beckwith (eds.) *The Abortion Controversy: 25 Years after Roe v. Wade* (Belmont: Wadsworth), p.408.

all members of the human family; one of them is the inherent dignity of the person. These characteristics define our human identity, our 'nature'. In biblical language, altering or destroying any of them would change the very image and likeness of God into which we have been created. I hope it became apparent from the analysis that most, even if not all, religious traditions see the danger not in stem cell research as such, but in the destruction of human life in its embryonic stage. Religious believers in general acknowledge this danger even if they do not necessarily attribute personal status to zygotes. The use of alternate sources, even if they yield only pluripotent stem cells, is therefore preferable.[16]

16. This not completely clear convergence of opinions becomes even stronger when it comes to the question of producing human zygotes for purpose of life-destroying experimentation. It is apparent for the careful reader of footnotes, that on the position of the various religious traditions, I have used a lot of information from the presentations made by the representatives of various religious traditions to the National Bioethics Advisory Commission of the President of the United States. See National Bioethics Advisory Commission (2000) *Ethical Issues in Human Stem Cell Research* (Rockville, MD: NBAC).

Moral Dilemmas of Gene Transfer Techniques[1]

Imre Szebik

1. Introduction

Interventions often cited as gene therapy have become a subject of serious debates in the past decade, although public discussion usually lags far behind the development of such techniques. Furthermore, the questions raised by gene transfer techniques regarding the scientific, psychological, legal, and social aspects are so complex and uncertain that the assessment of the moral implications of the procedure is extraordinarily difficult.

Given this complexity and the uncertainty of these interventions, the question arises: how can bioethics approach the moral meaning of this treatment without avoiding the trap of 'ethical nihilism' or 'relativism'? Are the classical approaches in bioethics (like principlism, casuistry, virtue ethics) sufficient to resolve the moral dilemmas of the case or do we need a paradigm change in bioethics to be able to deal with the moral issues genetics poses? Do we need a 'bioethics revolution' or simply more scientific knowledge?

Coming to the details, among the important questions that need to be answered we find the following: Should we allow changing our genetic material, and if so, what might be the requirements for it? Is it ethically permissible to change the genetic material of the germ-line cells of human beings, thus cause a change that might be manifested throughout many generations? What might be the aim of the genetic intervention? Is it in the best interest of mankind to promote genetic changes that lead to increased somatic intellectual capabilities, or should we restrict our intervention only to those cases where there is a medical need? The advancement of novel genetic and other biological techniques force use to raise these questions and examine the possible consequences.

Since our words are value-laden, the term 'gene therapy' suggests that we have a tool that can be used for therapeutic purposes. Until know, however, no single

1. *Acknowledgments:* This paper was supported by the Canadian Institutes of Health Research; the Canadian Stem Cell Genomics and Therapeutics Network; and the Széchenyi Program of the Hungarian Government (Development of New Technologies for the Therapy of High Mortality Diseases: Cell and Tissue Transplantation Combined with Gene Therapy).

treatment protocol has been proven to be safe and efficacious, therefore eligible to be approved as standard clinical therapy despite some promising preliminary results. Consequently, at this time, the word 'therapy' should be avoided and substituted by the term 'gene transfer technique'.

This paper discusses moral issues related to gene transfer techniques in the context of using these techniques for medical purposes and addresses some of the dilemmas concerning this approach. Somatic and germ-line techniques will be differentiated and the use for enhancement will be discussed in some detail.

If applied to human beings, gene transfer techniques might be used for preventing or curing certain medical conditions, diseases (therapeutic aim) and also for improving some of the properties or capabilities of the human body (enhancement aim). They may also be used for reproductive purposes, like for cloning; this latter function, however, will not be the subject of our reflections.

2. Somatic Cell Gene Transfer Techniques

Gene transfer techniques, if used for curing medical conditions by altering somatic cells (i.e., all cells except the reproductive cells) of the human body (also called 'somatic cell' or 'somatic-line gene therapy') usually do not raise specific moral questions. It is considered as a novel, very sophisticated tool to modify the genetic substance of some of the cells of the human body to correct their function.

Two claims might be worth mentioning here. Somatic cell gene transfer techniques might be unacceptable for those who reject medical interventions in general, like Amish people, since they believe these interventions are unnatural and grossly intervene into the course of nature. It is hard to argue this claim since it rejects many achievements of civilization.

Others might reject the somatic cell gene transfer techniques since they modify or manipulate the genetic substance of the human body and they believe the 'holy grail', the very essence of the human body, should be kept intact. In their view, only the manipulation of the human *body* is allowed, not its genetic substance. This claim is based on the assumption that the genetic substance of the human body has a unique status among the chemical substances. Although it would be difficult to contest the importance of the role of DNA and RNA in the life of the organism, the claim that they are unique and therefore should not be subject to intervention may rely on genetic determinism. If the argument goes further and claims that it is not the peculiar status of the DNA that should prevent us from any intervention but its complexity, then what could justify any intervention on the human central nervous system, the complexity of which is at least comparable to that of genetic substance. The *complex-*

ity argument is often claimed with regard to the germ-cell gene transfer technique and will be also discussed below.

We may conclude that these claims, although worth reflecting upon, do not pose any substantial argument against somatic cell gene transfer techniques.

3. Germ-Line Gene Transfer Technique

This intervention aims to alter or replace the genetic substance of the germ cells (ovum and sperm). The basic difference between the somatic and germ-cell gene transfer technique is that the application of the germ-cell gene transfer technique affects not only the person involved but future generations as well. Let us see now what arguments have been elaborated for and against this type of gene transfer.

One of the most powerful arguments for germ-line gene transfer techniques is that they are highly efficacious, since they cure a devastating genetic condition virtually 'forever', and not only for the given patient.[2,3] On the other hand, if a mistake were made, it would be very hard to 'undo'. There is an important counter-argument against the efficacy argument. Namely, if the birth of babies with genetic disorders may be prevented by the application of prenatal and pre-implantation genetic diagnostic measures, why should we waste our resources on this risky technique—given the fact that although the technique in itself might be very promising, failure of the technique cannot be excluded either. Certainly, this rather technical consideration would be valid only if all genetic defects could be detected before birth and if our society were willing to face the necessary consequences and accept a huge number of abortions.

While the efficacy argument certainly provides a strong support in favor of the germ-line gene transfer technique, we also have to look at other details and further consequences of the intervention.

4. The Moral Dilemma of Research on Human Beings

It is often mentioned in the (reproductive) cloning debate that although cloning in itself might be considered as an acceptable intervention, we may never achieve the

2. Kathleen Cranley Glass, Charles Weijer, Denis Cournoyer, Trudo Lemmens, Roberta M. Palmour, and Stanley H. Shapiro (1999) "Structuring the Review of Human Genetics Protocols, Part III: Gene Therapy Studies," *IRB: A Review of Human Subjects Research*, 21(2): 1–9.
3. LeRoy Walters (1986) "The Ethics of Human Gene Therapy," *Nature*, 320: 225–227.

'perfectness' of the cloning technique in an ethically acceptable way. This consideration argues that although we might be able to reach a point where the technique becomes perfect and fail-proof, we may not be able to accomplish this goal in an ethically acceptable way. Namely, during the research period it is almost unavoidable that the cloning technique causes various failures in the human cells. And it is quite possible that not all of these failures would be diagnosed during the first period of the development of the embryo, thus these embryos 'must be born' with birth defects. Since the questionable benefit of the technique will probably never justify the birth of a baby with more or less serious birth defects even in the research phase, the techniques of reproductive cloning in its prefect phase is ethically unacceptable due to the fact that it was developed in an unethical way.

Is this powerful argument valid for germ-cell gene transfer techniques as well? Probably it is, although the situation is somewhat different. While in the case of cloning the potential benefit of the intervention is quite questionable, the promising benefits of germ-cell gene transfer techniques are far less disputable, as the above-mentioned efficacy argument attests. On the other hand, we cannot maintain that germ-cell gene transfer techniques are the only way to treat or cure those individuals who suffer from a genetic disease, since somatic-cell gene therapy or other non-genetic medical therapeutic options might be available for these people. Certainly, one could claim that these latter therapeutic options also have side effects, therefore a similar positive risk–benefit ratio might be established even in the cases of germ-cell gene transfer research protocols.

This counter-argument sounds quite strong. But then again, while the conventional therapeutic options usually affect only individuals with the given genetic condition or disease, the germ-cell gene transfer techniques will also alter the genetic substance of future generations. And these generations would otherwise not necessarily be harmed by the side-effects of germ-cell gene transfer techniques, if conventional, non-germ-cell gene transfer techniques were applied. Although the discussion of this question without the practical details is somewhat difficult, we may still conclude that, in itself, the original consideration regarding the research phase of germ-cell gene transfer techniques might be acceptable for those cases when the only therapeutic option is a germ-cell gene transfer technique.

Although the 'playing God' argument sounds somewhat arbitrary, it is very often mentioned in the literature. It argues that an intervention into this very complex system is beyond what human beings are allowed to do. Again, it is hard to contest this respectable religious belief, but the argument in itself does not define where the borderline is between the jurisdiction of God and that of human beings, if this border exists at all. Transplantation of a heart or bone marrow seems to have at least as profound an effect on the human organism as a gene transfer intervention, not to mention the use of powerful psychotropic medicines on an everyday basis. If one accepts the medicine as a tool to help to relieve the suffering of our fellow, the quality of technique used for this purpose might be of secondary importance.

The classical 'slippery slope' argument is also frequently mentioned in the context of germ-cell gene transfers: once we develop this technique, there is virtually no way to prevent its application for purposes not beneficial for humankind. Although the slippery slope argument is a relevant one, it is hard to imagine a world where only those interventions are applied which may not possibly be used in a harmful way. Therefore, although the 'slippery slope' argument is valid in our case, it is too weak in itself to prevent the application of human gene transfer techniques on human beings.

Four more arguments deserve attention. First, the 'right to be born with an untinkered genetic patrimony' argument, which is frequently cited in European political discourses.[4] Even if we accept the existence of this right, it remains an open question whether it is better to be born with an 'intact' genetic substance that includes, for example, a dominant lethal mutation, or to be born with a corrected gene, even if this right is violated and the possibility of side effects of the correction cannot be excluded. Second, through germ-line gene transfer techniques we interfere with the natural process of evolution. This is very true, however, we have to see that almost all of our present medical activities can be characterized as denying natural selection, since medical interventions aim to help those who otherwise would not survive and reproduce. Third, distributive justice might be violated, since gene transfer techniques are extremely expensive and only the richer would profit from it. Again, true enough, one should never forget this aspect of research, but this argument is not strong enough in itself, if one considers the price we have to pay for other sophisticated medical technologies. Fourth, the 'future generations' argument claims that germ cell gene transfer techniques are unique in that sense that their effect is not restricted to the patient treated but may affect many generations. Therefore, germ-line gene transfer techniques must not be used since future generations may not consent to the intervention. Moreover, long-term effects pose an additional risk one has to consider.

But before analyzing the 'future generations' argument, and draw a conclusion about the moral status of germ-cell gene transfer techniques in general, let me describe a case where germ-cell genetic interventions are in fact done, although, this is not the intended effect of the procedure. The intervention is called *in vitro* ovum

4. Ronald Munson and Lawrence H. Davis (1992) "Germ-Line Gene Therapy and the Medical Imperative," *Kennedy Institute of Ethics Journal*, 2(2): 137–158.

5. Jacques Cohen, Richard Scott, Tim Schimmel, Jacob Levron, and Steen Willadsen (1997) "Birth of Infant after Transfer of Anucleate Donor Oocyte Cytoplasm into Recipient Eggs," *The Lancet*, 350: 186–187.

nuclear transplantation. It is used to treat a rare form of infertility.[5] The technical essence of the intervention is that the nucleus of the ovum of the infertile mother is transplanted into a donor's ovum, after the nucleus of the latter cell is destroyed. Thus, the donor nucleus regains its capability to become fertilized and may be used for *in vitro* fertilization purposes. Without discussing further technical details of the procedure, there is an important phenomenon that must be mentioned here: the host cell contains a genetic substance, the so-called *mitochondrial DNA* (mtDNA). Since this small amount of DNA is not eliminated from the host cell, as a result of the intervention, the offspring will harbor genetic substance from three persons: from her mother, her father, and the woman who donated her enucleated egg cell. This intervention does not use sophisticated methods to correct a tiny gene, it simply replaces (probably most of) the mtDNA of the mother with the one of the donor. How can we assess the ethics of this intervention? What are the risks and benefits associated with this intervention? Can we assess them at all?

The indication of the procedure is clear: to help infertile couples to get a genetically related offspring. While this motivation is usually considered highly laudable, when we consider the risk, that is, the potential harm and its magnitude, we encounter difficulties. Having three genetic parents might become a nightmare for lawyers. However, scientists argue that the mtDNA is only a tiny fraction of our genome, and the proteins it encodes have only basic physiological functions, therefore, the argument goes on, the whole issue has no scientific significance. Maybe so, but how can we be sure that this argument is really right? How can one prove that one gene is less relevant than another, after both of them express themselves and code a protein? May we claim that one given protein is of less importance in determining ourselves as human beings? Without exploring the still uncertain consequences of this scientific intervention, how could we determine whether the benefit of helping infertile couples outweighs the risk we cannot even estimate?

Could the 'precautionary principle' bring us closer to the solution? This principle states that if a research team plans to introduce a technique the potential harms of which are serious or irreversible and the uncertainties are vast, then they have to take anticipatory actions to prevent harm. Furthermore, the proponents have the obligation to examine a full range of alternatives, and the decisions must be open, informed, and democratic, and all parties affected must be included in the decision-making procedure.[6]

6. Stuart Newman and Carolyn Raffensperger (2000) "Genetic Manipulation and the Precautionary Principle," in Pamela A. Norton and Laura F. Steel (eds.) *Gene Transfer Methods* (Natick, MA: Eaton Publishing), p.242.

First, these questions bring us back to the 'future generations' argument. There are many complex interventions, unrelated to genetic engineering, in which similar decisions may deeply affect our physical and social environment or the education of forthcoming generations, and their well-being as well.[7] Thus, it is not unique to the germ-line gene transfer techniques that they could have a major effect on future generations. Although this parallelism does not diminish the weight the 'future generations' argument has with regard to germ-line gene transfer techniques, we have to see that the obligation to assess the consequences of our actions is not restricted to this kind of genetic intervention.

Second, even if we make our decision-making procedure as optimal as the precautionary principle demands, how can we cope with the vast scientific uncertainty, and the uncertainty of the answers we may have for our ethical dilemmas?

5. The Issue of Enhancement

Analyzing the possibilities of genetic enhancement is no less controversial. The question of normality and healthiness might be obvious in many cases, but as the following case of EPO illustrates, it might be very cumbersome. Erythropoetin (EPO) is a hormone that regulates the production of red blood cells in the human organism. It works on a basis of negative feed-back, inhibiting the overproduction of these cells. At the same time, synthetic EPO is an illicit doping drug since the increased amount of red blood cells has a major competitive advantage for sportsmen. However, it turned out that a sportsman, who was an Olympic champion in the 1960's, had a genetic mutation that resulted in overproduction of EPO, therefore in overproduction of red blood cells of the body affected. Can we state this champion is 'less normal' than others? Can we say that the artificial genetic realization of this phenomenon of enhancement otherwise existent in nature is morally impermissible since it compromises our basic values on nature, talent and diligence?

6. Coda

Instead of a conclusion, let me call attention to a controversy of distributive justice. When we discuss social, legal, theological, psychological, anthropological, ethical, and other aspects of contemporary genetics, I think it is worth remembering

7. Imre Szebik and Kathleen Cranley Glass (2001) "Ethical Issues of Human Germ-Cell Therapy: A Preparation for Public Discussion," *Academic Medicine*, 76: 32–38.

the warning thoughts of Solomon Benatar.[8] He emphasized that while we may be very clever in discussing the essence and consequences of various very sophisticated genetic techniques, we must not forget that the vast majority of our fellow human beings on this Earth are suffering from conditions and diseases that could be prevented or cured with tools presently available for only a couple cents. Moreover, these people will probably never benefit from any of the achievements genetics offers us, as these are unaffordable for them. Citing this warning, I certainly do not want to undermine the importance of gene transfer techniques, nor the significance of debating its social implications. Nevertheless, I believe that when we discuss ethical questions of a particular problem, we should not see it with an ethnocentric eye; the perspective of our less affluent fellows is at least as relevant as ours.

8. Solomon R. Benatar (1999) "A Perspective from Africa on Human Rights and Genetic Engineering," in Justine Burley (ed.) *The Genetic Revolution and Human Rights*, The Oxford Amnesty Lectures (Oxford and New York: Oxford University Press).

III.

Law in the Genetic Era

Seeking Common Ground on Genetic Issues: The UNESCO Declaration on the Human Genome

Roberto Andorno

1. Introduction

In 1997 the United Nations Educational, Scientific and Culture Organization (UNESCO) made a first important step towards the adoption of international legal standards for genetic issues with the approval of the Universal Declaration on the Human Genome and Human Rights.[1] This international instrument, which can be regarded as a "futuristic advance" in international law,[2] is at present the most complete global initiative aimed at protecting human dignity with regard to genetic technologies.

It is worth stressing that the UNESCO initiative in this area is perfectly in line with the objectives of this international organization. According to Article 1 of its Constitution, the main purpose of UNESCO is "to contribute to peace and security by promoting collaboration among the nations through education, science and culture in order to further universal respect for justice, for the rule of law and for the human rights and fundamental freedoms."[3] Now, it is clear that the increasing access to the human genome has profound implications for human rights and fundamental freedoms. While genetic advances promise amazing contributions to the well-being of individuals and humanity in general, they also raise significant ethical and legal concerns: How should genetic information be stored, shared and disclosed? How can we prevent the use of genetic testing for discriminatory purposes? Should patents on human genes be allowed? Do we have the right to clone people? Is human genetic engineering in conformity with human dignity?

Bearing in mind these among other questions and considering that the enterprise of laying down common standards on this sensitive area necessarily requires an open dialogue between cultures, it is understandable that UNESCO, where different cul-

1. The full text of the UNESCO Declaration can be found in the Appendix of this volume.
2. Noëlle Lenoir (1998) "La Déclaration sur le génome humain, avancée futuriste," *Le Monde*, December 9, 1998.
3. The official text of the UNESCO Constitution is available, among other sources, at http://www.unesco.org/confgen/en_constitution.shtml and http://www.icomos.org/unesco/unesco_constitution.html.

tural traditions are represented, appears as the ideal place for the discovery of such common criteria. In addition to this, the personal concern of the former UNESCO Director General, Federico Mayor for bioethical issues contributed decisively to the implication of the organization in this area. In 1993, Dr. Mayor established the UNESCO International Bioethics Committee (IBC), whose first task was to consider the opportunity of elaborating an international instrument on the ethical and legal implications of genetics. After four years of consultation and discussion between representatives of a large number of countries, the Legal Commission charged with the task of developing proposals concluded that UNESCO should gradually develop global political consensus for international norms on the human genome, first promoting the development of a non-binding document. The draft, finalized by a committee of governmental experts from eighty-one Member States, was adopted unanimously by the UNESCO General Conference on November 11, 1997. The following year, the United Nations General Assembly formally endorsed the Declaration.

Consequently, the instrument that was finally adopted is technically a 'Declaration', not a 'Convention', and therefore it is not binding for the Member States. The drafters of the document were clearly aware of the fact that if they had decided to prepare a binding instrument, the document would never have been approved. Nevertheless, this circumstance does not mean that the Declaration is a purely symbolic instrument. On the contrary, from now on it plays an important role helping to promote and harmonize legislative responses to new challenges raised by genetic advances. In fact, the chief merit of the Declaration consists in having laid down minimum standards to encourage—not to oblige—the member states in their commitment to adopt legally enforceable regulations at a national level. In any case, in order to ensure a major efficacy of the Declaration, a number of effective mechanisms will probably be needed in the future to overcome governments' lack of interest in implementing the common standards adopted.[4]

The Declaration consists of a Preamble and seven sections: A. Human dignity and the human genome; B. Rights of the persons concerned; C. Research on the human genome; D. Conditions for the exercise of scientific activity; E. Solidarity and international cooperation; F. Promotion of the principles set up in the Declaration; G. Implementation of the Declaration.

The purpose of this paper is to present and analyze the main provisions of the UNESCO Declaration.

4. Allyn L. Taylor (1999) "Globalization and Biotechnology: UNESCO and an International Strategy to Advance Human Rights and Public Health," *American Journal of Law and Medicine*, 25(4): 479–541, see p.512.

2. The Human Genome as a Common Heritage of Humanity

The general objective of the UNESCO Declaration is to ensure that the new possibilities opened by genetic developments will not be used in a way that would be contrary to human rights and human dignity. Within this context, one of the most serious concerns of the drafters of the Declaration was the need to protect the human genome from improper manipulations that may endanger the identity and physical integrity of future generations. To respond to this concern, the Declaration provides at the end of Article 1 that the human genome "is the heritage of humanity." This notion is inspired by the concept of 'common heritage of humanity', taken from international law, which tends to preserve some of the fundamental components of the environment as well as some precious cultural resources for the benefit of humankind as a whole. According to Ambassador Arvin Pardo, of Malta, who is considered the progenitor of the concept of common heritage of humankind, this notion is composed of five elements: first, the common heritage cannot be appropriated; second, it requires a system of management in which all users have a right to share; third, it implies an active sharing of benefits; fourth, it supposes reservation for peaceful purposes; and fifth, it implies reservation for future generations.[5]

Certainly, the final version of the UNESCO declaration does not take the complete expression 'common heritage of humankind' which was included in the drafts of the document. Instead, Article 1 just provides that "in a symbolic sense, it [the genome] is the heritage of humanity." This change is due to the concern of some Member States that the notion of 'common heritage of humankind' could be interpreted in crude commercial terms, that is, with exactly the opposite meaning of what was intended, allowing the appropriation of human gene sequences by multinational companies.[6] There was also the fear that the category of 'common heritage of humankind' could pose a threat to individual rights and be used as an entitlement for enhancing the human gene pool, that is, for eugenic purposes.[7] Therefore, the adjective 'common' was suppressed, and at the same time, it was added that the expression "heritage of humankind" should be applied to the genome only in a "symbolic sense."

5. Kemal Baslar (1998) *The Concept of the Common Heritage of Mankind in International Law* (The Hague: Nijhoff), p.81. The above mentioned elements were reiterated in many other UN documents and particularly, in the UN Declaration of Principles Governing the Seabed and Ocean Floor, passed on December 17, 1970.

6. Bartha Maria Knoppers (1999) *Le génome humain: patrimoine commun de l'Humanité?* (Montreal: Fides), p.35.

7. Ludger Honnefelder (1998) "Stellungnahme aus ethischer Perspektive zur 'Allgemeinen Erklärung über das menschliche Genom und die Menschenrechte' der UNESCO," in *Jahrbuch für Wissenschaft und Ethik, Band 3* (Berlin: de Gruyter), p.226.

This expression wants to stress the radical difference between the genetic structure of humankind and other 'external' goods that have been included in this category in the past (sea beds, outer space, the moon, etc.). By the same token, it means that genetic research engages the responsibility of the whole of humanity and that its results should benefit present and future generations.[8]

It is interesting to point out that, with the UNESCO Declaration, humanity as such is regarded for the first time as a common heritage to be protected.[9] Certainly the notion of 'crimes against humanity', the clearest example here being genocide, has been part of international law since the end of the Second World War. However, the category of crimes against humanity aims at avoiding inhumane acts committed against various ethnic, social or religious groups and therefore does not cover the protection of the human species as such, or indeed the preservation of its genetic structure, which is precisely the main purpose of the UNESCO Declaration.[10]

3. Human Dignity and the Human Genome

The most impressive feature of the UNESCO Declaration is perhaps the very central role it gives to the notion of human dignity, which is repeated fifteen times throughout the document.[11] No other international instrument had ever before focused so much on this notion. Dignity may indeed be considered the 'key concept' of the Declaration, which facilitates the understanding of the whole instrument.[12]

The reference to human dignity in international instruments is certainly not new. The Universal Declaration of Human Rights, adopted by the United Nations in 1948, already proclaimed in its Preamble the "inherent dignity . . . of all members of the human family" in order to form a basis for the list of rights recognized. According to its first Article, "all human beings are born free and equal in dignity and rights."[13]

8. Héctor Gros Espiell (1999) "Introduction," in *Genèse de la Déclaration universelle sur le génome humain et les droits de l'homme* (Paris: UNESCO), p.3.

9. Christian Byk (1998) "A Map to a New Treasure Island: The Human Genome and the Concept of Common Heritage," *The Journal of Medicine and Philosophy*, 23(3): 234–246, see p.235.

10. Bertrand Mathieu (2000) *Génome humain et droits fondamentaux* (Paris: Economica et Aix-en-Provence: Presses Universitaires d'Aix Marseille), p.92.

11. See Preamble, Articles 1, 2, 6, 10, 11, 12, 15, 21, and 24.

12. Roberto Andorno (2001) "La dignidad humana como noción clave en la Declaración de la UNESCO sobre el genoma humano," *Revista de Derecho y Genoma Humano (Law and the Human Genome Review)*, 14: 41–53.

13. For the official text of the United Nations Universal Declaration of Human Rights, see http://www.unesco.org/general/eng/legal/human-rights.shtml.

In fact, all the main international legal instruments on human rights refer to human dignity as a basic concept.

Even if not new to international law, the notion of 'human dignity' is becoming increasingly important in human rights documents, especially in the field of bioethics. The massive recourse to this notion to promote the respect for human rights in this area is not surprising: human dignity is one of the very few common values in our world of philosophical pluralism. This principle is universally accepted as the ground of democracy, and its reasonableness is not discussed at the political and legal level.

What does the notion of 'human dignity' mean? The usual understanding of this word refers to the idea that all human beings, irrespective of age, sex, physical or mental ability, religion, ethnic or social group, possess intrinsic worthiness and deserve unconditional respect. The word "inherent" is used to indicate that such a dignity does not rely on a particular feature or capacity of persons but only on their simple existence. This is why it cannot be gained or lost, and it does not admit of any degrees. Again, if we want to give a more precise account of the notion of dignity we can have recourse to the Kantian explanation of this concept, as a requirement of non-instrumentalization of persons. According to the German philosopher, persons should always be treated as an end in themselves and never as a means only.[14] Dignity is, in this sense, contrasted with 'price', the kind of value for which there can be an equivalent, whereas 'dignity' makes a person irreplaceable.

Articles 1 and 2 of the UNESCO Declaration adopt the notion of 'human dignity' as basis for their provisions, the first, with regard to humanity as such, and the second, with regard to individuals. According to Article 1, "the human genome underlies the fundamental unity of all members of the human family, as well as the recognition of their inherent dignity and diversity." The expression employed is certainly ambiguous and could be interpreted in a crude materialistic way. However, such interpretation was not the intent of the Declaration's drafters, as it clearly results from other provisions of the document. It seems evident that both the unity of the human family and the dignity of the human person, which are an expression of philosophical ideals, cannot be solely founded on the genetic information that characterizes our species. On the contrary, such fundamental principles are based on values and convictions that completely transcend biology. Human dignity is not the result of a particular combination of DNA, even a very complex one. In this respect, Article 2, Paragraph (b) points out in clear terms that "dignity makes it imperative not to reduce individuals to their genetic characteristics and to respect their unique-

14. Immanuel Kant (1911) *Grundlegung zur Metaphysik der Sitten, Band IV* (Berlin: Akademie-Ausgabe), p.428.

ness and diversity." Therefore, the theories of genetic determinism, which pretend to explain every aspect of the human personality and behavior on the basis of a specific gene, are discarded.

What is then the purpose of Article 1? This provision just aims to stress the fact that, from a genetic perspective, there is a deep interrelatedness between all members of the human species. Human genes are common to all past, present and future generations because the same genetic structure is inherited from one generation to the next.[15] Therefore, and independently of any changes that might evolve, which are recognized by Article 3, each individual inherits the same basic genetic structure of those that preceded him or her. This is why the human genome can be considered a 'common heritage of humanity' and therefore, fully deserves to be protected from irresponsible manipulations.

In addition to this, it is important to note that the 'genetic unity' of the human species is an additional argument in favor of the ethical and legal principle of equality between individuals. In this respect, Article 2, Paragraph (a) states that "everyone has a right to respect for their dignity and for their rights regardless of their genetic characteristics." This provision, which is a specific application of the general principle of equality, reminds us that it does not admit any exception due to genetic illness or predisposition to genetic illness that might be discovered among individuals with the development of new technologies. In other words, each human being, as a holder of intrinsic dignity, is entitled to inalienable rights, which are the same for all, regardless of their genetic make-up. This is why illness does not justify the consideration of some persons as being less valuable than others, or to deny them the exercise of their fundamental rights.

The scope of Article 2, Paragraph (a) is however larger than the prohibition of discriminative practices for genetic reasons, which are explicitly forbidden by Article 6. Its purpose is to reaffirm a vision of a society founded on the principle of human dignity. More precisely, it attempts to avoid the division of human society into 'genetically valid' and 'genetically non-valid' members, which would constitute a disastrous effect of scientific knowledge. Such a social stratification would weaken the proper foundation of democracy, whose basic principle is precisely the equality of all before the law.

In conformity with the idea of human dignity, Article 10 provides that "no research . . . should prevail over respect for human rights, fundamental freedoms and human dignity . . ." This is a classic principle of medical ethics that also appears with

15. Emmanuel Agius (1998) "Patenting Life: Our Responsibilities to Present and Future Generations," in Emmanuel Agius and Salvino Busuttil (eds.) *Germ-Line Intervention and our Responsibilities to Future Generations* (Dordrecht: Kluwer), p.76.

a very similar formulation in the Helsinki Declaration[16] on research on human subjects, adopted by the World Medical Association in 1964: "... considerations related to the well-being of the human subject should take precedence over the interests of science and society" (Article 5). In both cases, the idea is the same: individuals should not simply become instruments for the benefit of science, because science is not an absolute, but just a means at the service of persons. If in some cases non-therapeutic research is permitted, it is only within strict limits (for instance, free and informed consent of the subject; absence of serious risk for his or her life; and purpose of improving medical procedures).

Regarding the complex issue of the patenting of human genes, the UNESCO Declaration merely provides that "the human genome in its natural state shall not give rise to financial gains" (Article 4). Given that the ethical and legal problem is raised precisely by the patenting of human DNA sequences in something other than its natural form, and that this kind of patent is currently permitted in the United States and Europe, the Declaration gives the impression of having eluded the real problem. It was precisely in order to clarify this situation, that on September 14, 2001 the UNESCO International Bioethics Committee approved an "Advice on the Patentability of the Human Genome" according to which "there are strong ethical grounds for excluding the human genome from patentability." The Advice also recommends "that the World Trade Organization (WTO), in its review of the TRIPs Agreement,[17] clarify that, in accordance with the provision of Article 27(2), the human genome is not patentable on the basis of the public interest considerations set out therein, in particular, public order, morality and the protection of human life and health."

4. Informed Consent: The Right to Know and the Right Not to Know

The conditions for any genetic intervention, in particular, the informed consent of the individual and a prior assessment of potential risks and benefits are covered by Article 5 of the Declaration. This provision embodies a fundamental principle of medical ethics according to which a biomedical intervention may only be carried out after the patient has been informed of its purpose, nature, risks and consequences, and has freely consented to it. The principle of informed consent, which is the paradigmatic expression of patient's autonomy, has its origins in the Nuremberg Code of 1947, having thereafter been included in the Helsinki Declaration of 1964.

16. For the official text, see http://www.wma.net/e/policy/b3.htm.
17. The 'TRIPs Agreement' means Agreement on Trade-Related Aspects of Intellectual Property Rights.

At present, virtually all medical ethics guidelines and research codes hold that physicians and researchers must obtain the informed consent of patients and subjects prior to any substantial intervention.

For an informed consent to be valid, the patient must be considered competent to make the decision. If the individual is unable to consent, as in the case of minors, the intervention may only be carried out for his or her direct health benefit and subject to the authorization prescribed by law. Exceptionally an intervention without direct benefit can be carried out on a person unable to consent, one condition being that the person is only exposed to "a minimal risk and a minimal burden" (Article 5, Paragraph d). In this respect, it is interesting to point out that the European Convention on Human Rights and Biomedicine employs the same formulation regarding non-therapeutic research on persons unable to consent (Article 17[2]).

Article 5, Paragraph (c) recognizes both the right to know and the right not to know the results of genetic tests. Regarding the right not to know, the UNESCO Declaration is in conformity with other international instruments, like the European Convention on Human Rights and Biomedicine (Article 10[2]); the World Medical Association Declaration of the Rights of the Patients (Article 7[2]), and the WHO Guidelines on Ethical Issues in Medical Genetics and the Provision of Genetic Services of 1997 (Table 7). It should be stressed however that the UNESCO Declaration does not specify the conditions for the exercise of this right. In addition to this, the right not to know has been criticized at the theoretical level as being in contradiction with patient's autonomy, with doctors' duty to inform patients and with solidarity with family members.[18] This happens especially when nondisclosure poses a risk of serious harm to the patient's relatives who, without that vital information, could be deprived of preventive or therapeutic measures. In spite of these criticisms, there seem to be strong reasons for recognizing the right not to know. Individuals may indeed have a legitimate interest in not knowing their genetic make-up to avoid serious psychological consequences. This interest, far from being contrary to autonomy, may constitute an enhancement of autonomy, because the decision to know or not to know is not taken out of the hands of the patient by the doctor. But the right not to know, as in fact most rights, is not an absolute but a relative right, in the sense that it may be restricted when disclosure to the patient is necessary in order to avoid a risk of serious harm to family members, which means that some form of prevention or treatment is

18. John Harris and Kirsty Keywood (2001) "Ignorance, Information, and Autonomy," *Theoretical Medicine and Bioethics*, 22(5): 415–436.

available. Let us remember that, according to Article 9 of the Declaration, the principles governing informed consent and confidentiality, far from being absolute, can be limited by law, "for compelling reasons within the bounds of public international law and the international law of human rights."

5. Confidentiality of Genetic Data

Article 7 guarantees the confidentiality of genetic data associated with an identifiable person. It is worth remembering that health care professionals have a duty of confidentiality towards their patients and that they are in principle not allowed to disclose that information to third parties without patient consent. The purpose of Article 7 is to stress that this general duty also includes genetic information. This solution is fully justified: as a result of the widespread access to genetic tests, an increasing volume of genetic information about individuals becomes more and more available. While this information has the potential to provide health benefits, it may also be a source of harm to people. One of the major concerns is the potential misuse of genetic information resulting in any kind of discrimination or stigmatization. A regulation of the duty of confidentiality in this specific area is needed, because of the uniqueness of genetic information: first, genetic tests have implications not only for the individuals who have been tested, but also for their blood relatives; second, this information has an almost uniquely identifying nature, including its capacity to confirm, deny or reveal family relationships; third, genetic tests not only concern the past, but also the future of people and therefore, may have a significant predictive value; fourth, genetic information could be obtained from a very small amount of material, possibly secured without the donor's consent.[19]

To the doctor's duty of confidentiality of genetic information corresponds, from the patient's point of view, the 'right to genetic privacy'. This right may be considered as deriving from the more general 'right to privacy', recognized by the Universal Declaration of Human Rights (Article 12) and the International Covenant on Civil and Political Rights (Article 17). The difficult question is to determine what practice may constitute a violation of the right to genetic privacy. Article 7 does not provide any guidance in this respect. However, in 2001 a Working Group of the UNESCO International Bioethics Committee considered this issue. According to its Draft Report, human genetic data may only be collected and stored for the provision of

19. John K. Mason, R. Alexander McCall Smith, and Graeme T. Laurie (1999) *Law and Medical Ethics*, Fifth edition (London: Butterworths), p.167; UK Human Genetics Commission (2002) *Inside Information. Balancing interests in the use of personal genetic data*, May 2002 (London: HGC), pp.28–32.

health care, research on the human genome, and forensic purposes. The collection of genetic data for purposes of discrimination is therefore precluded.[20]

In 2001, the Director General of UNESCO, Mr. Koïchiro Matsuura, entrusted the International Bioethics Committee with the drafting of an international instrument on genetic data. This document, which probably will take the form of a new Declaration, has great relevance, because it will be the first international instrument prepared by the IBC since the adoption of the Declaration on the human genome in 1997. The drafting of this document should be undertaken in the 2002–2003 biennium, focusing in particular on the collection, processing, storage, and use of genetic data. To this end, the IBC set up a Drafting Group, which has already met three times, in February, April, and July 2002. The preliminary draft that has already been prepared covers, among others, the following topics: purposes for the storage and use of genetic data; informed consent; genetic discrimination; data confidentiality; population-based genetic studies; fate of material; anonymization of genetic data; historical or archived samples; and collection of samples in developing countries.

6. Human Reproductive Cloning

The UNESCO Declaration did not confine itself to laying down very general principles, but also dared to adopt some exceptionally precise provisions. This is the case with Articles 11 and 24 that declare two concrete practices as "contrary to human dignity": human reproductive cloning and germ-line interventions, respectively.

With regard to 'cloning', it is important to clarify that this term is here used to indicate the production of an entire individual (that is, an embryo) by 'nuclear transfer'. This technology involves transferring the nucleus—which contains all of the individual's genetic information—from a cell of an adult person into an oocyte of another whose own nucleus has been removed. If successful, this procedure leads to the formation of an embryo, genetically identical to the cell's donor. Depending on the destination of the embryo, a distinction is usually made between 'reproduc-

20. Draft Report on Collection, Treatment, Storage and Use of Genetic Data, September 3, 2001, Rapporteurs: Sylvia Rumball and Alexander McCall Smith. This document can be read at the website www.unesco.org/bioethics. The International Declaration on Human Genetic Data, adopted by the 32nd Session of the UNESCO General Conference in Paris in October 2003, expressed this position forcefully as follows: "Every effort should be made to ensure that human genetic data and human proteomic data are not used for purposes that discriminate in a way that is intended to infringe, or has the effect of infringing human rights, fundamental freedoms or human dignity of an individual or for purposes that lead to the stigmatization of an individual, a family, or a group or communities." (The Editor's note; for the official text, see the Appendix to this volume.)

tive cloning' and 'therapeutic cloning'. In the first case, the embryo is transferred to a woman's uterus, beginning a process that may lead eventually to the birth of a baby. In the second case, the embryo's inner mass is harvested and grown in culture for subsequent derivation of embryonic stem cells, which may have therapeutic applications in the treatment of serious degenerative disorders, like Alzheimer or Parkinson's diseases. Although the ethical and legal debate has predominantly focused on reproductive cloning, the acceptability of therapeutic cloning is also controversial, because it relates to the debate on the use of human embryos for research purposes.

Since the announcement of the successful cloning of the sheep Dolly from a cell of an adult animal in 1997, there is a widespread concern over the possibility that the same technique could be applied to produce genetically identical human beings. The main objections raised against this practice may be summarized as follows: First, the procedure gives those who create the clone unjustifiable powers over that individual, who would be deliberately produced to resemble an existing individual (or even a dead one) just to satisfy the desires of third persons. In this way, cloning would become a new and radical form of instrumentalization of people.[21] Second, although human beings are not reducible to their genes, the fact is that, given their physical similarity to the 'original' and between themselves, clones would seem like replacements. This situation would undermine their quest for individuality and make them appear, to themselves and to society, as replaceable 'copies', rather than irreplaceable originals.[22] A third objection is that the cloned child would be produced without genetic parents and would therefore be irreversibly deprived of the possibility to relate his or her existence to a 'father' and a 'mother' in the normal sense of these terms.[23] Finally, even on purely scientific grounds, human reproductive cloning is a dangerous procedure: data on the reproductive cloning of animals demonstrate that only a small percentage of attempts are successful; that many of the clones die during gestation; that newborn clones are often abnormal or die; and that the procedures may carry serious risks for the mother. Now, such devastating consequences in humans make the procedure unacceptable.[24]

21. Leon R. Kass (1998) "The Wisdom of Repugnance," in Leon R. Kass and James Q. Wilson, *The Ethics of Human Cloning* (Washington: The AEI Press), pp.3–59.

22. Henri Atlan (1999) "Possibilités biologiques, impossibilités sociales," in Henri Atlan *et al.* (eds.), *Le clonage humain* (Paris: Seuil), p.29.

23. George J. Annas (1998) "Why We Should Ban Human Cloning," *The New England Journal of Medicine*, 339(2): 122–125.

24. US National Academy of Sciences (2002) *Scientific and Medical Aspects of Human Reproductive Cloning* (Washington: National Academies Press).

The position of the UNESCO Declaration regarding human cloning is in conformity with most international and national instruments that cover this issue. At the international level, the most recent initiative aimed at preventing human cloning was taken in December 2001 by the United Nations General Assembly that established an "Ad Hoc Committee on an International Convention against the Reproductive Cloning of Human Beings." Other important international bans on—mainly reproductive—human cloning, adopted at both the global and the regional level, are the following: Charter of Fundamental Rights of the European Union of 2000 (Article 3[2]); Additional Protocol on Human Cloning to the European Convention on Human Rights and Biomedicine of 1998; various Resolutions of the European Parliament; WHO Resolutions WHA50.37 (1997) and WHA51.10 (1998); World Medical Association Resolution on Cloning of 1997.

At the national level, many countries have already adopted specific provisions that ban human cloning: UK Human Reproductive Cloning Act, 2001; German Embryo Protection Act, 1990, Article 6; Swiss Constitution, Article 119(2)(a); Spanish Penal Code, Article 161(2); Japanese Act No. 146 of 2000 on Human Cloning; Peruvian Penal Code, Article 324. Although in the United States there is still no federal legislation on the issue, on July 31, 2001, the House of Representatives passed a bill (the Human Cloning Prohibition Act) to ban both reproductive and therapeutic cloning. In addition, a number of states (California, Michigan, Louisiana, Virginia, and Rhode Island) have already outlawed human cloning.

7. Germ-Line Interventions

The UNESCO Declaration's ban on germ-line interventions is also in conformity with most ethical and legal guidelines on the topic. The objections raised against this practice stem from the fact that, unlike the alteration of the genes in the somatic cells, which affects only the individual treated, any alteration in the germ cells (gametes) or in the early embryo before the stage of differentiation would be passed to the next generation. The arguments against germ-line interventions are of different nature, depending on the purpose of the procedure.

First, in the case of germ-line interventions for *therapeutic* purposes, that is, for preventing the transmission of diseases, if we leave aside the controversy on embryo research, the objection is not based on intrinsic ethical arguments, but on the risks of irreversible harm to future generations. We are increasingly aware that the relationship between genes, environment and diseases is terribly complex. Even genes that are clearly associated with diseases in one context can be beneficial in others. Thus, the most elementary prudence requires that germ-line gene modification, which at present risks bringing more damage than benefits, should not be undertaken.[25] This solution is nothing but an application of the 'precautionary principle', which deals with scientific uncertainty and potential risks of new technologies. In addition to this, it is important to recognize that, in practice, any presumed beneficial effects of germ-line interventions would affect only a relatively few number of cases. This is why the idea of eliminating 'harmful' genes from the entire human population is more utopian than real. Such a global result, if ever, could only be realized over a time scale of thousands of years and with recourse to massive coercive programs, which would be morally unacceptable.

Second, in the case of germ-line interventions for *enhancement* purposes, the objections are more philosophical. The most fundamental argument is based on the idea that we do not have the right to predetermine the characteristics of future individuals. That means that people should be free to develop their potentialities without being biologically conditioned by the particular conceptions of 'good' and 'bad' human traits that were dominant at the time of those who preceded them. In other words, genetics should not become the instrument for an intergenerational tyranny.[26] A second objection is that the procedure would profoundly affect our own self-perception as 'persons', that is, as autonomous beings, leading to consider ourselves as mere 'things' or biological artifacts designed by others.[27] A third objection is that even if social agreement on the 'ideal' human being is reached, this will inevitably reinforce discrimination and stigmatization of those who do not fall into the accepted standards of genetically desirable traits.

As it has been already pointed out, most ethical and legal regulations covering this issue strongly discourage or frankly prohibit germ-line interventions. For instance, the European Convention on Human Rights and Biomedicine provides that "an intervention seeking to modify the human genome may only be undertaken for preventive, diagnostic or therapeutic purposes and only if its aim is not to introduce any modification in the genome of any descendants" (Article 13). At the national level, some legal provisions and guidelines already aim at preventing germ-line interventions: UK Human Fertilisation and Embryology Act, 1990, Schedule 2, Paragraph 1(4); Australian National Health Medical Research Council (NHMRC), Guidelines

25. The Council for International Organizations of Medical Sciences (CIOMS) (1990) *Declaration of Inuyama on Human Genome Mapping, Genetic Screening and Gene Therapy,* full text available at http://www.cioms.ch/1990_texts_of_guidelines.htm; British Medical Association (1998) *Human Genetics. Choice and Responsibility* (Oxford: Oxford University Press), p.198.

26. Hans Jonas (1985) *Technik, Medizin und Ethik* (Frankfurt am Main: Insel), pp.162–203.

27. Council for Responsible Genetics (CRG) (2000 [1992]) *Paper on Human Germline Manipulation,* available at http://www.gene-watch.org/programs/cloning/germline-position.html.

for Ethical Review of Research Proposals for Human Somatic Cell Gene Therapy and Related Therapies, 1999; Austrian Gene Technology Act of 1994, Article 76; Brazilian Act No. 8974 of 1995 on Genetically Modified Organisms, Article 13(1); French Civil Code, Article 16(4); German Embryo Protection Act of 1990, Article 5; Swiss Constitution, Article 119(a); US National Institutes of Health (NIH), Guidelines for Research Involving Recombinant DNA Molecules, 1998, Appendix M.

8. Solidarity and International Cooperation

Articles 17 to 19 deal with the issue of solidarity and international cooperation in the field of genetics. Such solidarity should take place on two different levels. First, solidarity should be advocated inside each country, where governments should help individuals, families, and groups particularly vulnerable to genetic diseases. Second, at the international level, solidarity should be especially promoted to help developing countries to benefit from the results of genetic research. This issue needs to be addressed without delay because, without solidarity, genetic advances threaten to widen rather than narrow the gap between rich and poor people and between developed and developing countries. Therefore, the hope behind the UNESCO Declaration is that all humanity will benefit from scientific advances in this area. This means, in particular, that industrialized countries will share the results of genetic research in a free and open manner with the world's areas most in need of such scientific assistance.

Article 17 stipulates that States should take the appropriate measures to ensure the promotion of active solidarity towards individuals, families and population groups who are particularly vulnerable to genetic diseases or disabilities. In particular, States should encourage research on identification, prevention and treatment of genetic diseases, especially rare diseases, as well as endemic diseases affecting a significant number of world's population.

Article 18 provides that States should "make every effort . . . to continue fostering the international dissemination of scientific knowledge" in three specific areas: first, the human genome as a matter of general knowledge; second, human diversity; and third, in genetic research. This Article also stresses the need for scientific and cultural cooperation in this area between industrialized and developing countries.

Article 19, Paragraph (a), enumerates the measures that States should adopt in the context of international cooperation with developing countries. The aims of such measures are: to assess the risks and benefits pertaining to research on the human genome to be carried out and abuse to be prevented; to strengthen the capacity of developing countries to carry out research in biology and genetics, taken into account their specific problems; to allow that the results and benefits of research, which is primarily carried out in the industrialized countries, could be used to promote economic and social progress for all; to favor the free exchange of scientific knowledge and information in the areas of biology, genetics and medicine.

9. The Central Role of the States

The Declaration strongly stresses the responsibility for the implementation of the common principles adopted lies principally with the States. This is obvious if one considers that the document is not binding and therefore inevitably requires the initiative of governments to become effective at the national level. In this respect, one of the drafters of the Declaration has pointed out that this instrument is ultimately no more than "a call to attention" to the community of States, so as to make them more conscious of the many ethical dilemmas that research on the genome presents.[28]

The Declaration stresses that the governmental initiative is not restricted to the adoption of adequate legislation, but includes "all appropriate measures" aimed at the implementation of the principles adopted (Articles 20 and 22). Among the tasks assigned to States are:

- "to foster the intellectual and material conditions favorable to freedom in the conduct of research on the human genome and to consider the ethical, legal, social and economic implications of such research" (Article 14);
- "to provide the framework for the exercise of research on the human genome" (Article 15);
- to facilitate "the establishment of independent, multidisciplinary and pluralist ethics committees" (Article 16);
- "to promote the practice of solidarity towards individuals, families and population groups who are particularly vulnerable" to genetic diseases (Article 17);
- to foster "the international dissemination of scientific knowledge concerning the human genome" (Article 18);
- to adopt measures enabling developing countries to carry out genetic research and to benefit from the achievements of such research, so that their use "can be to the benefit of all" (Article 19);
- to promote "the education in bioethics at all levels" (Article 20);
- to facilitate an open discussion on these issues, ensuring the free expression of various socio-cultural, religious and philosophical opinions (Article 21).

28. Ernst Benda, interview in *Frankfurter Rundschau*, November 1, 1997. See also this interview in the UNESCO Press Review on the Universal Declaration on the Human Genome and Human Rights, 1998.

10. The UNESCO International Bioethics Committee (IBC)

In Article 24, the UNESCO Declaration institutionalizes the International Bioethics Committee, which had worked on an ad hoc basis since its creation in 1993. In 1997, the 29/C Resolution 17, adopted at the 29th session of the UNESCO General Conference invited the Director General to convene, as soon as possible, an *ad hoc* working group to redefine the integration and tasks of the Committee. The working group, composed by representatives from several Member States, met in Paris between March 25 and 27 in 1998. As a result of that meeting, the working group drew up the Draft Statutes of the IBC, which were then adopted by the UNESCO Executive Board.

It is interesting to note that the IBC is the only body at the global level that engages in ethical reflections on research in biology and genetics. According to Article 24, the IBC is responsible for following up the principles set out in the Declaration and for "the further examination of issues raised by their applications and by the evolution of the technologies in question." To accomplish this task, the IBC is entitled to make recommendations, to organize appropriate consultations with parties concerned and to give advice concerning the follow up of the Declaration, "in particular regarding the identification of practices that can be contrary to human dignity."

Since the approval of its Statutes in 1998, the IBC consists of 36 experts from different countries and various disciplines, such as medicine, genetics, law, and philosophy, who are appointed by the Director General. The IBC members are in a peculiar situation because they are proposed by the States they come from, but they do not officially represent their governments and therefore, they do not engage them with their positions. Once appointed, the IBC members act independently and on their personal capacity and conscience. However, the Committee's composition is based on the assumption that its members in some way represent the socio-cultural values of the countries they come from. Precisely for this reason the Statutes provide that the Director General, when making his choice, shall take into account "cultural diversity" and "balanced geographical representation" of the IBC members (Article 3).

The IBC members meet once a year for a few days. Nevertheless, the main substantial activity is developed through various Working Groups, which deal with specific issues and elaborate drafts reports, opinions and recommendations, to be then submitted to the plenary of the IBC. Among the last reports adopted by the IBC, the following can be mentioned: "Draft Report on the Solidarity and International Cooperation between Developed and Developing Countries concerning the Human Genome"; "Ethical Aspects of Human Embryonic Stem Cell Research"; "Follow-Up of the International Symposium on Ethics, Intellectual Property and Genomics"; "Draft Report on Collection, Treatment, Storage and Use of Genetic Data."

The IBC Statutes adopted in 1998 created a parallel body, the Intergovernmental Bioethics Committee (IGBC), which comprises 36 representatives of Member States, who are elected by UNESCO General Conference (Article 11). This body aims at reflecting in some way the position of the Member States on bioethical issues. The interest of States in being directly implicated in the discussion of bioethical issues is perfectly understandable, if one considers that, after all, UNESCO is an intergovernmental organization.

According to the Statutes, the main task of the IGBC consists in the examination of the opinions and recommendations of the IBC, particularly those relating to the follow-up of the Universal Declaration. The IBC and the Director-General are then informed of the IGBC's point of view so that the Director-General may transmit this information, along with the opinions and recommendations of the IBC, to the Member States, the Executive Board and the General Conference of UNESCO. The IGBC may also make proposals for appropriate action regarding the IBC's opinions and recommendations. The IGBC meets at least once every two years.[29]

11. Implementation of the Declaration

Although the main responsibility for the implementation of the Declaration lies with the States, the IBC has considered it necessary to suggest some concrete ways of action. For that purpose, it drew up the "Guidelines for the Implementation of the Universal Declaration on the Human Genome and Human Rights", which were approved by the Intergovernmental Bioethics Committee (IGBC) and endorsed on November 16, 1999 by the General Conference of UNESCO.[30]

The Guidelines take into account the fact that "the implementation of the Declaration is all the more urgent since scientific progress in genetics and biology is accelerating and both giving humankind hope and creating ethical dilemmas." On that basis, they "seek to identify not only the tasks devolving on the different actors in the implementation of the Declaration but also modalities of action for their achievement."

29. The IGBC met for the first time in Rabat, Morocco, from October 7 to 13 in 1999, and for the second time in Paris, from May 14 to 16 in 2001.

30. In 2002, the UNESCO evaluated both the results obtained through the Guidelines and the impact of the Declaration. This evaluation will be examined at a joint session of IBC and IGBC and will be submitted by the Director General in 2003 to the statutory bodies of the Organization along with any relevant recommendations.

Among the various objectives and activities proposed, the following can be mentioned:

1. The dissemination of the principles set forth in the Declaration. It supposes, in particular, the translation of the Declaration into the largest possible number of languages, as well as the organization of seminars, symposia and conferences on topics relating to the Declaration at the international and regional level.
2. The raising of consciousness, especially through education, about the importance of the ethical issues at stake in genetics and biology. For this purpose, the Guidelines suggest various possible activities: the drafting of a commentary on each of the articles of the Declaration; the publishing of books on the subject, both for a non-specialist public and for the professional groups concerned; the preparation of programs of education and training in bioethics at the secondary and university level, as well as for special programs for teachers and trainers; the preparation of information kits and audiovisual materials on specific subjects of bioethics; the organization of multimedia exhibitions especially designed for young people.
3. The promotion of an exchange of ideas and analyses regarding bioethical issues at the international and regional level, in order to identify practices that could be contrary to human dignity. In this respect, the Guidelines suggests, as the Declaration itself provides, the creation of independent, pluralist, and multidisciplinary ethics committees.
4. The encouragement of dialogue and public debates between industrialists, members of civil society, vulnerable groups, scientists and political leaders.
5. The support for freedom of research, especially in genetics and biology, and the promotion of scientific and cultural cooperation between the countries of the North and the South.
6. The preparation of examples of legislation and regulation that embody the principles set out in the Declaration.
7. The effective cooperation with other international organizations that also deal with the issues covered by the Declaration. For this aim, the Guidelines suggest the creation of an interagency committee within the United Nations system open to other intergovernmental organizations and responsible for the coordination of activities related to bioethics.

12. Conclusion

The UNESCO Declaration is the first global response to the ethical, social, and legal concerns raised by genetic advances. It represents a significant effort of the international community in the search for common legal standards in this fundamental area, where the interests of present and future generations are at stake. Probably the main originality of this instrument is the fact that, for the first time, the human genetic structure is regarded as a common heritage to be protected. In addition, various fundamental issues regarding human rights in the context of genetic developments are covered.

Nevertheless, the major task in the search for an international genetics law is still to be faced. First, because most of the principles included in the Declaration are still very general and need to be clarified through additional discussion and consultation with Member States. Second, because the Declaration is not legally binding and therefore, its principles largely depend on the States' initiative to become effective at the national level.

Who Owns Genetic Information?

David M. R. Townend[1]

1. Introduction

There are, undoubtedly, enormous opportunities for social and medical advances and for massive commercial benefit in the exploitation and use of genetic information. Equally, however, there is massive potential harm that accompanies this genetic revolution. There are clear possibilities of discrimination against individuals and the stratification of citizens across the world order on the basis of genetically unequal life chances: life chances that individuals have always had, but that no longer seem random in the light of this revolution in understanding and knowledge. So now is the time to ensure that equal protection is available for all citizens. This is especially true in the current period where raw genetic information is becoming increasingly available, but where the possibilities for medical treatment of the individual through that information are much more limited than may be the case in the future. The possible harm from the use of this largely unknown commodity of raw information is now therefore at its most dangerous.

Given the potential harm, regulation of the control and use of genetic information is imperative. One would expect that, as with other forms of information, property rights would form the basis of the regulations over genetic information. This is not the case. As with the law in the United Kingdom on body parts,[2] there is a traditional reluctance to create rights of ownership over genetic information for the subjects of that information. Rather, the traditional protections have come from privacy and anti-discrimination law. What property rights there are, have been almost exclusively[3] in patents granted for the invention of products through the use of genetic information. This chapter, in seeking to identify the property in genetic

1. Acknowledgements: I am grateful to Professors Deryck Beyleveld, Roger Brownsword and John Merrills, and to Mr. Mark Taylor and Ms. Jessica Wright, for discussions on aspects of this paper, and to the participants in the CEU workshop.

2. See John K. Mason, R. Alexander McCall Smith, and Graeme T. Laurie (1999) *Law and Medical Ethics*, Fifth edition (London: Butterworths), Chapter 20.

3. For an exception, see the 1995 Oregon Senate Bill 276.

information, first explores the traditional regulation of genetic information, concluding that genetic information is not 'owned' in its own right and that its protection is not adequate. In the second part of the chapter, the question of "who should own genetic information?" is examined through the property theory of Professors Beyleveld and Brownsword. From that starting point a property right in genetic information is established and its implications for the protection of the subject (the 'owner', it will be argued) of genetic information is explored.

2. Can Information Be the Subject of Property Rights?

There is, underpinning the question 'Who owns genetic information?', an implied question: 'Can information itself be the subject of a property right?' This should be briefly answered before considering the present legal regulation of genetic information. That facts are commodities is very clear. However, a claim to a property right includes some form of exclusive possession[4] and the justification for such a right hangs on the need for marshaling the exclusive nature of the property through morally justifiable monopolies for particular individuals against others about things. This need for a property right stems largely from issues of social control because of scarcity of the resource, where access to the thing is limited and such access damages or extinguishes the commodity. Information as a subject of exclusive possession because of the need for social control of scarce resources would appear difficult to assert as any number of individuals can enjoy possession of the information at the same time without damaging or exhausting the commodity. This would be a standard argument against the granting of property rights in information or in intellectual property. However, such an argument misses another key function of property: the reward and catalyst for creation. This is central to the justifications for the creation of property rights presented by Locke, Nozick, and Mill.[5] In granting the monopoly over property, the State grants monopoly rights over the exclusive part of information—that is

4. See, for example, the claims analyzed as "substantive rights" in Deryck Beyleveld and Roger Brownsword (2001) *Human Dignity in Bioethics and Biolaw* (Oxford and New York: Oxford University Press), pp.173–175.

5. Locke justifies the property monopoly as resulting from the individual's ownership of the self expressed through labor creating the thing; Nozick extends Locke's justification to allow an individual's claim only for their additional contribution to the state of the art; Mill justifies a property monopoly as the most effective balance of competing claims to control of the thing to produce the greatest benefit. See Edwin C. Hettinger (1989) "Justifying Intellectual Property," *Philosophy and Public Affairs*, 18(1): 31–52.

its controlled release, which has a scarcity value. Thus, while there is some confusion as to the extent of the property in information,[6] there is equally a clear body of cases that establishes the principle that information can be property.[7] From this, genetic information is clearly in the realm of things that could be the subject of property rights, and so the question can proceed as to who owns that property.

3. How Is Genetic Information Protected in Law at the Present Time?

There is a tendency to use the language of property already in this regard. One hears comments about '*their* genetic information', and the like, implying ownership. When the television producer Steve Bing's genetic information was taken from the dental floss he had thrown away in his rubbish bin and was used as evidence in another couple's child maintenance suit, the UK Human Genetics Commissioner, Lady Kennedy, was reported to talk in terms of criminalizing the theft of genetic information.[8] English law has had difficulty with the concept of theft of information,[9] as has already been mentioned; the language from this prominent lawyer, however, has clear property tones. Despite this use of language, genetic information is not regulated currently though property law, with the very specific and limited exception of patents gained over inventions using genetic information.[10] Further, it is very difficult to answer the general question 'Who owns genetic information?' because the control and use of that information is not regulated through property rights and the ownership of

6. See in English law, for example, *Oxford v. Moss* (1978) 68 Cr App Rep 183, where the theft of the information contained in examination papers was not recognized. See also The Law Commission (1998) *Legislating the Criminal Code: Misuse of Trade Secrets*, the Law Commission Consultation Paper No. 150 (London: The Law Commission), available at http://www.lawcom.gov.uk/351.htm, last accessed on March 2, 2003.

7. See, for example, *Faccenda Chicken v. Fowler* (1987) 1 Ch 117. For a detailed exposition of the law in this area, see Lionel Bently and Brad Sherman (2001) *Intellectual Property Law* (Oxford: Oxford University Press), pp.923ff.

8. See Martin Bentham and Lorraine Fraser (2002) "Move to Outlaw Secret DNA Testing by Fathers," *The Daily Telegraph*, May 19, 2002, available at http://www.telegraph.co.uk/news/main.jhtml?xml=/news/2002/05/19/ndna19.xml, last accessed on March 2, 2003. The case coincided with the publication of the Human Genetics Commission (2002) *Inside Information: Balancing Interests in the Use of Personal Genetic Data* (London: Human Genetics Commission), see at http://www.hgc.gov.uk/insideinformation/index.htm.

9. See footnote 6.

10. That these can encompass ownership of genetic information through genetic material will be explained later in the chapter.

the information is left outside the law of property. This is seen when one examines the regulation of genetic information by its use. The uses can be grouped as follows:

1. Private Benefits in the Private Interest;
2. Public Benefits in the Public Interest;
3. Commercial Benefits in the Private (or arguably Public) Interest.

The most interesting legal position relates to the first of these uses: private benefits in the private interest, where one would assume a strong property right for the individual subject of genetic information. The users of genetic information under this head are the individual subjects of the genetic information—those about whom the genetic information relates. This use then concerns individual identity and its protection and exploitation, for example, in medical diagnosis and treatment. There is a private benefit in the genetic information (as the information is personal to the claimant) and is based on the legitimate private interest of the claimant. The protection here, however, is afforded traditionally not by property rights but through two other human rights: the right to privacy and the right against discrimination.[11]

The second group of public benefits in the public interest occurs, for example, in the uses of genetic data for research, for issues of public health, and for civil and criminal forensic uses. Here use is said to be in the public interest and for the public benefit; for example, through the sharing of research results or the protection of the population either against epidemic or against crime. Further duties, for example in relation to civil genetic testing in determining paternity, could arguably also be in the public interest in maintaining civil society. The crucial element in this intervention over the privacy and anti-discrimination rights of the individual is that the authority to make the intervention comes with an appeal to the 'public interest'. This is extraordinarily difficult to define and makes the rights of the individual particularly vulnerable to manipulation by the mass, and from other interests masquerading as the public interest. In the United Kingdom, Sir John Donaldson, the Master of the Rolls, gave the following famous description of the issue:

> The media [newspapers, radio and television] are an essential foundation of any democracy. In exposing crime, anti-social behaviour and hypocrisy and

11. For example, Articles 8 and 14 of the Convention for the Protection of Human Rights and Fundamental Freedoms (hereinafter: European Convention on Human Rights). There is a human right to property established in Article 17 of the Universal Declaration of Human Rights and in Article 1 of the First Protocol of the European Convention on Human Rights. However this is recognition of one's property and the right to own property, it does not of itself prescribe the limits and types of property. As genetic information is not seen traditionally as property, the human right to property protection does not create a property interest in genetic information.

in campaigning for reform and propagating the views of minorities, they perform an invaluable function. However, they are peculiarly vulnerable to the error of confusing the public interest with their own interest.[12]

The right to public interest intervention is perhaps justified by utilitarian principles of maximizing benefit for the many at the expense of the few. Alternatively, there are community property theories[13] that could equally be used to justify the ownership of genetic information not in the individual subject but in the general community. While Utilitarianism is very popular in the minds of legislators and the judiciary in determining matters of the public interest, community property theories have proved, beyond the realm of common land, to be very unpopular.

The third grouping of claims, commercial benefits in the private interest, is more problematic, but contains the only truly established area of property ownership over genetic information. Here the claimants derive commercial benefit from the exploitation of property in products and processes derived from genetic information through the patent system. The link is not direct, as will be shown, as the invention over which the patent is granted includes additional value beyond the simple genetic information. This is in most legal systems the gateway for property claims relating to genetic information as it fits most conventionally with the activity of commodifying information as property generally. Here the justification for this private commercial benefit is largely private interest: that the patent developer is rewarded for their inventive activity. However, a further veneer of public interest on the creation of a private monopoly is often argued through the claim that the invention afforded through the reward of the patent advances society.[14]

In this third group of commercial activity there are three other possible claimants for a use *as of right* of genetic information (which implies a control and use right over the information and, therefore, elements of a property right). These interests are claimed by the insurance industry, by employers, and most controversially by the entertainment industry. The insurance industry claims that it can legitimately demand genetic information known to the client for risk assessment when forming insurance contracts. Likewise, employers increasingly claim a right to know genetic information concerning subjects in their workforce, ostensibly for making health and safety judgments in the interests of the employees, but also potentially for making

12. *Francome and another v. Mirror Group Newspapers Ltd and others* (1984) 2 All ER 408, at p.413.

13. See Crawford B. Macpherson (1975) "Capitalism and the Changing Concept of Property," in Eugene Kamenka and R.S. Neale (eds.) *Feudalism, Capitalism and Beyond* (London: Edward Arnold), pp.104–124.

14. Again, the justifications are found in Locke, Nozick and Mill, see Hettinger, *op. cit.*

decisions generally against the employees' interests. Again, the treatment of Steve Bing shows the potential interest in genetic information in the news and entertainment industries. The step to exposure of similar circumstances by the media themselves 'in the public interest' could easily be a future development, and here the appeal to the public interest perhaps veils the media's commercial interests.[15]

4. Does the Regime of Anti-Discrimination and Privacy Rights and Patents Offer Adequate Protection to the Subjects of Genetic Information?

One route of protection used, then, comes from *rights against discrimination.*[16] Article 2 of the Universal Declaration of Human Rights requires that the other rights and freedoms of the Declaration should be enjoyed without discrimination. This is taken into Article 14 of the European Convention on Human Rights that establishes a similar right to enjoy the other Convention rights without discrimination. This, in the case of genetic information, could give a right to the subject of the information against discriminatory treatment by a public body because of their genetic status through an appeal to the Article 8 on right to privacy. Protocol No. 12 to the European Convention on Human Rights[17] has recently been negotiated. This protocol, if accepted by a State, requires a legal obligation on the part of that State not to discriminate. It remains to be seen how many States will sign this obligation. Article 26 of the International Covenant on Civil and Political Rights gives a stronger right against discrimination. The weakness in the ICCPR right is that the direct access to a hearing in individual cases is available only where a State has accepted to be bound by the optional protocol giving a right to that hearing. Where the optional protocol is not signed, protection is afforded through the reporting process of the Human Rights Committee. Further, in all these cases the definition of 'discrimination' is problematic.

15. Underlining Sir John Donaldson's point, see footnote 12 above.
16. See Conor A. Gearty (1999) "The Internal and External 'Other' in the Union Legal Order: Racism, Religious Intolerance and Xenophobia in Europe," in Philip Alston, Mara Bustelo, and James Heenan (eds.) *The EU and Human Rights* (Oxford: Oxford University Press), pp.327–356. Specific anti-discrimination legislation relating to genetic status is increasingly used in the US with difficulties of defining 'discrimination' and 'genetic information'. I am particularly grateful to Professor John Merrills and Ms Jessica Wright respectively for their guidance in these two areas.
17. Protocol No. 12 to the Convention for the Protection of Human Rights and Fundamental Freedoms (ETS-177); signed in Rome, November 4, 2000. For the official text, see http://conventions.coe.int/treaty/en/Treaties/Html/177.htm.

The traditional route for protection of genetic information is *privacy*. While the route concerns the control of information, the respect afforded to information through this route is not based in a property construction of the right (although that is not precluded in privacy arguments[18]): rather it is based upon a human rights argument. Privacy rights in relation to genetic information are generally protected by human rights conventions[19] and specifically protected through measures such as data protection law. Whether or not this is effective is in issue. Arguably, in relation to the use of genetic information, there are fundamental safeguards in data protection law, for example in European law, that must be in place to require the notification of the purpose of the collection and use of the information, and the consent to the use by the data subject. There is some reluctance, however, to accept that genetic information is covered by data protection regimes.

There are a number of concerns about the effectiveness of privacy protection. These are illustrated here through an analysis of some problems in the European Union and the United Kingdom. In Europe, the Directive on Data Protection[20] requires a coherent and adequate regime of protection for data subjects whose personal information is used in medical research. Undoubtedly, in some jurisdictions data protection certainly does work.[21] However, in other jurisdictions there will clearly be a failure to implement the Directive adequately. Indeed, in the UK one can have serious concerns about the protection of genetic information that is offered through the Data Protection Act of 1998. For example, at the outset of the collection of genetic information, if the information is intended solely or in part to be used in research, then the Directive clearly indicates that the data subject *must* be informed of that intention.[22] In implementing this, the UK has diluted the requirement inserting 'if practicable' into the Act. The Directive itself allows for the use of information that has been collected for one purpose (for example for treatment or a first specific

18. Although, to get to this argument one is rather close to the argument that all rights are property, as Narveson suggests and Beyleveld and Brownsword discuss in *Human Dignity in Bioethics and Biolaw*, *op. cit.*, p.173; removing the usefulness of the concept of property relationships as distinct from other relationships.

19. For example, Article 8 of the European Convention on Human Rights.

20. Directive 95/46/EC of the European Parliament and of the Council of 24 October 1995 on the protection of individuals with regard to the processing of personal data and on the free movement of such data, *Official Journal L 281, 23.11.1995*, pp.31–50.

21. Currently this is being tested in a EC Fifth Framework funded *'Quality of Life' Concerted Action* project evaluating the implementation of the Directive in relation to medical research across all the Member States, Norway (as a representative of the European Economic Area), and all the accession countries. The project, PRIVIREAL, can be found at www.privireal.org.

22. Article 10 of Directive 95/46/EC.

research purpose) at a later date to be used for a different purpose (for example an unrelated research project) not initially disclosed to the data subject. However, it is very possible that the data was collected with consent to specific purposes and then anonymized. Denying a use for a new purpose once the data subjects can no longer be identified is judged too harsh and for the Directive one must inform the data subject if it is practicable to do so.[23] This may offend the data subject's sensibilities and is, arguably, a bad enough invasion of privacy. The UK position, in extending the Directive's treatment of a later purpose to the situation when one is collecting data at the outset from the data subject, is much more worrying.

The use of data without informed consent is another example of weakness in the UK law. By Article 8 of the European Convention on Human Rights, the right to privacy—that "[e]veryone has the right to respect for his private and family life, his home and his correspondence"—in paragraph one is only supervened by the conditions outlined in paragraph two:

> There shall be no interference by a public authority with the exercise of this right except such as is in accordance with the law and is necessary in a democratic society in the interests of national security, public safety or the economic well-being of the country, for the prevention of disorder or crime, for the protection of health and morals, or for the protection of the rights and freedoms of others.

These supervening circumstances, while establishing that privacy is not an absolute right, indicate relatively extreme circumstances to removing the right. One removal of the right, in relation to the use of personal (for example genetic) information about an individual, would be using the information without the consent of the data subject. Article 8(2) must require relatively extreme circumstances before the data subject's consent is not required for the use of genetic information. This has considerable impact on the Data Protection Directive and the UK's Data Protection Act of 1998. Under the 1998 Act, the requirements listed in the Article 8 of the ECHR are not spelt out. However, the UK has been a signatory to the Convention since its creation[24] and implemented it directly into UK law in the Human Rights Act of 1998. Thus, whereas the Data Protection Act of 1998 suggests that, when considering the particular justification for the use of personal data about an individual, consent by that individual to the use is simply equal to a number of other justifica-

23. Article 11 of Directive 95/46/EC.

24. See John Wadham and Helen Mountfield (1999) *Blackstone's Guide to the Human Rights Act 1998* (London: Blackstone Press Limited).

tions on a list,[25] reading the Data Protection legislation through the Human Rights Act, Article 8 requires a different interpretation. Then, following Article 8, Schedule 2 and Schedule 3 conditions must, in each case, be interpreted such that the first on the list, *consent*, is the ordinary requirement and that the other justifications on the list should only be used when consent would not be required in line with the Article 8(2) exceptions in the European Convention on Human Rights.

It remains to be seen whether this interpretation will be followed in English law. One of the justifications is "circumstances specified by order of the Secretary of State."[26]

25. The lists appear in Schedules 2 and 3 of the Data Protection Act of 1998:

"Schedule 2

(1) The data subject has consented to the processing.

(2) The processing is necessary in performing a contract to which the data subject is party or in taking steps to create such a contract.

(3) Beyond contractual obligations, the data controller is compelled to process the data because of another legal obligation.

(4) Processing the data protects the 'vital interests of the data subject'.

(5) The administration of justice, government or public body, or duties conferred by statute, require the data to be processed.

(6) The data controller or third party to whom the data is disclosed require the processing of the data for their legitimate purposes which are not prejudicial to the 'rights and freedoms' of the data subject.

Schedule 3

(1) The explicit consent of the data subject to the processing has been given.

(2) The data controller has a legal right to process the data through employment.

(3) The processing of the data is in the data subject's vital interests and consent cannot be gained, or processing is in a third party's vital interests and the data subject unreasonably withholds his consent.

(4) The processing is carried out by a non-profit making organisation, or a body that 'exists for political, philosophical, religious or trade-union purposes'.

(5) The data subject has taken deliberate steps which have resulted in the data becoming public information.

(6) The processing is necessary for work relating to legal proceedings, legal advice, or defending legal rights.

(7) The administration of justice or government, or duties conferred by statute, require the data to be processed.

(8) The processing is necessary for medical purposes and is conducted by a health professional or other person who owes the same duty of confidentiality.

(9) The processing enables a furthering of racial equality and respects the freedoms and rights of the data subjects.

(10) The personal data are processed in circumstances specified in an order made by the Secretary of State for the purpose of this paragraph."

26. Schedule 3, paragraph 10.

This has been exercised in relation to medical research.[27] Here the new law indicates that use of personal data (including sensitive personal data, which would include genetic information) can be used without the subject's consent without further justification than that it is used for medical research. Clearly to satisfy Article 8(2) of the ECHR this requires a particular reading of the public benefit of research in relation to the breach of the individual's rights for the advancing of medical research. It also requires a particular view of the difficulty in obtaining consent from the individual to the use of the information when obtaining other consents from the subject.[28] It also requires a particular view of the science required to undertake medical research. Here, there seems to be an assumption, especially in the cancer research communities, that unless samples are automatically taken and made available to researchers, the research is endangered.[29]

Thus, privacy protection for the subject of genetic information can be undermined in law. It can also be undermined in practice in the use and protection of genetic information in medical research if research ethics committees do not make decisions about the legality of proposed research protocols. To suggest that individual data subjects have the protection of the courts to safeguard their legal privacy rights is to overlook the fact that medical research is often undertaken without notice to the individual data subjects. Thus, if the data subject does not know that a breach of his or her privacy is possible, he or she will not be able to protect themselves in the courts. The research ethics committees are created to give protection to the data subjects precisely because of their vulnerability and it is research ethics committees that must safeguard the data protection and other rights of the subjects of medical research.

In conclusion then, privacy and anti-discrimination rights arguably do not give sufficient protection to the subjects of genetic information, and they certainly do not give 'ownership' rights.

27. Health and Social Care Act of 2001 and Health Services (Control of Patient Information) Regulations (SI 2002 No. 1438).

28. The data subject will, in the majority of cases, be a patient and therefore will be giving consent to treatment at the point of giving the information.

29. See for example the report on the impact of the Data Protection Act of 1998 on medical research in *The Times*, May 15, 2001. This prompted Lord Campbell of Croy to ask a question to the Lord Chancellor in the House of Lords asking for assurance that the requirements of the 1998 Act did not "restrict or inhibit medical research likely to save lives" and whether the legislation can be "misread" to support the assertion that "the law is killing people" (Lord Campbell reporting the assertion in the newspaper article): House of Lords Debate, 16 July 2001, Hansard, Column 1261 (London: HMSO).

5. What Property Rights are Granted Through Patents? What Is the Impact of Such Rights on the Subjects of Genetic Information?

Genetic information is made the subject of property rights through patents, but this would seem to be contrary to the agreed position in respect of the protection of genetic information through the European Biotechnology Directive.[30] The anomaly has occurred through the application of patent law's own internal logic.

Patents are traditionally not granted for discovery. Rather, they are traditionally granted in line with a Lockean principle of added value through the human labor of invention. Something that did not exist before in nature is invented and the inventor is rewarded for his contribution (and the disclosure of the underpinning information into the public domain) by the grant of a patent. There is a second line of justification that is alternatively used to justify the creation of patent monopolies through utilitarian argument: the comparative detriment to society in granting a monopoly over the exploitation of information by one individual is outweighed by the greater social benefit of making the invention available in society and disclosing the information behind the invention to competitors.

So, the process of isolation of things found in the natural world can, in these justifications, be the subject of patents. However, it follows that the natural things themselves cannot be the subject of patents (and therefore property rights), being discoveries of natural things. This is reflected in the European Biotechnology Directive: "[the] human body at the various forms of its formation and development, and the simple discovery of one of its elements, including the sequence or partial sequence of a gene, cannot constitute a patentable discovery."[31] However, this is not the position adopted in patent law. It is not the case that there are only patents available for the process of isolation. Patent law has accepted the argument that by isolating a naturally occurring substance from its natural surroundings it is no longer a naturally occurring substance since it is not natural in its novel isolation. Further, because that isolation results from an action of human intervention requiring effort and skill, the resulting substance is not discovered and is therefore patentable. It is, in this argument, 'invented' in its isolation. However, this has the effect of rewarding the actual invention twice—once for the invention of the process of isolation and a second time for the product of operating that process.

30. Directive 98/44/EC of the European Parliament and of the Council of 6 July 1998 on the Legal Protection of Biotechnological Inventions, *Official Journal of the European Communities L213, 30.07.1998*, pp.13–21. For the full text, see the Appendix of this book, pp.323–337.
31. Article 5(1) of Directive 98/44/EC.

When this patent reasoning is applied to human tissue and genetic material, raw data, once isolated, becomes property, and body parts are owned where traditionally they were considered as not capable of being property. However, the owner of this property is not the subject of the genetic material or the donor of the tissue. Through the added-value thesis it is the 'inventor' of the isolated material who owns the property. The John Moore case[32] shows very clearly that ownership is not related to the human from whom the substance was taken: the property in the substance is created in the act of isolating the substance. In the case, Moore's tissue was taken in the course of medical treatment. In analyzing the tissue, it was found that it had remarkable qualities and the medical team treating Moore was able to patent their work and reap considerable financial benefits from their research. Moore sought to share in that financial benefit as 'owner' of the tissue. However, his property-based claim failed. This relates to body parts, but equally it extends from the tissue into the genetic information that is contained in the tissue. Those conducting research can, through an added-value argument, create property in genetic information; those from whom the genetic information was initially taken cannot enjoy such property rights.

The traditional and current position then in answer to 'Who owns genetic information?' should be clear: no one should 'own' genetic information. Individual subjects of genetic information should not own genetic information and, through the European Biotechnology Directive, neither should they be the object of ownership in patents. That patent law has found a route around the Directive and has created property rights over genetic information perhaps shows the commercial power and influence that is exerted over patent policy development rather than the development of a coherent policy to extend property rights over genetic information. The position is therefore that individuals gain some protection over their genetic information through human rights protections, whereas certain commercial interests gain property interests in their added-value inventions over genetic information. This poses first a policy question: is it appropriate to exclude the subjects of genetic information from ownership of 'their' genetic information? This would be a theoretical question if it were not for the major practical problem that is also posed. For this current regime to offer adequate protection to the individual subject of genetic information, human rights and property rights must have equal status, and they do not. Arguably, courts and legislators, with their roots firmly in economic interests, are much more comfortable in free market economies dealing with property rights than with creating hard individually protective and powerful human rights to compete with specific property rights. Institutionally, human rights are not designed to serve such a confrontational

32. *Moore v. Regents of University of California* (1990) 793 P 2d 479 and (1991) 111 S Ct 1388.

purpose against specific property rights. To protect genetic information subjects adequately, the question remains, 'Who should own genetic information?' The work of Professors Beyleveld and Brownsword addresses this point squarely first by identifying the formulation of traditional property rights and then by reformulating that right on the basis of *human dignity*.[33]

6. Creating a Property Right in Genetic Information

Clearly, as has been seen above, in some situations 'ownership' is recognized over genetic information, but there is a reluctance to extend the right of ownership in body parts and in genetic information to the individual subject. However, this is not a position that is necessary. Indeed, when one analyses, as Beyleveld and Brownsword do, the characteristics of a property right, then the opposite position may be required for human dignity to be observed. The problem, for Beyleveld and Brownsword, is in the characteristics of property that are usually observed in the formation of the concept of property.

Beyleveld and Brownsword point to the concepts of property presented by a number of writers. Each writer appeals to 'control' as the defining concept of property, and each indicates a number of essential elements of control necessary to establish a claim of 'property' in a certain thing. At one extreme, they point to Honoré,[34] who requires, in making a claim for 'property' in a thing, eleven separate rights of control.[35] Clearly, examining this list in detail, Honoré is setting a very high level of requirements that challenge some of the commonly understood 'property' rights, for example in requiring no fixed term over a property right[36] property in leases and fixed

33. Beyleveld and Brownsword, *Human Dignity in Bioethics and Biolaw*, *op. cit.*, Chapter 8: "Living with Dignity I: Ownership and Commodification of the Human Body and Its Parts," particularly pp.173–179, "The Concept of Property."

34. A. M. Honoré (1961) "Ownership," in A. G. Guest (ed.) *Oxford Essays in Jurisprudence* (Oxford: Clarendon Press), pp.107–147.

35. Honoré's conditions of ownership: "rights (1) to have exclusive physical control over an object; (2) to personally enjoy and use it; (3) to decide how and by whom it shall be used; (4) to derive income from it; (5) to alienate it and consume, waste, or destroy it; (6) to immunity from expropriation; (7) to transfer rights over it to others; (8) to no fixed term being set on the duration of the rights while the holder lives; all of which are (9) subject to the limitation that use harmful to others should not be made of the object; and (10) to liability to removal to pay debts or as compensation or penalty. Finally it is a characteristic of property (11) that, on abandonment, characteristics (1) to (10) will be invested in another." See Beyleveld and Brownsword, *op. cit.*, p.173.

36. Condition 8, see footnote 35.

term security rights is challenged. Beyleveld and Brownsword also point to the less extensive set of conditions for ownership required by Gaus.[37] For Gaus, the necessary rights conditional for making a property claim are the rights of 'use', 'exclusion', 'transfer', and 'compensation'.[38]

7. Is There a Better Understanding of Property Rights?

Beyleveld and Brownsword find that while there are considerable differences between the various characteristics, there is one element in common. It is that these views are all reduced to some description of the extent of 'control' necessary before 'property' in a thing can be observed: in that description they are "substantive claims."[39] Beyleveld and Brownsword indicate that there is a second requirement in conceptualizing property, that there is a "functional claim" being made:

> Specification along this dimension directs our attention to the fact that claims to property are not merely claims to some bundle of rights of control, but function to justify rights of control. "Why should I be able to exert control over P?" or even more pointedly, "Why do I have a right to exclusive use of P?" can be answered by saying, "Because P is my property!"[40]

Beyleveld and Brownsword go on to show that making a claim to a right based upon 'property' in the thing is fundamentally different to making that claim on the basis of other rights, 'personal integrity', 'freedom of religion', and the like. In the latter cases one makes the rights claim in the face of a challenge to one's 'personal integrity' or 'freedom of religion'. Beyleveld and Brownsword's first major contribution in the concept of property is to show that when one makes the claim to the right on the basis of 'property' in the thing, then simply making this statement means no any further justification of one's claim is required, and places the onus upon the other party to show evidence that the claim is falsely made or that the right should be suspended for another valid reason[41]: "'Because it's mine!' functionally means 'It is so related to me that I do not have to say why I should be able to use it or why I may exclude others from using it!'"[42]

37. Gerald F. Gaus (1994) "Property Rights and Freedom," in Ellen Frankel Paul, Fred D. Miller Jr., and Jeffrey Paul (eds.) *Property Rights* (Cambridge: Cambridge University Press), pp.209–240.
38. See Beyleveld and Brownsword, *op. cit.*, p.175.
39. *Ibid.*, p.176.
40. *Ibid.*, p.176.
41. *Ibid.*, pp.177–178.
42. *Ibid.*, p.178.

Having shown that claim to property is "preclusionary," Beyleveld and Brownsword then show that it is "rule-preclusionary" because the functional claim to property is also a moral claim to a relationship that is necessary to support primary rights.[43] Property is a secondary right, relating to the individual's insuperable primary right, the right to agency,[44] and as such the right can both conflict with another's primary right to agency and can be subordinated to that superior claim.[45] They use this "rule-preclusionary" concept of property to show that it can well be argued that, as one's agency is fundamentally dependent upon one's body, one should be able to make a claim to 'property' in one's own body.[46]

8. What are the Implications for the Ownership of Genetic Information Generally?

Beyleveld and Brownsword's concept is a most exciting concept of property. In this conceptualization, 'property' can become a radical concept. Beyleveld and Brownsword use the theory to suggest that the sale of tissue and the like, as in the Moore case, does not necessarily pose a threat to morality. Indeed, to deny the right to property in one's body would pose a greater threat, indicating that the market would find an appropriate value for the tissue sample (which, without the added value of the research and medical expertise would be worthless in its discarded form to the owner).[47] This market development of the theory is one clear possibility. However, the conceptualization also opens a 'social welfare' challenge to the traditional industrial view that property is a primary right in itself.[48]

There is a very strong argument that Beyleveld and Brownsword's conceptualization requires a reappraisal of how property rights have been allocated globally and whether the assumptions of the market are morally justifiable. Because as a matter of morals all property rights are held in relation to the agency of individuals,[49] the

43. *Ibid.*, p.178.
44. Beyleveld and Brownsword have developed their work (especially in *Human Dignity in Bioethics and Biolaw*) from the seminal work of Professor Gewirth: Alan Gewirth (1978) *Reason and Morality* (Chicago: Chicago University Press).
45. Beyleveld and Brownsword, *op. cit.*, p.178.
46. *Ibid.*, pp.179–194.
47. *Ibid.*, p.XXX.
48. See the discussions of property in Charles A. Reich (1964) "The New Property," *Yale Law Journal*, 73(5): 733–787; cf. for example, Macpherson, "Capitalism and the Changing Concept of Property," *op.cit.*
49. See Beyleveld and Brownsword, *op.cit.*, p.178.

issue is brought into sharp focus by the conceptualization, asking why institutions are maintained allowing property rights to prevail in the face of clear and urgent primary needs of agents: the conceptualization reminds us that it is not the starving who must make the case for the property that is not needed by those who have plenty, but that it is for the rich to justify withholding their surplus from the poor.[50] The key to this argument is that the claim is to the surplus. All agents have rights to property in things to the extent that they are needed for their own agency. Where one agent has surplus and another's primary needs are challenged, then the former must surrender his or her property to meet the needs of the latter. When resources are so scarce that demand universally outstrips supply, then the sharing of things must be on an equal basis. Beyond that, all agents also have rights in things to the extent that the ownership of a particular thing by one agent does not challenge the agency of another. Thus, when supply outweighs demand between particular agents, then the property rights of the one do not challenge the primary rights of the other: an individual has a right to property because of its necessary relationship with his or her primary rights unless enforcing the right places the primary rights of any other individual in jeopardy.[51]

9. What are the Particular Implications for Owning Genetic Information?

The case for the ownership of genetic information is made particularly interesting when one applies Beyleveld and Brownsword's concept of property to this commodity. One can see how the relationship between an individual's body parts and his or her self is fundamental to that individual's primary rights, and therefore property in one's body parts must be acknowledged. The extension of that argument to one's genetic information is logical and should not present difficulty. However, making the statement that one must have ownership of one's genetic information is not as easy as making the claim for ownership of one's arm or one's tissue sample, even though all such items have one's genetic information written through them. The problem is one of sameness.

50. For a further development of this work, see David Townend, "Social Welfare Challenges to Intellectual Property Traditions," forthcoming.
51. Equally, no one can place the property owner's primary rights in jeopardy by bringing a claim to secure their own primary rights in competition with the property owner's primary rights. This leads to the conclusion either that individuals have a right to safeguard their own primary rights, as the respect of another's agency only flows from a necessary respect for one's own agency first (which has implications for social control), or that in competing primary rights situations, resource sharing is required.
52. Applying Beyleveld and Brownsword, *op. cit.*

Much is made of the unique nature of genetic information; for example, forensic science allows juries to be presented with astonishing odds of the strength of DNA evidence. And in that individuality is the very reason for the need for the 'property' relationship to extend between oneself and one's genetic information: one's individuality is an essential part of one's primary rights. However, the justification is one of protection from harm. One makes the claim to property to prevent harm against one's primary rights, and the secondary, property right that one can exercise is subordinated to the primary rights of others.[52] In that justification there are, arguably, two surprising conclusions: first, one cannot exploit one's genetic information, one can only use one's property rights in genetic information as a shield; and second, it may be possible for others to use one's genetic information without one's consent. These conclusions can be explored under two questions: 'Can one exploit one's genetic information property?' and 'Who can own and use genetic information?'

10. Can One Exploit One's Own Genetic Information Property?

The individuality of genetic information offers the potential for its commodification. However, the difficulty arises through the sameness of genetic information. As has been widely discussed in the literature, human genetic information is closely related to that of other species.[53] Further, when one considers the genetic information not only within the species but also specifically within a family, then the difficulty in the exploitation of the property in genetic information becomes immediately clear. If one has an identical twin, then one's genetic information is shared and identical, making disclosure of genetic information by the one a full disclosure of the genetic information of the other. This is rare but the same problem occurs in the commonplace. One's genetic information contains vast amounts of the genetic information of one's parents, one's siblings, and one's children. This makes the disclosure of one's own genetic information a disclosure of significant amounts of the genetic information of many other people. Indeed, in this respect we are barely individuals. To exploit the property in one's genetic information without the consent of one's own would be to remove their property in their genetic information. The problem that must prevent the exploitation by an individual of their genetic information is of defining 'one's own' for the purpose of consent.

53. For example, comparing three million letters of the chimpanzee genetic code with the human genome draft, there is only a 1.3 percent difference between the two.

It is not, therefore, that genetic information is incapable of being owned. It is clear that the individual subject of genetic information has the strongest claim to ownership of the information as such a right is essential, if they are to exercise their primary rights effectively. However, because that secondary right cannot be exploited without causing damage to the primary rights of others (and, indeed, to their secondary rights to the property in their own genetic information), the exploitation of the property in genetic information looks very much to be severely restricted, if not impossible. Put crudely, if I sell my genetic information, then I also sell much of my extended family's genetic information, thus I endanger their primary rights by the sale and therefore my action cannot be morally justified. The individual's property right in genetic information can therefore only be a protection—a *shield*. It is not property that one can exploit. What then is left for the use of genetic information?

11. Who Can Use and Own Genetic Information?

Applying Beyleveld and Brownsword's concept of property, ownership of genetic information clearly originates in the subject of that information. This is a radical change from the traditional position (explained at the beginning of this paper) that any ownership over genetic information comes largely through the patent system, and is granted to inventors for the skill and labor added to the raw information. Further, the use of genetic information in the traditional view is justified through various public interest and privacy balances. The question remains: if legislators accept the subject-based property protection argument advanced above, how useful will such a strengthening of the property right in genetic information be in protecting the individual subject of the genetic information? How will they benefit from owning their genetic information?

12. Applying a Subject-Based Property Protection of Genetic Information

First, in relation to patents created by the use of genetic information, the added value of scientific intervention can no longer be seen as the originating factor of property in genetic information. Arguably, the position is not the same as Beyleveld and Brownsword's outline for the exploitation of body parts. The issue could not be one of finding the price for the information on the open market because of the harm that would necessarily be occasioned upon other subjects with critically similar genetic information. There is, however, another route to the possible justification of property in genetic information patents. As has been shown, property in genetic information is a secondary right, subordinate not simply to the primary rights of the individual subject of the genetic information but also to the primary rights of all individuals. Therefore, if withholding the use of the property in some specific genetic information challenges the primary rights of another individual, without jeopardizing the rights of the owner, then such an action (of withholding the use of the information) cannot be justified. Where the patent is for products and processes that safeguard the primary rights of individuals, arguably it is not possible to withhold the necessary license to allow the use of one's genetic information. Here, the possible harm to the common-owner-group by the disclosure will be subordinated to the primary rights of the beneficiary of the product or process. This argument applies forcefully to require participation in medical research over one's genetic information generally.

Perhaps the most important contribution to be made by accepting that there is a property for the individual in his or her genetic information, even with the caveat that its exploitation is severely constrained by the property rights of others with similar genetic information, is in changing expectations in relation to other claims about privacy and discrimination. In both cases, the claims of commercial and other interests to rights to control, use or access an individual's genetic information are brought onto an equal level: all the claimants appeal to a property right. The construction of that right shows that the individual in whom the genetic information originates cannot have that right removed without their license (going beyond full and informed consent and reflecting their property in the information that is at issue) without an appeal to the jeopardy of another individual's primary rights. This sets the level of protection for the individual much higher than that which has been afforded when the protection has been through privacy or anti-discrimination rights.

This strengthening of the genetic information subject's rights through the recognition of his or her property in the information has a strong impact on that individual's rights in relation to claims for access to that information made by employers, the media, or insurance companies. In all three cases, the property concept requires that the commercial claimant, in seeking to require the owner of the genetic information to disclose that information against their will (i.e., require them to relinquish their property rights over the genetic data) must appeal to overriding harm to a third party's primary rights. In employment it is perhaps only possible to see such appeals arising in limited numbers of health and safety-at-work issues (although this situation is particularly hard to envisage). It is impossible to see how the entertainment media could ever be in a position to override the property right of the individual.

The insurance industry position is very difficult. Insurance companies have a long tradition of discrimination between individuals on the basis of actuarial assessments made from various facts required from the individual. This has extended into requiring known information owned by the data subject relating to sensitive personal information such as medical conditions—for example, the HIV and AIDS status of the applicant for insurance. This is arguably not irrelevant discrimination but it has a harmful effect on the individual. The information clearly belongs to the individual seeking to enter a contract with the insurance company. However, commercially the company can determine whether or not to enter into the contract and what information it will require before making the offer through setting the cost and terms of the policy. The impact of genetic testing upon risk assessment is extraordinary, especially at the present time when the interpretation and implications of genetic information are relatively unsophisticated. There are perhaps only two responses to the genetic impact on the insurance industry: the first is to move to a flat-rate insurance cover; the second is for the state to maintain the industry as it is, offering low-cost insurance to those who cannot be insured within the ordinary operation of insurance premiums because of their genetic status.

This leveling of life-chances would equally apply to the provision of education opportunities and other rights of citizenship that are challenged by the increased knowledge of the genetic information revolution. Strengthening the protection from discrimination for individuals by lifting the status of their relationship with their genetic information to that of a property right within the concept of property outlined above may help; however, such rights must be supplemented by the State. Why should States take on this leveling role? Because while medical science cannot yet offer a balancing for genetic disorders, society can and must maintain equal citizenship.

The Genetic Testing of Children

Sheila A. M. McLean

Advances in the understanding of the genetic basis of many diseases and conditions have so far failed to produce much in the way of cures or palliative therapy, but they have spawned increased interest and capacity in detection through individual genetic testing or population screening. Indeed, Stone and Stewart have said that, "[i]t seems that we are being subjected to a second wave of evangelistic screening fervor generated, at least in part, by dramatic advances in diagnostic technology."[1] Although not directly referring to genetic screening or testing, it is clear that their intuition applies equally in this arena, driven by the capacity to detect more and more genes which influence health (and even behavioral) characteristics.

It is, of course, incontrovertible that screening plays a significant role in public health strategy, and is a popular tool in the armory of epidemiology. Testing, on the other hand is generally the term used when the intention is a direct one-on-one diagnosis. Thus, it is possible to differentiate testing and screening, but there are also similarities between them which will be highlighted where relevant in this chapter. In any case, as the House of Commons Select Committee on Science and Technology pointed out, "[a] 'genetic test' may be either for diagnostic or screening purposes."[2]

Additionally, concerns about screening, such as the individual or public understanding of the results derived from it, are as significant in individual or familial testing as they are in population genetics. Indeed, arguably the private and the public dimensions cannot easily be separated. As the Nuffield Council on Bioethics said:

> A broad public understanding of the scientific basis of medical genetics is essential if informed public policy decisions are to be taken about the introduction of genetic screening programmes. Such programmes . . . have both an individual and a public dimension.[3]

1. D. Stone and S. Stewart (eds.) (1994) *Towards a Screening Strategy for Scotland* (Glasgow: Scottish Forum for Public Health Medicine), p.45.
2. The House of Commons Select Committee on Science and Technology (1995) *Human Genetics: The Science and Its Consequences*, Volume 41–1 (London: Her Majesty's Stationery Office).
3. Nuffield Council on Bioehtics (1993) *Genetic Screening: Ethical Issues* (London: Nuffield Council on Bioethics), p.75.

Both screening and testing—but perhaps more significantly, testing—have supporters and detractors. Gannon and Villiers suggest that:

> Genetic testing promises both benefits and problems. The potential benefits may include improved quality of life . . . At the same time there are risks involved in the activity, such as public health errors, and unfair commercial practices that could impede progress in the field.[4]

Generally speaking, where a therapy is available neither testing nor screening of a target population are usually perceived as being particularly controversial, although Hubbard and Wald caution that "[t]he mind-set behind genetic testing rests on a societal view of disabilities that should not go unchallenged."[5] And Hepburn says that "[t]he notion that early diagnosis and intervention is important therapeutically is almost an article of faith in medicine."[6] This early detection and intervention is the sustained and overt goal of genetic testing, unlike screening which is often anonymized and conducted primarily or solely for epidemiological reasons.

Medicine as a discipline is fundamentally interventionist and therefore genetic testing, in a sense, is roughly equivalent to other forms of diagnostic intervention. In other words, where there is a cure or successful treatment available then the rationale for the test or screening program is self-evident. However, as has already been noted, in the majority of conditions currently being tested for, or where testing is possible, there remains a significant shortfall between diagnostic and therapeutic capacity. It is also true to say that the implications of a genetic test may be less straightforward than the more standard medical diagnosis. As Hubbard and Wald has said:

> Genetic predictions, whether they involve testing or screening, are based on the assumption that there is a relatively straightforward relationship between genes and traits. However, genetic conditions involve a largely unpredictable interplay of many factors and processes.[7]

The combination of enthusiasm for screening and the uncertainties associated with its outcome, as well as the fact that many tests can reveal susceptibility but little more than that, have generated considerable interest in, even disquiet about, the use

4. Pippa Gannon and Charlotte Villiers (2000) "Regulation of Genetic Testing," in Laura McGregor, Tony Posser, and Charlotte Villiers (eds.) *Regulation and Markets Beyond 2000* (Aldershot: Ashgate), p.127.

5. Ruth Hubbard and Elijah Wald (1993) *Exploding the Gene Myth* (Boston: Beacon Press), p.28.

6. Elizabeth R. Hepburn (1996) "Genetic Testing and Early Diagnosis and Intervention: Boon or Burden?" *Journal of Medical Ethics*, 22(2): 105–113, p.105.

7. Hubbard and Wald, *op. cit.*, p.36.

of genetic testing. Yet, ". . . in modern Western societies it is often relatively simple to encourage people to seek diagnostic testing. Culturally, the ascendancy of science has engendered a growing desire to exert control over events and to reduce uncertainty."[8] Taken together, these observations compel us to inquire into a number of questions surrounding the use and value of genetic testing. As Szabo warns:

> Genetic testing performed today carries particularly high stakes for patients. The tests not only detect current conditions, but they can foresee future medical conditions. If the sensitive issues surrounding these tests are not properly handled, a host of ethical and social problems are likely.[9]

Of course, these issues arise concerning the testing of both adults and children, but the position of children merits separate consideration since there are some matters, derived from their social and legal situation, which merit special attention.

In this chapter, therefore, it is intended to ask three main questions while interrogating the issue of the genetic testing of children. The first question that must be answered is: why would we wish to do such testing? Secondly, it must be asked on whose authority any such testing can be carried out. And finally, we must ask whether or not there are limitations on the kinds of outcomes that we are seeking; in other words what should we test for?

1. Why Test?

Arguably, the discovery of a genetic condition where such a condition is incurable or untreatable serves no good and may cause harm. Of course, some genetic conditions can be treated or strategies developed to minimize or obviate their impact. However, no matter the circumstances, the current hype surrounding genetic discoveries may lead the individual concerned to have a different attitude to these, as opposed to other, medical problems. The Danish Council of Ethics, for example, has suggested that the results of genetic tests may lead the individual to revise his or her opinion of him or herself, may make them feel unwell even though otherwise asymptomatic, and they may be treated differently by others. As the Council says:

> Just as persons found through screening to have a particular gene or chromosome composition may happen to feel abnormal or outright ill . . . so may other react to the persons involved by giving them a wide berth. The detection of certain genetic traits can thus form the basis for branding certain persons

8. Hepburn, *op. cit.*, p.107.
9. Joan Szabo (2001) "Regulating Genetic Testing," *Physicians Financial News*, 19(2): 5, 8.

and groups among the population, with the possibility of discrimination proper as a result.[10]

Although these concerns are clearly not specific to children, it may be that they are more acute in a family situation, not least because when a test has been conducted on a child it is highly unlikely that other family members—particularly parents—would be unaware of the diagnosis. Any incentive or impetus to treat people differently on the basis of a genetic test result would likely be magnified in that situation. Thus, despite the values generally ascribed to autonomy and confidentiality, testing of one member of a family seems *de facto* likely to result in the information being shared. If a child or young person agrees to sharing the information, then arguably this is non-controversial. However, young children are unlikely to be able to provide such an undertaking; consequently their capacity to manage such highly sensitive information is removed or reduced for all time. This is, of course, particularly acute where the testing is conducted for a condition which is unlikely to arise until later in life, of which the child currently shows no symptoms, and where no therapy is available. Although some of the reasons for testing (outlined in more depth below) might apply in this situation, the Advisory Committee on Genetic Testing (ACGT, UK)[11] concluded that "[p]resymptomatic genetic testing of young children for disorders not currently influenced by therapy, and where onset is normally in adult life, is not recommended."[12]

In 1994, a working party of the UK Clinical Genetics Society (CGS) traced the arguments concerning the genetic testing of children.[13] Broadly, they concluded that there were arguments both for and against genetic testing of children. However, they also conceded that genetic testing of children was different from other medical interventions because it very often had no immediate clinical goal. To this might be added the argument that genetic information is in any event different from other medical information because there are real, and potentially harmful, consequences which flow simply from its availability. Further, as the British Medical Association (BMA) pointed out:

10. The Danish Council of Ethics (1993) *Ethics and Mapping the Human Genome* (Copenhagen: The Danish Council of Ethics), p.60.

11. Advisory Committee on Genetic Testing (1998) *Report on Genetic Testing for Late Onset Disorders* (London: ACGT), available at http://www.doh.gov.uk/genetics/lodrep.htm.

12. *Ibid.*, p.5.

13. Working Party of the Clinical Genetics Society (CGS) (1994) "The Genetic Testing of Children," *Journal of Medical Genetics*, 31: 785–797.

> Opposition to testing children often rests on the proposition that the parents are denying the child, and later the adult, the option of not knowing his or her genetic status. It is also suggested that testing at the request of a child's parents, and giving the results to them, denies the child, and later the adult, the confidentiality which he or she would expect as an adult.[14]

Looking at the arguments against testing, the Clinical Genetics Society felt that these roughly revolved around the impact of tests on the individual child him or herself. Even ignoring the wider possibility of discrimination flowing from the holding of genetic information, the working party concluded that there may be detrimental consequences for the individual child flowing directly from the test. For example, there may be a loss of self esteem; it may affect the way the child is treated within the family or the wider community; and the provision of testing to young children in particular might breach current United Kingdom policy on the need for counseling before or in tandem with screening.[15]

Arguments in favor of genetic testing on the other hand seem to relate to the interests of third parties rather than to those of the child. For example, one benefit of the genetic testing of a particular child might be that the family would learn to come to terms with the child's illness. Equally, since genes are shared in families, testing could be offered to other members of the family in order to identify whether or not they too were at risk from a genetic condition. Indeed, the BMA have indicated that "[i]t appears that there is a strong desire for such testing among some parents of children affected by genetic disorders."[16]

Thus, achieving a balance in respect of genetic testing very much depends on where emphasis is placed. If the interests of the individual child are to be prioritized, then it would appear that the arguments against genetic testing carry more weight than those in favor. However, if the well-being of the family as a unit (or the 'rights' of the parents) is prioritized, then there would seem to be reasonably good arguments in favor of testing—although it must be borne in mind that the same problems of self-esteem and possible discrimination apply to the third parties just as much as they would to the individual child being tested. Therefore, it cannot safely be concluded that even the interests of third parties in the genetic testing of a particular child within the family will be served by the test actually being carried out.

14. British Medical Association (BMA) (1998) *Human Genetics: Choice and Responsibility* (Oxford: Oxford University Press), p.81.

15. For a fuller analysis of the report, see Sheila A.M. McLean (1995) "Genetic Screening of Children: The UK Position," *Journal of Contemporary Health Law and Policy*, 12: 113–130.

16. BMA, *op. cit.*, p.80.

In response to the Clinical Genetics Society report, the Genetic Interest Group's (GIG) response took issue with some of their conclusions. GIG describes itself as ". . . the national umbrella organization for the charities, voluntary organizations, and support groups which serve those affected by specific genetic disorders and their families."[17] Not least, the GIG response suggested that:

> The report is overly preoccupied with psychological considerations, and the harm that knowledge of genetic disorders can cause within families. With little evidence, this seems to reflect more the fears of doctors that they will be held responsible for negative reactions, rather than the needs of families.[18]

Indeed, and perhaps unsurprisingly given its own description of itself, GIG was rather more concerned with the integrity of the family unit, rather than the 'rights' or interests of the children themselves. This emphasis is demonstrated in its own characterization of the "focus" of its response which was ". . . to argue for the right of parents to have their children tested, except in the case of adult onset conditions."[19]

For the children themselves, of course, it is often argued that the obtaining of genetic information, where no treatment is available, cannot be seen as anything other than a negative, in that it reduces the child's capacity for future autonomy. In other words, especially where no treatment follows the test, the child is disadvantaged for the future by being unable to make that choice him or herself and burdened (in some cases at least) with information which they would not have wanted.[20] The response of GIG to the CGS Working Party Report shows both how complex and how deeply entrenched attitudes can be in this arena.

Of course, the situation with regard to children is compounded by the fact that they are in some cases unable legally to agree to or reject testing, as they are regarded as legally incompetent. Even where competence is asserted, the decision to agree to or refuse testing may yet be doubted. The majority of problems or tensions will emerge in two situations, which will be considered in the next section. Both relate to the issue of who has the authority to make decisions about testing, but raise slightly different questions. The first arises when children are too young and/or immature to be legally allowed to make healthcare or health related decisions, and the second arises when the child is likely to be sufficiently mature for their competence to be potentially vindicable.

17. Genetic Interest Group (1995) "GIG Response to the Clinical Genetics Society Report: 'The Genetic Testing of Children'." Available at http://www.gig.org.uk/docs/gig_testingchildren.pdf.
18. *Ibid.*, p.2, para.1.4.
19. *Ibid.*, p.3, para.1.6.
20. See below for further discussion of this issue.

2. On Whose Authority?

When children are very young, it is clear that they are unable to agree even to ordinary medical intervention on their own behalf. Thus, parents or guardians are generally entrusted with taking such decisions for them. However, the authority of a parent to make decisions of this sort is limited in scope and effect. Any such decision is lawful only where it can be said to be in the best interests of the child. Testing for phenylketonuria (PKU), for example, might easily be seen as satisfying that test, since there is a clinical strategy which can be applied to the condition and the child clearly benefits from its application.

However, given the nature of the arguments in favor of testing, where the link between diagnosis and therapy is more tenuous or even absent, it is likely that any benefits of testing accrue (at least in the short term) to others—for example the parents or the family unit—rather than to the individual child. In other words, it must be questionable whether or not the best interests of the child can be served by agreement to testing in these circumstances, and it may therefore be that any purported consent from the parents is technically of limited legal value. The same might be said about pre-symptomatic diagnosis of childhood onset conditions, although there are arguably more reasons to regard this as relatively aligned to other medical interventions of a non-genetic type. Irrespective of the type of test being contemplated, it is generally accepted in common and statute law that the issue of primary concern when third party decisions are made on behalf of a child is the welfare or best interests of the child; not those of others.

As is often the case in these sensitive and complex areas, arguments about what is in the 'best interests' of any person can be mounted, apparently convincingly, from a variety of perspectives. Thus, it would be a mistake to imagine that, simply because a test is laid down in law, it is not open to a wide range of diverse interpretations. In fact, the 'best interests' test is notoriously subject to different interpretations and should be regarded with some caution. Nonetheless, it remains a guide (albeit a vague one) as to what renders a parental decision for a child lawful.

This is not to say that the 'best interests' test has not been interpreted over the years, yet its parameters are by no means clear-cut. For example, it has traditionally been assumed that in this context the best interest of the child relates to his or her *medical* best interests. It is for this reason that concentration generally focuses on the possibility of palliation or cure following a diagnosis. However, a recent judgement in the Supreme Court of Ireland[21] preferred the concept of family autonomy to the

21. *North Western Health Board v. W (H)* [2001] IESC 70.

medical best interests of a child.[22] And in at least one UK case, the wishes of the parents of a disabled newborn were also allowed to prevent treatment which could have been life-saving from being provided, in part because of their personal objections and not because of the direct welfare interests of the child.[23] Additionally, some recent UK cases have extended the concept of best interests in ways that, for some at least, seem to be difficult to reconcile with an expectation that the welfare of the individual concerned will be prioritized. In one recent case, for example, a mentally disabled woman was the subject of a non-consensual bone marrow removal in order to help her seriously ill sister.[24]

Clearly the mentally disabled woman could not agree to the intervention, but the court nonetheless authorized the operation to go ahead on the basis that the young woman's emotional welfare would suffer were she not to provide a donation for her sister. In particular, the Court was concerned that the mother, who had a close relationship with the mentally disabled young woman, would be less able to maintain that relationship should the other sibling die, thereby causing harm to the interests of the young woman concerned.

This seems to stretch the concept of best interests considerably, even if intuitively it is a decision with which one has sympathy.[25] In terms of genetic testing, the welfare or interests of the individual being tested are restated as being the priority. The Declaration of Inuyama on Human Gene Mapping, Genetic Screening and Gene Therapy, for example, says, "[t]he central objective of genetic screening and diagnosis should always be to safeguard the welfare of the person tested."[26]

However, as with some of the cases mentioned above, it is by no means easy to correlate the non-consensual acquisition of genetic information with welfare, and this is particularly true where the collection of the information has no obvious or direct therapeutic intention or outcome. For example, this will apply when testing only provides information with no hope of treatment, or in most—if not all—screening programs. Indeed, research has tended to show that screening of all sorts can raise anxiety

22. For a commentary, see Graeme T. Laurie (2002) "Better to Hesitate at the Threshold of Compulsion: PKU Testing and the Concept of Family Autonomy in Eire," *Journal of Medical Ethics*, 28(3): 136–138.
23. *Re T (a minor) (wardship: medical treatment)* [1997] 1 All ER 906.
24. *Re Y (adult patient) (mental incapacity: bone marrow transplant)* [1996] 35 BMLR 111.
25. For an interesting analysis of the 'best interests' test, see John K. Mason and R. Alexander McCall Smith (1999) *Law and Medical Ethics*, Fifth edition (London: Butterworth), particularly pp.254–255.
26. The Council for International Organizations of Medical Sciences (CIOMS) (1990) *Declaration of Inuyama on Human Genome Mapping, Genetic Screening and Gene Therapy*, Article IV. Full text available at http://www.cioms.ch/1990_texts_of_guidelines.htm.

levels,[27] even if only temporarily.[28] Moreover, although some would argue that there is in reality no difference between genetic and other health-related information,[29] for many there is a critical difference in that genetic information is more sensitive and arguably more open to abuse. Considerable ink has been spilled in identifying the wide range of areas of life in which the mere holding of genetic information, whether or not properly understood, might adversely affect the individual.[30]

Despite these concerns, the House of Commons Select Committee on Science and Technology[31] concluded that "[p]arents should . . . be able to ask for genetic diagnosis to be carried out on a particular child where a family history suggests this may be appropriate."[32] Indeed, they identified a further ostensible argument in favor of testing children, which they believed would become more common place. This additional argument was that it "will bring benefits by avoiding multiple referrals due to misdiagnosis."[33]

This approach is interesting, at least in part because it seems in some sense to contradict their general position that mass genetic screening for public health reasons should not be undertaken unless "it is for an important condition, it offers a clear diagnosis and prognosis, and it informs choice and may offer opportunities to prevent the problem arising."[34] In specific reference to children, they continue, "there should be no mass screening for public health reasons in childhood unless a treatment for the disorder exists."[35]

No such caveats appear to apply in the case of parents seeking such tests on their children. In respect of childhood screening for late onset conditions their conclusion was, however, that "[w]e do not believe that children should have genetic diagnosis for late onset disorders. Such diagnosis is only justifiable, if those requesting it have fully considered all its implications."[36] Again, the apparently unequivocal commitment of the first sentence (which seems designed to protect the young) is tempered

27. Cf. BMA, *op. cit.*, especially p.103.
28. Hilary Bekker, *et al.* (1994) "The Impact of Population Based Screening for Carriers of Cystic Fibrosis," *Journal of Medical Genetics*, 31(5): 364–368.
29. Ron L. Zimmern (1999) "Genetic Testing: A Conceptual Exploration," *Journal of Medical Ethics*, 25(2): 151–156.
30. For a range of areas of concern, see the discussion in Roger Brownsword, William Cornish, and Margaret Llewelyn (eds.) (1998) *Law and Human Genetics: Regulating a Revolution* (Oxford: Hart Publishing).
31. The House of Commons Select Committee on Science and Technology, *op. cit.*
32. *Ibid.*, paragraph 92, pp. xli-xlii.
33. *Ibid.*
34. *Ibid.*, paragraph 83, p.xxxix.
35. *Ibid.*, paragraph 92, pp.xli-xlii.
36. *Ibid.*, paragraph 80, p.xxxviii.

by the second which seems to presume that the understanding of a third party is, or should be, relevant.

The Clinical Genetics Society Working Party identified an interesting division on this particular point. Its report noted that geneticists were less likely to be happy to accept parental authority as validating genetic testing of children than were pediatricians[37]—perhaps reflecting the different nature of the knowledge possessed by each group as to the intricacies of genetic information or indeed the nature of the relationship between the parties and the individual children.

Of course, a different situation may obtain in the case of older children. Following the case of *Gillick v. West Norfolk and Wisbech Hospital Authority*[38], it is now clear that older or more mature children may be permitted to engage in medical interventions on their own authority. However, the situation differs between the jurisdictions in Scotland and that in England and Wales.[39] In Scotland, it has long been recognized at common law that children under the age of 16 can enter into certain agreements on their own behalf, and indeed that when they reach the level of maturity which permits this, the parents' role is reduced if not obliterated. More recently, legislation in Scotland allows for children to agree to participate in medical treatment or *any medical procedure* on their own behalf.[40] Thus, even if genetic testing is not seen as a standard therapy, but is rather seen as a medical 'procedure', then the Gillick-competent child in Scotland seems unquestionably to be permitted by law to agree to testing on their own behalf without the involvement of their parents. The position of the child in Scotland, therefore, seems to be very much in line with the view of the UK Advisory Committee on Genetic Testing, which proposed (in this case, even regarding the most extreme of all situations—namely testing for pre-symptomatic testing for late-onset conditions) that:

> Requests from adolescents . . . should be considered, taking into account the circumstances of the individual and their family. If an adolescent is competent to give consent but there are no clear benefits of testing, some might suggest that testing should be deferred until the age of majority is reached. However, the competent adolescent is entitled to make a personal decision on this matter after a full discussion and exploration of the issues.[41]

37. CGS, *op. cit.*, p.787.

38. *Gillick v. West Norfolk and Wisbech Hospital Authority* [1982] 1 All ER 402.

39. For an excellent discussion of this, see Sarah Elliston (1996) "If You Know What's Good for You . . .: A Consideration of Refusal of Consent to Medical Treatment by Children," in Sheila A.M. McLean (ed.) *Contemporary Issues in Law, Medicine, and Ethics* (Brookfield, VT and Aldershot: Dartmouth).

40. Age of Legal Capacity (Scotland) Act of 1991, Art. 1(1)(b); Children (Scotland) Act of 1995.

41. Advisory Committee on Genetic Testing, *op. cit.*, p.5.

In England and Wales the situation is more complicated. As in Scotland, children may become Gillick-competent under the age of 16 but it is not clear that their consent to 'procedures' as opposed to treatment would necessarily be valid. What is clear is that where a therapeutic medical act is contemplated, children can indeed agree to this on their own behalf. However, and again unlike the situation in Scotland, the child's refusal of an intervention—even if it qualifies as 'treatment'—may be overridden by their parents. In fact, this right of veto extends even to young people between the ages of 16 and 18.[42] It remains a matter of speculation whether or not the English Courts would treat a genetic test on the same basis as they have dealt with standard medical interventions, but it seems likely that—given that they are not purely and simply 'treatment'—the views of parents would be given even greater dominance over the views of the young people concerned.

3. What Should We Test For?

The toll exacted by genetic conditions cannot be underestimated: for neither individuals nor communities. Writing as long ago as 1985, Kevles noted that:

> Genetic and chromosomal illnesses or malformations are reported to account for between twenty and thirty percent of all pediatric hospital admissions. Twelve percent of all adult hospital admissions are said to involve illnesses with a significant genetic component. At least fifteen percent of all diagnoses for mental retardation report it as unambiguously hereditary.[43]

In 1992, the British Medical Association claimed that "[g]enetic and pre-genetic diseases affect one in every twenty people by the age of 25 and perhaps as many as two in three people during their lifetime."[44] Clearly these figures demonstrate the real connection between ill health and genetic predisposition, and as such they will likely fuel the rapid development of genetic tests and, hopefully, therapies.

It has already been suggested that testing for conditions that are treatable may be relatively uncontroversial. Indeed, they may represent little more than the standard medical inquiry. However, it must be borne in mind that genetic information may

42. Family Law Reform Act of 1969, as interpreted in cases such as *Re W (a minor) (medical treatment)* [1992] 4 All ER 627; *Re R* (1992) 7 BMLR 147.
43. These figures presumably relate to the United States. See Daniel J. Kevles (1995) *In the Name of Eugenics: Genetics and the Uses of Human Heredity* (Cambridge, MA: Harvard University Press), p.291.
44. British Medical Association (1992) *Our Genetic Future: The Science and Ethics of Genetic Technology* (Oxford: Oxford University Press), p.1.

have consequences which do not routinely arise where the diagnostic intervention is for non-genetic conditions. However, genetic testing may also be used in two other situations.

First, it may be felt desirable to *test for carrier status*. At present, this is generally only available pre-natally, but the House of Commons Select Committee on Science and Technology considered the possibility that it might be made more widely accessible. The Committee concluded that:

If carrier screening becomes routine, serious consideration should be given to offering it outside the context of antenatal care. If such screening is offered, there should be sufficient educational support to ensure that no stigma attaches to unaffected carrier status.[45]

It might, arguably, be evident why families would wish to have this information and the BMA notes that:

> [a] small study of parents' attitudes to testing the siblings of patients with cystic fibrosis, for example, found that 90 per cent would like to know the carrier status of their unaffected children and 91 per cent believed that they had a right to the information.[46]

It is less clear, however, why children would want to have such information. Certainly carrier status becomes relevant when an individual is faced with making reproductive decisions, but there is little evident value in children (or their families) having this knowledge before that stage in life is reached.[47] In any event, of course, an adult could refuse to undergo such testing and make uninformed reproductive decisions perfectly lawfully, even if we felt that it is questionable morally. Any moral criticism of such decisions would be based in the concept of 'intergenerational justice',[48] which requires current generations to pay careful attention to the well being of those generations to come.

More controversial still, arguably, is *predictive testing for late-onset conditions*. While there may be reasons—both for the families and for the state—why this infor-

45. The House of Commons Select Committee on Science and Technology, *op. cit.*, paragraph 95, pp.xlii-xliii.

46. British Medical Association, *Human Genetics . . ., op. cit.*, p.80.

47. It should be noted that the GIG would not entirely agree with this assertion. In its view, "After suitable counseling, parents have the right to make an informed choice about whether or not to have their children tested for carrier status. Ideally, children should only be tested when of an age to be involved in the decision." See Genetic Interest Group, *op. cit.*, p.5, paragraph 2.2.

48. For further discussion, see Sheila A. M. McLean (2000) "Intergenerational Justice: A Brave New World or a Leap in the Dark?" *Genetics Law Monitor*, 1(1): 1–3.

mation could be valuable, yet again it is unclear how it benefits the individual child. As Kinderlerer and Longley say:

> Notwithstanding discrimination, genetic testing is likely to be an important tool in identifying individuals susceptible to disease, not only for attempting to protect them (or their offspring) from the consequences of the disease, but for the future consideration of determining priorities and allocation of resources within health services.[49]

The standard arguments against such testing include those already referred to, but also revolve around the fact that obtaining this knowledge where a child has not sought it precludes the future exercise of autonomy by that child when he or she reaches an age at which he or she could have agreed to or refused such a test on his or her own behalf. Indeed, the UNESCO Universal Declaration on the Human Genome and Human Rights appears explicitly to recognize this 'right not to know', declaring that: "The right of every individual to decide whether or not to be informed of the results of genetic examination and the resulting consequences should be respected."[50]

In a recent article, Robertson and Savulescu note, however, that there is some evidence that suggests that families where certain genetic conditions, such as bipolar disorder, are present, would agree to have their children tested. They seek to argue that—although clearly there are problems with genetic testing for late onset conditions in children—there are also conceivable benefits from it. As they say, "[j]ust as authors have listed potential harms of genetic testing, there are conceivable potential benefits as well."[51] Thus, they conclude, "the objection to childhood testing based on a violation of future autonomy fails."[52]

This conclusion is by no means unarguable. In any event, although it may well be that families report that they would wish tests to be carried out, two objections remain. First, it seems clear that the benefits which they envisage as accruing from such tests are for the family rather than the individual child—something which it has already been suggested is insufficient to justify the potential harm to the child. Second, even where people have expressed enthusiasm for the use of tests, once they become

49. Julian Kinderlerer and Diane Longley (1998) "Human Genetics: The New Panacea," in Brownsword *et al.* (eds.), *op. cit.*, pp.11–28, see especially pp.19–20.

50. See Article 5(c). The full text of the UNESCO Declaration can be found in the Appendix of this volume.

51. John A. Robertson and Julian Savulescu (2001) "Is There a Case in Favour of Predictive Genetic Testing in Young Children?" *Bioethics*, 15(1): 26–49, see p.42.

52. *Ibid.*

available, their enthusiasm can be considerably dimmed. The House of Commons Select Committee on Science and Technology, for example, noted that:

> One unexpected consequence of the availability of a test for Huntington's and the intensive counselling given before the test has been a fall in the perceived demand for such a test. Before such a test existed 80 per cent of those with a family history of the disease thought they would like to know their status; when the test became available and those coming forward had been counselled, only 10 per cent did so.[53]

Again, however, the rationale for testing of children would generally relate more to others than to the child. The place of third parties in the scale of values must be regarded as dubious for so long as our commitment to children is to protect them rather than ignoring or reducing their interests in the actual or purported interests of others, or on the basis of some speculative benefit. In any event, both the Clinical Genetics Society Working Party and the House of Commons Select Committee concluded that there should be a presumption against such testing.

4. Conclusions

The rapidly evolving science of genetic screening and testing seems likely to be in the forefront of medical development in this century. Fletcher and Wertz indeed predict that:

> Advances in genetic knowledge . . . will gradually affect every person with or without the capacity to choose genetic screening and testing. Information about the molecular contents of virtually every gene harmful to human health can be—and gradually will be—used for testing. Human beings will be genetically laid bare and vulnerable as never before. Approaches to diagnosis, treatment and prevention of genetic and multifactorial diseases will grow and proliferate.[54]

53. The House of Commons Select Committee on Science and Technology, *op. cit.*, paragraph 78, p.xxxviii.

54. John C. Fletcher and Dorothy C. Wertz (1990) "Proposed: An International Code of Ethics in Medical Genetics before the Human Genome is Mapped," in Zbigniew Bankowski and Alexander M. Capron (eds.), *Genetics, Ethics, and Human Values: Human Genome Mapping, Genetic Screening, and Therapy* (Geneva: CIOMS), p.97.

These developments, which can realistically be expected to be rapid, will undoubtedly "create unique dilemmas for parents, their children and their health care providers."[55] This leads logically enough to the question of how we control testing generally, specifically with regards to children—for control or regulation seems likely to be the only (or the major) way of offering a balance between knowledge and abuse. Berg[56] suggests that the health benefits, which it can be anticipated will arise from the "responsible use of genetic tests," will likely mandate their continued development and expanded use. But he argues that "a prerequisite would be to establish rules and practices to secure that information about, or results from, genetic tests are not being used to the tested individual's disadvantage."[57]

Finally, there are matters of specific concern for children. It is accepted that, where a child is young, information about his or her clinical status is made available to those who authorized the discovery of such information—generally the parents. This is routinely uncontroversial because first, the parent has the authority to ask for diagnosis and treatment and second, treatment is generally seen to be in the child's 'best interests'. However, where what is being sought is a diagnosis with no therapy then there may be a subtle shift in the applicable values or interests. Equally, where the information being sought both leads to no treatment and its very existence might be damaging for the child, now and in the future, then it might be argued that the circumstances which would otherwise justify the passing on of information about one person to a third party no longer exist, even although there is a clear familial bond between the parties.

Where the child is older, and more likely to be considered a patient on his or her own right, it is unclear how the provision of otherwise confidential information to third parties (in this case parents) could be justified. This would seem to suggest that no testing should be carried out where the harm (which may result from the mere availability of the information) outweighs the benefit (generally to third parties). Of course, if we adopted the approach that families are units rather than a collection of individuals, then it might possibly be argued that the unit has interests that supersede those of the individual. However, adopting this approach would be to move our legal tradition away from seeing children as adults in waiting, worthy of legal protection,

55. Diane E. Hoffman and Eric A. Wulfsberg (1995) "Testing Children for Genetic Predispositions: Is It in Their Best Interest?" *Journal of Law, Medicine, and Ethics*, 23(4): 331–344, see p.341.
56. Kåre Berg (1994) "Confidentiality Issues in Medical Genetics" and "The Need for Laws, Rules, and Good Practices to Secure Optimal Disease Control," in *Ethics and Human Genetics: Proceedings of the Second Symposium of the Council of Europe on Bioethics, Strasbourg, November 30–December 2, 1993* (Strasbourg: CDBI, Council of Europe). CDBI-SY-SP (93) 3.
57. *Ibid.*, p.7.

towards the perception that what occurs in the family is not to be scrutinized where it serves the purposes of that family.

It may well be that the so-called 'genetics revolution' will ultimately lead us in that general direction, but we should be cautious about permitting this to happen unless and until we have contemplated all of the possible consequences. Genetics may well point us to a more communitarian approach, and reduce the power that the concept of individuality or autonomy currently has, but we should arguably be wary about conceding this unless all of the implications have been thoroughly evaluated. The cult of autonomy may have gone too far for some, but the rationale for its pre-eminence in the aftermath of the Second World War still has power even in the 21st century. For children, the value of anticipated autonomy, coupled with the right to protection from harm, should serve as significant barriers to the over-emphasis on their genetic make-up at the expense of their humanity.

A Disability Rights Approach to Genetic Discrimination

Gregor Wolbring

1. Introduction

It has become a global concern that the advancement in the ability of genetic manipulation and prediction might lead to genetic discriminations. The World Medical Association Declaration on the Human Genome Project states:

> Genetic discrimination in private insurance and employment: here is a conflict between the increasing potential of new technologies to reveal genetic heterogeneity and the criterion for private insurance and employment. It may be desirable, regarding genetic factors, to adopt the same tacit consensus, which prohibits the use of race discrimination in employment or insurance. Genetic mapping may become a source of stigmatization and social discrimination, and the 'risky population' may turn into a 'defective population'.[1]

Article 6 of the Universal Declaration on the Human Genome and Human Rights of UNESCO states that "No one shall be subject to discrimination based on genetic characteristics that is intended to infringe or has the effect of infringing human rights, fundamental freedoms and human dignity."[2] The European Convention for the Protection of Human Rights and Dignity of the Human Being declares in its Article 11 on Non-Discrimination that "Any form of discrimination against a person on grounds of his or her genetic heritage is prohibited."[3] And finally the Draft World Health Organization (WHO) guidelines on bioethics state: "Genetic information

1. World Medical Association Declaration on the Human Genome Project. Adopted by the 44th World Medical Assembly in Marbella, Spain, September 1992. For the complete text, see http://www.wma.net/e/policy/g6.htm.
2. The UNESCO Universal Declaration on the Human Genome and Human Rights (hereinafter the 'UNESCO Declaration') was adopted in 1997. For the complete text, see www.unesco.org/ibc/en/genome/projet/index.htm, and the Appendix to this volume.
3. The European Convention for the Protection of Human Rights and Dignity of the Human Being with regard to the Application of Biology and Medicine: Convention on Human Rights and Biomedicine (hereinafter the 'Convention on Human Rights and Biomedicine') was adopted in Oviedo, on April 4, 1997. For the complete text, see http://conventions.coe.int/treaty/en/Treaties/Html/164.htm, and the

should not be used as the basis for refusing employment or insurance. Exceptions would have to be legally defined."[4]

This concern on the international level is also reflected on the national level by the fact that many countries are developing legal instruments to prevent the occurrence of various forms of genetic discrimination. On the surface, this seems to be a reasonable right. The prohibition or prevention of genetic discrimination seems to be a logical extension of the prohibition of sexual, racial and other discrimination.

This paper looks at how the debate around the issue of genetic discrimination and the introduction of laws aiming at prohibiting various forms of genetic discrimination relates to disabled people, their rights, and their self-perception. Normally people do not think about disabled people when they think about genetic discrimination. From a disability rights perspective, three questions arise:

1. What is the meaning of the term 'genetic discrimination'?
2. What do laws prohibiting genetic discrimination cover? and
3. Whom do laws prohibiting genetic discrimination cover?

Before I try to answer these three questions, I have to give a little background as to what is meant by the term 'disability' or the term 'disabled people'.

2. What Is a Disability?

There are two main interpretations of the term 'disability'. These may be expressed in a dialectic of the 'medical model' (mM) versus the 'social model' (sM).[5] Within the *medical individualistic model*, disability is viewed as a defect, a problem inherent in the person, directly caused by disease, trauma, other health conditions, or a deviation from certain norms and leading to a low quality of life for the person and their relatives. Management of the disability of the disabled person or person-to-be is aimed at cure, prevention, or adaptation of the person (for example, by various normalizing assisting devices) to ensure a functioning or existence as normative as possible. Medical

Appendix to this volume. See also the Council of Europe Committee of Ministers Recommendation No. R (92) 3 of the Committee of Ministers to Member States on Genetic Testing and Screening for Health Care Purposes (Adopted by the Committee of Ministers on February 10, 1992), at the website http://cm.coe.int/ta/rec/1992/92r3.htm.

4. The World Health Organization (WHO) *Draft Guidelines on Bioethics* was presented to the World Health Assembly in May, 1999. For the complete text, see http://www.nature.com/wcs/b23a.html.

5. See also Jonathan Penney (2002) "A Constitution for the Disabled or a Disabled Constitution? Toward a New Approach to Disability for the Purposes of Section 15(1)," *Journal of Law and Equality*, 1(1): 83–115, available at http://www.jle.ca/files/v1n1/JLEv1n1art3.pdf.

care, preventive medicine, and rehabilitation towards the norm are viewed as the primary issues, and at the political level, the principal response is that of modifying or reforming health care policy and making preventive medicine more efficient. At the emotional level, the principal response is one of pity and rejection. Disabled people are seen as a suffering entity and as having a low quality of life.[6]

However, the medical model is in contradiction to studies that show how disabled people (by the social model) rate their own quality of life as equal to or higher than their non-disabled counterparts,[7] and do not see themselves as a medical condition.[8] This view of disabled people is reflected in the *social model* (sM) of disability, which sees disability mainly as a socially created problem and as a matter of the full integration of individuals with different biological realities and abilities into society.

6. See Royal College of Physicians (1989) *Prenatal Diagnosis and Genetic Screening* (London: RCP); Arthur L. Caplan (1992) "If Gene Therapy Is The Cure, What Is the Disease?" in George J. Annas and Sherman Elias (eds.) *Gene Mapping: Using Law and Ethics as Guides* (Oxford and New York: Oxford University Press), pp.128–141; and Diane Coleman (1996) *Testimony Before the Constitution Subcommittee of the Judiciary Committee of The House of Representatives*, April 29, 1996, available at http://www.mcil.org/mcil/mcil/testimon.htm.

7. Gregor Wolbring (2001) "Folgen der Anwendung genetischer Diagnostik für behinderte Menschen" (Consequences of the Application of Genetic Diagnostics for Disabled People), Expert Opinion for the Study Commission on the Law and Ethics of Modern Medicine of the German Bundestag, pp.10–17; available in electronic form: see the German language text at the webpage http://www.bundestag.de/gremien/medi/medi_gut_wol.pdf. See also Kenneth A. Gerhart, Jane Koziol-McLain, Steven R. Lowenstein, and Gale G. Whiteneck (1994) "Quality of Life Following Spinal Cord Injury: Knowledge and Attitudes of Emergency Care Providers," *Annals of Emergency Medicine*, 23(4): 807–812; C. Ray and J. West (1984) "Social, Sexual, and Personal Implications of Paraplegia," *Paraplegia*, 22(2): 75–86; Myron G. Eisenberg and Constance C. Saltz (1991) "Quality of Life Among Aging Spinal-Cord Injured Persons: Long-Term Rehabilitation Outcomes," *Paraplegia*, 29(8): 514–520; Saroj Saigal, David Feeny, Peter L. Rosenbaum, William Furlong, Elizabeth Burrows, Barbara L. Stoskopf, and Lorraine Hoult (1996) "Self-Perceived Health Status and Health-Related Quality of Life of Extremely Low-Birth-Weight Infants at Adolescence," *JAMA, The Journal of the American Medical Association*, 276(6): 453–459, at p.453; Laura A. Cushman and Marcel P. Dijkers (1990) "Depressed Mood in Spinal Cord Injured Patients: Staff Perceptions and Patient Realities," *Archives of Physical Medicine and Rehabilitation*, 71: 191–196; Gale G. Whiteneck *et al.* (1985) *The Rocky Mountain Spinal Cord Injury System*, Report to the National Institute of Handicapped Research (Washington, DC: National Institute of Handicapped Research), pp.29–33.

8. Adrienne Asch (1999) "Prenatal Diagnosis and Selective Abortion: A Challenge to Practice and Policy," *American Journal of Public Health*, 89(11): 1649–1657; Marque-Luisa Miringoff (1991) *The Social Costs of Genetic Welfare* (New Brunswick, NJ: Rutgers University Press); Ruth Hubbard (1990) *The Politics of Women's Biology* (New Brunswick, NJ: Rutgers University Press), Chapters 12 to 14; Abby Lippman (1991) "Prenatal Genetic Testing and Screening: Constructing Needs and Reinforcing Inequities," *American Journal of Law and Medicine*, 17(1–2): 15–50; Martha A. Field (1993) "Killing 'the Handicapped'—Before and After Birth," *Harvard Women's Law Journal*, 16: 79–138; Michelle Fine and Adrienne Asch (1982) "The Question of Disability: No Easy Answers for the Women's Movement,"

Disability is not seen as an attribute or defect of an individual, but as caused by the reaction of society towards the biological reality of the individual. Disability is a complex collection of disadvantages, many of which are created by the environment, particularly the social environment and the socially mediated aspects of the physical environment. In this view, the management of the problem requires social action, and it is the collective responsibility of society at large to make the environmental and emotional modifications necessary for the full participation of people with different biological realities in all areas of social life. The issue is therefore an attitudinal or ideological one requiring social change, which at the political level becomes a question of human rights. The social model allows 'ableism'[9] (discrimination based on the lack of expected abilities) to be seen in the same light as racism or sexism.

The social model of disability does not negate that a disabled person has a certain biological reality (like having no legs) that makes that person different in her/his abilities, which make her/him not fit the norm. However, it views the 'need to fit a norm' as the disability itself and questions whether many deviations from the norm need a medical solution (adherence to the norm) or a social solution (change or elimination of norm). Unfortunately, the social model and the self-perception of disabled people is often ignored by non-disabled people or even denounced as irrelevant.[10] The social model also does not negate the existence of symptomatic acute medical problems, which should be treated. However, in the same way as a 'non-disabled' who is ill

Reproductive Rights Newsletter, 4(3): 19–20; Marsha Saxton (1984) "Born and Unborn: The Implications of Reproductive Technologies for People with Disabilities," in Rita Arditti, Renate Duelli-Klein, and Shelley Minden (eds.) *Test-Tube Women: What Future for Motherhood?* (Boston, MA: Pandora), pp.298–312; Anne Finger (1990) *Past Due: A Story of Disability, Pregnancy, and Birth* (Seattle, WA: Seal Press); Deborah Kaplan (1994) "Prenatal Screening and Diagnosis: The Impact on Persons with Disabilities," in Karen H. Rothenberg and Elizabeth J. Thompson (eds.) *Women and Prenatal Testing: Facing the Challenges of Genetic Technology* (Columbus: Ohio State University Press), pp.49–61; Adrienne Asch (1989) "Reproductive Technology and Disability," in Sherrill Cohen and Nadine Taub (eds.) *Reproductive Laws for the 1990s* (Clifton, NJ: Humana Press), pp.59–101; Adrienne Asch and Gail Geller (1996) "Feminism, Bioethics, and Genetics," in Susan M. Wolf (ed.) *Feminism and Bioethics: Beyond Reproduction* (Oxford and New York: Oxford University Press), pp.318–350.

9. 'Ableism': a discourse of beliefs, processes, and practices that produce a particular kind of self and body (the corporeal standard) that is projected as perfect and species-typical, and therefore essentially and fully human. In this view, then, disability is seen as a diminished state of being human. See Fiona A.K. Campbell (2001) "Inciting Legal Fictions: 'Disability's' Date with Ontology and the Ableist Body of the Law," *Griffith Law Review*, 10(1): 42.

10. At the Tenth "Genetic Technology and Public Policy in the New Millennium" symposium, Daniel W. Brock, a bioethicist at the National Institutes of Health, gave a presentation entitled "Genetic Testing and Selection: A Response to the Disability Movement's Critique." In the presentation, Brock told the audience that "Our notion of how good a person's life is fully determined by their own subjective self-assessment." Quoted in Roberto Rivera y Carlo (2002) "Targeting the Disabled," *Boundless*, December 5,

with a flu and who is treated for this acute illness with medicine is *not defined* by this concrete medical problem, a 'disabled' can have acute medical periods without being solely seen as a medical problem.

Furthermore, due to the technological developments on the horizon that will make the non-genetic and genetic enhancements of human beings possible, a third, so-called 'transhumanist model of disability' is appearing. The *transhumanist model of disability* (tM) does not see the 'disabled person's body' as subnormal. Rather, it sees the human body in general as deficient, as something in need of improvement. Therefore it believes in improving *any* human body if possible whether it's the body of a so called 'disabled' or a so-called 'non-disabled'. In this model, in essence, everyone is disabled (mM, tM), everyone has defects in need of fixing. At the societal and political level, the principal response is the support for the development of *any* new technology that could potentially improve the human body and the norm.[11]

3. Who Is 'Disabled'?

Laws prohibiting disability discrimination protect persons who either

- presently have a disability (symptomatic);
- were disabled in the past (past-symptomatic) (such as laws in Australia, Canada, Hong Kong, New Zealand, the Philippines, United Kingdom, United States);
- may be disabled in the future (asymptomatic) (laws in Australia, Hong Kong, Sweden, United States); or
- who are regarded as being disabled (laws in Australia, Hong Kong, New Zealand, the Philippines, United States).[12]

2002, available at http://www.boundless.org/2002_2003/features/a0000685.html. Compare also John Harris (2000) "Is There A Coherent Social Conception of Disability?" *Journal of Medical Ethics*, 26(2): 95–100; Solveig Magnus Reindal (2002) "Disability, Gene Therapy, and Eugenics: A Challenge to John Harris," *Journal of Medical Ethics*, 26(2): 89–94.

11. Gregor Wolbring (2003) "NBIC, NGOs, and the Three Types of Disabled People," paper written for the conference "Within and Beyond the Limit of Human Nature," held between October 12 and 16, 2003 in Berlin, Germany. (NBIC stands for Nanotechnology, Biotechnology, Information Technology and Cognitive Sciences). Available at the International Center for Bioethics Culture and Disability webpage http://www.bioethicsanddisability.org/boell.html.

12. For the list of countries that have adopted anti-discrimination laws for persons with disabilities, see Theresia Degener and Gerard Quinn (2000) "A Survey of International, Comparative and Regional Disability Law Reform," especially the footnotes 88 to 114. Available in English language at http://www.dredf.org/symposium/degener1.html. For an extensive list of links to national legal norms prohibiting disability discrimination, see http://www.bioethicsanddisability.org/dislawstatistic.html.

The question of who is covered by laws prohibiting disability discrimination plays a role in the discriminative approach of laws and law proposals prohibiting genetic discrimination, and these will be discussed below.

However, there is another problem lurking on the horizon. With the increase in capabilities to change the genetic and non-genetic compositions of human beings, the question arises whether someone is still seen as disabled (sM) and covered by laws prohibiting disability discrimination if it is perceived that there is a medical or technological cure available for the disability (mM). The answer seems to be 'no' if we look at recent rulings of the United States Supreme Court. These, in essence, state that if a person does not experience a substantial limitation in a major life activity when using a mitigating measure such as medication, glasses, or other corrective devices, then that person is not considered disabled under the law.[13]

This interpretation has huge implications for disabled people and their parents. It would seem that a disabled adult or parents on the behalf of their disabled child or child-to-be have to take any cure—genetic or non-genetic—if available because the disabled adult or the disabled child would not be covered by laws prohibiting disability discrimination anymore and could be discriminated against without any legal recourse. That makes it of course hard for the parents of a disabled child (mM) or

13. *Sutton et al. v. United Air Lines, Inc.* (1999) 527 U.S. 471 (Case No. 97–1943, Argued April 28, 1999, Decided June 22, 1999). *At Issue:* Whether nearsightedness that can be corrected with glasses or contact lenses is a 'disability' under the Americans with Disabilities Act. *Decision:* The court ruled that the Americans with Disabilities Act does not cover people whose disabilities can be sufficiently corrected with medicine, eyeglasses or other measures. See *United States Reports*, 527: 471–515, at http://www.supremecourtus.gov/opinions/boundvolumes/527bv.pdf and for a syllabus, http://laws.findlaw.com/us/000/97-1943.html.

Murphy v. United Parcel Service, Inc. (1999) 527 U.S. 516 (Case No. 97–1992, Argued April 27, 1999, Decided June 22, 1999). *At Issue:* Whether medication or other mitigating measures should be considered in assessing if an individual is 'disabled' under the terms of the Americans with Disabilities Act. *Decision:* The justices ruled that the condition the petitioner suffers and controls with medication is not considered 'disabling' under the Americans with Disabilities Act. The condition and medication preclude him from performing only a particular job and do not substantially limit him in employment. See *United States Reports*, 527: 516–525, at http://www.supremecourtus.gov/opinions/boundvolumes/527bv.pdf and for a syllabus, http://laws.findlaw.com/us/000/97-1992.html.

Albertson's, Inc. v. Kirkingburg (1999) 527 U.S. 555 (Case No. 98–591, Argued April 28, 1999, Decided June 22, 1999). *At Issue:* Whether a truck driver blind in one eye who received a waiver from Department of Transportation vision standards is 'disabled' for the purposes of the Americans with Disabilities Act. *Decision:* The justices voted unanimously that the plaintiff in this case was not protected by the Americans with Disabilities Act. They ruled that employers who set job qualifications based on federal safety standards do not have to overlook those standards when a worker obtains a waiver from the federal agency. See *United States Reports*, 527: 555–580, at http://www.supremecourtus.gov/opinions/boundvolumes/527bv.pdf and for a syllabus, http://laws.findlaw.com/us/000/98-591.html.

the disabled adult (mM) to refuse these medical or technological 'cures' even if they go against their self-perception because they would have no legal recourse against the behavior of the non-disabled towards them. In essence, they are forced to fix themselves medically or technologically in order to prevent being further disadvantaged by society. Furthermore, if a disabled child (mM) or a disabled adult (mM) has to be cured by medical or technological means, the disabled adult or the parents on behalf of their disabled child might as well follow the transhumanist model of disability (tM), and decide to use not only a medical or technological 'cure' to measure up to the norm, but also a medical or technological 'solution' to *surpass* the norm.

4. Laws Prohibiting Genetic Discrimination

4.1 What Do Laws Prohibiting Genetic Discrimination Cover?

At first glance, most laws and law proposals prohibiting genetic discrimination cover health insurance and employment discriminations. It is, however, problematic from a disability rights perspective that they do not cover disability and life insurance discriminations.

In order to look at the connection between disability and genetic discrimination laws more deeply, this paper would have to look at every country separately as every country has different laws relating to disabled people and different laws relating to genetic discrimination. This is far beyond the scope of this article. Instead, this paper uses the situation in the United States as an example to outline the dynamics between disability and genetic discrimination. Two articles—one by DeBris[14] and one by Lemmens[15]—contain a lot of information in regards to the legal situation of genetic discrimination in Europe and elsewhere. From the discussion of the problems in the US people can extrapolate the problems for disabled people in their countries by looking at existing laws and legal proposals with regards to genetic discrimination and disability discrimination.

14. Sonia Le Bris (2001) "Give Me Your DNA, and I'll Tell You Who You Are . . . Or Who You'll Become: Questions Surrounding the Use of Genetic Information in Europe," *Isuma: The Canadian Journal of Policy Research*, 2(3): 90–101. Available at http://www.isuma.net/v02n03/lebris/lebris_e.pdf.

15. Trudo Lemmens (2000) "Selective Justice, Genetic Discrimination, and Insurance: Should We Single Out Genes in Our Laws?" *McGill Law Journal*, 45(2): 347–412. Also available at http://www.journal.law.mcgill.ca/arts/452lemme.pdf.

4.2 Current Laws Prohibiting Genetic Discrimination in the USA

The situation is fairly complicated in the United States as there are a few federal laws that could be seen as already covering genetic discrimination. Namely,

1. *The Americans with Disabilities Act:* an anti-disability discrimination law[16], which has been interpreted by the Equal Employment Opportunity Commission (EEOC) to prohibit some forms of genetic discrimination. In 1995, the EEOC issued enforcement guidance advising that an employer who takes adverse action against an individual on the basis of genetic information may regard the individual as having a disability and, therefore, may be violating the ADA.[17] The EEOC recently settled its first court action challenging an employer's use of genetic testing and also issued a finding of cause in an administrative complaint filed by a woman whose work contract was terminated based on a genetic test result.[18]

   In Australia, Hong Kong, Sweden, and the United States laws prohibiting disability discrimination cover people who might develop a disability *in the future* (presently asymptomatic disabled); and in Australia, Hong Kong, New Zealand, the Philippines, and the United States[19] laws prohibiting disability discrimination cover those who are *regarded as being disabled.* The laws prohibiting disability discrimination in any of these countries could be used to fight genetic discrimination. Furthermore, it is interesting that the disability rights commission in the United Kingdom asked that asymptomatic disability be added to the definition of disability within the British laws prohibiting disability discrimination to protect them from genetic discrimination.[20] The same demand was voiced by 84.8 percent of human geneticists in Germany[21];

16. Americans with Disabilities Act of 1990 (ADA). Public Law 336 of the 101st Congress, enacted on July 26, 1990, codified in the *United States Code*, Volume 42, Chapter 126, *Equal Opportunity for Individuals with Disabilities*, as Sections 12101–12213. For the official text, see http://www.usdoj.gov/crt/ada/pubs/ada.txt.

17. Department of Labor, Department of Health and Human Services, Equal Employment Opportunity Commission, Department of Justice (1998) *Genetic Information and the Workplace* (Joint Government Report of January 20, 1998), p.6; available at http://www.dol.gov/asp/programs/history/herman/reports/genetics.htm. See also EEOC (1995) *Compliance Manual*, Vol. 2, Section 902, Order 915.002, 902-45.

18. *EEOC v. Burlington Northern Santa Fe Railroad*, Civ. No. 01-4013 MWB (N.D. Iowa Apr. 23, 2001) (Agreed Order).

19. See footnote 12 above.

20. The Disability Rights Commission (DRC, United Kingdom) calls for ban on bosses and insurers wanting genetic tests. See DRC (2001) "Responses to the 'Whose Hands on Your Genes?' Consultation," available at http://www.hgc.gov.uk/whoygconsultation_responses/drc.htm.

21. See Table 13 in Gerhard Wolff, Dorothy C. Wertz, and Irmgard Nippert (1999) "Ethik und Genetik: Ergebnisse der Umfrage zu Problemaspekten angewandter Humangenetik," *Medizinische Genetik*, 11(3): 308–318.

2. *The Health Insurance Portability and Accountability Act (HIPAA):* "which prohibits group health plans from using any health status-related factor, including genetic information, as a basis for denying or limiting coverage or for charging an individual more for coverage"[22]; and
3. *Title VII of the Civil Rights Act of 1964:* "which prohibits employment discrimination based on race or gender, may prohibit employers from denying employment based on genetic markers linked to race, such as that for sickle cell anemia, or linked to gender, such as those for breast or ovarian cancer."[23]

However, none of the above laws are very clear or perceived as potent in preventing genetic discrimination. Furthermore, laws on the state level are needed regardless of existing federal laws, as important areas, such as the insurance industry in general, are not federally regulated.[24] An overview of state-based genetic discrimination laws can be found at the webpage of the Council for Responsible Genetics.[25] In short, most of the state statutes deal exclusively with employment discrimination and health insurance contracts to the exclusion of life and disability insurance.[26]

4.3 Whom Do the Laws Prohibiting Genetic Discrimination in the USA Cover?

Two strategies could have been employed to form new policies with regard to laws prohibiting genetic discrimination in the United States: strengthening the existing laws, such as the ADA, or developing new anti-discrimination laws. It seems that the focus is currently on the development of new laws and not the strengthening of existing laws. Why is this? The answer becomes obvious when we look at whom these

22. National Council on Disability (2002) *Position Paper on Genetic Discrimination Legislation*, March 4, 2002 (Washington, DC: The National Council on Disability). Available at http://www.ncd.gov/newsroom/publications/geneticdiscrimination_positionpaper.html.
23. *Ibid.*
24. Sophia Kolehmainen, Peter Shorett, Sara Gambin, and Paul Billings (2002) "Comments to the National Conference of Insurance Legislators on the Proposed Genetic Discrimination Act," in *Genetic Testing, Discrimination, and Privacy Program* (Cambridge, MA: Council for Responsible Genetics), available at http://www.gene-watch.org/programs/privacy/insurance.html.
25. Council for Responsible Genetics (2001b) "Genetic Discrimination Legislation in the United States," in *Genetic Testing, Discrimination, and Privacy Program* (Cambridge, MA: Council for Responsible Genetics), available at http://www.gene-watch.org/programs/privacy/summary2001.html.
26. See, for example, the New Jersey Genetic Privacy Act of 1996. Tillie Young (2002) "The New Jersey Genetic Privacy Act (1996): Some Key Points," in *Genes at Work* (Newark: University of Medicine and Dentistry of New Jersey), available at http://www.umdnj.edu/genesatwork/clinicalserv/08_nj.htm; and National Conference of State Legislatures (2003) "Genetics and Life, Disability, and Long-Term Care Insurance," in *Genetic Technologies Project* (Denver and Washington: NCSL), available at http://www.ncsl.org/programs/health/genetics/ndislife.htm.

laws protect. There are a variety of ways these laws could be shaped. These laws could prohibit genetic discrimination against, for example, people who *have the gene* for Huntington Chorea, whether or not they exhibit symptoms related to Huntington Chorea. Or these laws could prohibit discrimination against people who have the gene for Huntington Chorea *until they develop the symptoms* related to Huntington Chorea. This means that it would be legal to discriminate against someone with the gene for Huntington Chorea as soon as that person exhibits the symptoms related to Huntington Chorea. In other words, one is protected from discrimination as long as one is viewed as asymptomatic (in the framework of the ADA, this would be an 'asymptomatic disabled') but as soon as one is viewed as symptomatic (a symptomatic disabled), one is not protected anymore.

It seems that all the law proposals and laws in place are aimed at preventing genetic discrimination of individuals based on their genetic predisposition to certain conditions as long as, in the medical sense, they have not developed the symptoms yet.[27] This means that the law proposals in the United States prohibiting genetic discrimination cover asymptomatic disabled people but not symptomatic disabled people. This has huge implications for symptomatic disabled people because to accept symptoms as a reason for discriminatory treatment not only questions the current laws prohibiting disability discrimination, which are based on the assumption that symptomatic disabilities are not a reason for discrimination but it seems to further remove symptomatic disabled people from being seen within a human rights and social justice framework. The fear that the symptomatic disabled would be further removed from this framework seems to be justified if we look at the fact that the distinction between symptomatic and asymptomatic disabled is endorsed by the American Civil Liberties Union (ACLU). The ACLU states in its campaign for a law prohibiting genetic discrimination "that Congress should take immediate steps

27. Paul R. Billings (1992) "The Scientific Basis of the 'Genetic Revolution': A Selective Review," in *The Genome, Ethics and the Law: Issues in Genetic Testing* (Washington, DC: American Association for the Advancement of Science), pp.23–45, at p.26. Joseph S. Alper, Lisa N. Geller, Carol I. Barash, Paul R. Billings, Vicki Laden, and Marvin R. Natowicz (1994) "Genetic Discrimination and Screening for Hemochromatosis," *Journal of Public Health Policy*, 15(3): 345–359; Larry Gostin (1991) "Genetic Discrimination: The Use of Genetically Based Diagnostic and Prognostic Tests by Employers and Insurers," *American Journal of Law and Medicine*, 17(1–2): 109–144; Carol Lee (1993) "Creating a Genetic Underclass: The Potential for Genetic Discrimination by the Health Insurance Industry," *Pace Law Review*, 13(1): 189–228; Mark A. Rothstein (1995) "Genetic Discrimination in Employment: Ethics, Policy and Comparative Law," in Swiss Institute of Comparative Law (ed.) *Human Genetic Analysis and the Protection of Personality and Privacy*. Proceedings of an International Colloquium (Zürich: Schultheiss Polygraphischer Verlag); Michael S. Yesley (1998) "Protecting Genetic Difference," *Berkeley Technology Law Journal*, 13(2): 653–665, at 662;. and Lemmens, *op. cit.*

to protect genetic privacy . . . [because] it is inherently unfair to discriminate against someone based on immutable characteristics that do not limit their abilities . . ."[28]

This focus on giving the asymptomatic disabled a separate law that protects them might reflect the fact that asymptomatic people do not feel protected by the ADA. It might also reflect the fact that asymptomatic people do not see themselves as disabled and do not want to be covered under the ADA. Human rights groups, the academia, and the policy makers agree that there is an ethical, moral, and political difference between discrimination against symptomatic and asymptomatic disabled people. Still, these new laws further marginalize the symptomatic disabled people from the human rights movement and further entrench a discriminatory approach of action that is solely justified by values based on a medical view of disability and a societal view of other characteristics.

4.4 Principles for Future Legislation Preventing Genetic Discrimination

Of course, disabled people in the United States do not agree with the fact that laws prohibiting genetic discrimination cover only asymptomatic disabled people. The National Council on Disability has looked at the issue of genetic discrimination numerous times and their *Principles for Genetic Discrimination Legislation* states that "it is crucial for any proposed legislation addressing genetic discrimination to reflect the following principles"[29] in various domains of social life. Thus, in the field of *employment*, "any proposed legislation must provide effective prohibitions against discrimination by employers based on genetic information."[30]

Recommendations for laws prohibiting genetic discrimination in the field of *insurance* and for regulating the collection and disclosure of genetic information by *insurance companies:*

> Any proposed legislation should contain comprehensive protection against genetic discrimination by providers of health, life, disability, and other types of insurance. Legislation must bar insurers from making decisions about enrollment based on genetic information. It must also prohibit insurers from using genetic information in determining premium or contribution rates, or other terms or conditions of coverage. Finally, it must bar insurers from requesting or requiring an individual to undergo genetic testing.

28. American Civil Liberties Union (2002) "ACLU Renews Calls for Congress to Ban Genetic Discrimination," *Workplace Rights, Press Release*, February 13, 2002 (New York: ACLU), available at http://www.aclu.org/WorkplaceRights/WorkplaceRights.cfm?ID=9688&c=180.
29. National Council on Disability, *op. cit.*
30. *Ibid.*

Any proposed legislation must prohibit insurers from requiring, requesting, collecting, or buying genetic information about individuals who are covered or seeking coverage.[31]

Recommendations for prohibiting genetic discrimination in the *health care system* and for regulating the collection and disclosure of genetic information by *health care providers*:

Any proposed legislation should protect against genetic discrimination by health care providers. Health care providers must not be permitted to refuse treatment to individuals, or treat them differently, based on genetic information. For example, 'futile care' policies, under which medically indicated treatments may be denied based on determinations that such treatments would be 'futile' in light of an individual's expected quality of life, should be prohibited to the extent that they result in denials of treatment based on genetic information.

Any proposed legislation must prohibit health care providers from requiring, requesting, or collecting genetic information about individuals who are seeking treatment. Providers may only collect this information for the purpose of providing genetic testing or other genetic services.[32]

The principles to be taken into account in any future legislation on *education and technical assistance*:

Any proposed legislation should include funding to permit education and technical assistance to be provided by appropriate organizations in order to ensure that individuals affected by the legislation are aware of its requirements.[33]

At the end of this position paper, The National Council on Disability criticizes the various courts for interpreting the Americans with Disabilities Act in a number of ways that severely restrict the number of people protected:

They have interpreted ADA to protect only individuals who are so limited in major life activities that they are extremely debilitated, and to exclude protection for many individuals who take mitigating measures to control the effects of their impairments. They have made it extraordinarily difficult for indivi-

31. *Ibid.*
32. *Ibid.*
33. *Ibid.*

duals to establish that an employer regards them as substantially limited in working. These interpretations of ADA have resulted in far more limited protection than Congress envisioned when it passed ADA.

The same interpretations that make ADA difficult to enforce by individuals with genetic markers make it difficult to enforce by individuals with a range of health conditions. When Congress considers protections for individuals who experience discrimination based on genetic information, it should also ensure that if these individuals are eventually diagnosed with medical conditions, they will be protected against discrimination.[34]

The American College of Medical Genetics seems to follow the sentiment of the National Council on Disability, when they state in their Position Statement, among other things, that:

All Americans should be protected by genetic nondiscrimination legislation addressing both health insurance and employment.

Protection against genetic discrimination in health insurance is needed for individuals covered by all public and private programs, whether through group or individual plans, and regardless of the mechanism by which the program is regulated.

Protection against discrimination in employment must cover the many participants in the process including employers, employment agencies, labor organizations, and training programs, and at all steps of the employment process.

[. . .] Like all health information, the privacy of genetic information must be adequately protected. Protection against unfair discrimination based on genetic risk for disease is achieved only by strategies that restrict use of genetic information in enrollment and rate setting.

[. . .] Definitions must avoid creating arbitrary distinctions that unfairly exclude some individuals from protections afforded to others.

Protected genetic information must include that based on evaluation, testing, and family histories of individuals and their family members.

Legislation should include enforcement to ensure compliance.[35]

34. *Ibid.*

35. Michael S. Watson, and Carol L. Greene (2001) "Points to Consider in Preventing Unfair Discrimination Based on Genetic Disease Risk: A Position Statement of the American College of Medical Genetics," *Genetics in Medicine*, 3(6): 436–437. Available at http://www.acmg.net/resources/policies/pol-025.pdf.

However, the problem of disabled people goes further than the fact that they are not covered by laws prohibiting genetic discrimination and that they are discriminated against with regards to life and disability insurance, which many cannot obtain.[36] The issue of genetic discrimination is a much broader one than just insurance and employment. Below I try to point out some forms of genetic discrimination that are rarely mentioned in the debates on genetic discrimination.

5. What Is Genetic Discrimination?

5.1 Pre-Birth Genetic Discrimination

There are many problems in the wording of documents demanding the prohibition of genetic discrimination. For example, one of the problems with the wording in Article 11 of the Convention on Human Rights and Biomedicine[37] is that it uses the term 'person'. The term 'person' is normally used for living human beings, although even this might change in the future.[38] Nevertheless, whatever the possible redefinition of the term 'person' in the future, the current use of the term 'person' in connection with the prohibition of genetic discrimination is already problematic. It is problematic because it leaves the impression that the act of discrimination only occurs on the level of the person, meaning the living human being, which is not the case. The act of discrimination is not associated with a cell, a zygote, an embryo, a fetus, a person, or a human being. Discrimination is associated with a *characteristic* of a cell, a zygote, an embryo, a fetus, and a human being. There are many ways to select or deselect the genotype or phenotype of one's offspring or of living members of society through, for example, pre-fertilization or pre-implantation diagnostics, prenatal or postnatal testing, and infanticide. The key question is how a society or

36. See footnote 26.

37. As already mentioned, Article 11 on Non-Discrimination in the Convention on Human Rights and Biomedicine states that "Any form of discrimination against a person on grounds of his or her genetic heritage is prohibited." See footnote 3.

38. Moves are underway to link the term 'person' to certain abilities of the living human being such as cognitive abilities. These abilities would question the use of the term person for a newborn human being, a human being with a cognitive ability below a certain level of 'IQ' or awareness of their environment (people with Alzheimer, comatose people, and so on). Joseph Fletcher (1979) *Humanhood: Essays in Biomedical Ethics* (Buffalo, NY: Prometheus Books). For more definitions of what makes a being a person, see James Park (2002) "Personhood Bibliography," at http://www.tc.umn.edu/~parkx032/B-PERSON.html. For a collection of Internet links on 'personhood', see the International Center for Bioethics Culture and Disability webpage at http://www.bioethicsanddisability.org/personhood.html.

a person decides which characteristics can be selected for or deselected against in an offspring or offspring-to-be.

Many of the genetic discriminations that would affect the characteristics labeled as 'disabilities' (mM) will happen through the use of prenatal predictive testing and selective abortion or pre-implantation genetic diagnostics and selective embryo implantation into the wombs of women through *in vitro* fertilization. And indeed many countries and international documents developed disability discriminatory prebirth strategies by allowing deselection based on the genetic characteristics labeled as 'disabilities' and by prohibiting deselection based on the genetic characteristics labeled as 'sex'. Genetic discrimination is even more clear in the distinction between prohibiting sex selection for 'social reasons' (gender balancing and wish of parents) and allowing sex selection for 'medical reasons' (preventing, for example, so-called X-linked diseases). (See the laws and legal practices of the State of Pennsylvania[39], India[40], China[41], Germany[42], Turkey[43], Great Britain[44], and Norway[45]). Some countries, such as Canada[46], intend to adopt acts that would outlaw sex selection for 'social reasons'.

39. Pennsylvania Consolidated Statutes Annotated, Title 18, [189]3204 (c), as amended November 17, 1989, P.L. 592, No. 68, [189] 2.

40. The Prenatal Diagnostic Techniques (Regulation and Prevention of Misuse) Act of 1994 bans sex selection. For more, see point 277 in India's First Report on the Convention on Elimination of All Forms of Discrimination Against Women (CEDAW), at http://wcd.nic.in/CEDAW4.htm.

41. Article 8 of The Act on Maternal and Infant Health Care (China), as it was published in *Renmin Ribao* (Peoples Daily), October 28, 1994. See also Xin Mao and Dorothy C. Wertz (1997) "China's Genetics Services Providers' Attitudes Toward Several Ethical Issues: A Cross-Cultural Survey," *Clinical Genetics*, 52: 100–109.

42. *Gesetz zum Schutz von Embryonen (Embryonenschutzgesetz, ESchG), vom 13. Dezember 1990.* BGBl. I 1990 S. 2746-2748 (BGBl III 453-19). See Article 3 on Banning Sex Selection (Verbotene Geschlechtswahl), http://www.bmgesundheit.de/rechts/genfpm/embryo/embryo.htm.

43. In Turkey, the problem reached such proportions that the Ministry of Health has restricted clinical application of sex selection methods when there is no medical indication. See Çağri Kalaça and Ayşe Akin (1995) "The Issue of Sex Selection in Turkey." *Human Reproduction*, 10(7): 1631–1632.

44. Human Fertilisation and Embryology Authority (1993) *Sex Selection*, A Public Consultation Document (London: HFEA), p.8, paragraph 38.

45. *Lov om medisinsk bruk av bioteknologi.* Act No. 56 of 1994 on the Medical Use of Biotechnology, Chapter 4: Preimplantation Diagnosis, Section 4(3): Sex Selection, and Chapter 5: Prenatal Diagnosis, Section 5(4): Information about the Gender of the Fetus. Text available in Norwegian at http://www.lovdata.no/all/hl-19940805-056.html.

46. Bill C-13 for an Act respecting Assisted Human Reproduction. First reading, October 9, 2002. Available at http://www.parl.gc.ca/37/2/parlbus/chambus/house/bills/government/C-13/C-13_1/C-13_cover-E.html. See also an earlier attempt to draft an act on human reproduction in Canada, which did not complete the legislative process: Bill C-47, for an Act respecting Human Reproductive Technologies and Commercial Transactions relating to Human Reproduction, 1996–97. Available at http://www.parl.gc.ca/bills/government/C-47/C-47_1/17946bE.html.

Furthermore, at least three international documents demand the prohibition of sex selection for 'non-medical reasons'. First, the European Convention on Human Rights and Biomedicine, which states in Article 14 on the Non-Selection of Sex: "The use of techniques of medically assisted procreation shall not be allowed for the purpose of choosing a future child's sex, except where serious hereditary sex-related disease is to be avoided."[47] Second, the WHO Draft Guidelines for Bioethics, which asserts in Article 21 that "Sex is not a disease. Except for severe sex-linked genetic disorders, the use of genetic services for the purpose of sex-selection is not acceptable."[48] Third, the Report of the International Bioethics Committee of the UNESCO on Pre-Implantation Genetic Diagnosis and Germ-Line Intervention, which states in paragraph 68 that "Destruction of embryos for non-medical reasons or termination of pregnancies because of a specific gender are not 'counterbalanced' by avoiding later suffering by a severe disease. Sex selection by PGD or PD is therefore considered to be unethical."[49]

Bioethicists and genetic professionals often share the opinion that "sex selection lowers the status of women in general and perpetuates the situation that gave rise to it" and therefore sex selection should be prohibited.[50] Many ethicists and geneticists are against sex selection for a variety of other reasons as well, as an international sur-

47. Article 14 of the Convention on Human Rights and Biomedicine. See footnote 3 and the Appendix to this volume.

48. Article 21 of the World Health Organization's *Draft Guidelines on Bioethics*. See footnote 4.

49. PGD stands for 'Pre-Implantaion Genetic Diagnosis'. See International Bioethics Committee of the UNESCO (2003) *Report of the IBC on Pre-Implantation Genetic Diagnosis and Germ-Line Intervention*, SHS-EST/02/CIB-9/2 (Rev. 3), April 24, 2003 (Paris: United Nations Educational, Scientific, and Cultural Organization), p.9. paragraph 68. Available at http://unesdoc.unesco.org/images/0013/001302/130248e.pdf.

50. For example, in her excellent summary of feminist views, Susan Sherwin argues that "[u]nless there is evidence that women would agree to this practice if they were free of patriarchal coercion, we cannot treat it as an acceptable local custom, even if the majority of citizens in areas where it is customarily practiced now approves of it." See Susan Sherwin (1992) *No Longer Patient: Feminist Ethics and Health Care* (Philadelphia: Temple University Press), p.74. "According to Sherwin's criteria, sex selection in Asian cultures is wrong because it results from and contributes to the oppression of women. Sex selection is wrong even if the woman herself requests it without direct coercion from her husband or partner, because her request emanates from a coercive culture. Therefore, in this case the majority culture should not respect requests for sex selection and should not consider itself 'ethnocentric' for doing so. Rejecting requests for sex selection does not denigrate an entire culture, only its oppressive aspects." Sherwin's argument is recited in Dorothy C. Wertz and John C. Fletcher (1993) "A Critique of Some Feminist Challenges to Prenatal Diagnosis," *Journal of Women's Health*, 2(2): 173–188, see p.181. Also available at http://www.umassmed.edu/shriver/research/socialscience/staff/wertz/critique.cfm.

vey testifies.[51] However, they do not accept the same argument if used by disabled people demanding a stop to the deselection based on abilities.[52,53] All of these laws and the international documents in the end justify the distinction between disability deselection and sex selection by labeling disability as a medical problem as outlined above.

However, if they accepted the social model of disability, which sees 'ableism'[54] on par with sexism, this selective prohibition based on selective characteristics would not be justifiable and Articles 11 and 14 of the European Convention on Human Rights and Biomedicine would directly collide with each other. Furthermore, as prohibition of sex selection can only work on the non-person pre-birth state of a human being, as that is where sex selection takes place, we have the case that Article 14 demands the prohibition of genetic discrimination (sex is a genetic condition) against a non-person. And if genetic discrimination against a non-person is to be outlawed than it should follow that it has to include the outlawing of genetic discrimination against any non-person independent of which characteristic is targeted in the non-person. In this sense Articles 11 and 14 of the European Convention collide.

6. Discrimination in Wrongful Life Suits and Wrongful Birth Suits[55]

The term *wrongful life* was originally used to define a case in which the child or someone on behalf of the child sues for compensation for general damages, namely for the pain and suffering incurred as a result of the harm or injury of being born

51. See Dorothy C. Wertz and John C. Fletcher (1998) "Ethical and Social Issues in Prenatal Sex Selection: A Survey of Geneticists in 37 Nations," *Social Science and Medicine*, 46(2): 255–273, and especially Table 10 at p.256. Also available at http://www.umassmed.edu/shriver/research/socialscience/staff/wertz/sexselecttable.htm#Table10
52. Edward Stein (1998) "Choosing the Sexual Orientation of Children?" *Bioethics*, 12(1): 1–24. See also Gregor Wolbring (2003a) "Disability Rights Approach to Bioethics," *Journal of Disability Policy Studies*, 14(4): 154–189; and Gregor Wolbring (2000) *Science and the Disadvantaged.* Occasional Papers of the Edmonds Institute (Washington, DC: The Edmonds Institute). Also available at http://www.edmonds-institute.org/wolbring.html.
53. As Andrew Brown of Amnesty International says, in discussing tests for Down's syndrome, "if society regards the presence of such disease as an acceptable reason for aborting a foetus, this makes it harder to preserve equality of respect for those already born. One might argue that their human worth, if not their human rights, have been diminished." See "Amnesty's Latest Fear: How Our Genes May Determine Our Fate," *The Independent*, February 18, 1998, p.19.
54. See footnote 9.
55. See the International Center for Bioethics Culture and Disability webpage "Wrongful Birth and Wrongful Life Suits," at http://www.bioethicsanddisability.org/wrongfulbirth.html.

rather than not being born.[56] In the original case in 1963, a child sued for being born out of wedlock.[57] The first wrongful life suit in which someone sued on behalf of a disabled child was decided in 1967.[58] In wrongful life suits involving 'disabled children', the plaintiff (the disabled child or someone on behalf of the disabled child) seeks compensation for general damages, namely for the pain and suffering incurred as a result of the harm or injury of being born *impaired* rather than of not being born. In these suits, the disabled child also sues for compensation for *special damages*, namely for the extraordinary medical, hospital, institutional, educational, and other expenses that are necessary to properly manage and treat the congenital or genetic disorder until the child reaches the adult age or even for his or her whole life. In wrongful life lawsuits involving disabled children, it is charged that having the disabled child in question was a legally bad act because having to live a life with the disability in question is worse than not existing at all.[59] The disability discrimination of wrongful life suits is evident in the fact that courts entertain wrongful life suits based on the claim of a disabled child,[60] but refuse to entertain wrongful life cases for other reasons, such as illegitimacy, poverty, or race.[61] This is by now also reflected

56. Jeffrey R. Botkin (1988) "The Legal Concept of Wrongful Life," *JAMA (The Journal of the American Medical Association)*, 259(10): 1541–1545. For a list of USA wrongful life suits see, Sven Tintemann (2000) "Thema: 'Wrongful Life'. Vergleich der Argumentation US-amerikanischer und deutscher höchstrichterlicher Entscheidungen," avaliable at http://userpage.fu-berlin.de/~stsurfer/ha/methosemho.doc.
57. *Zepeda v. Zepeda* (1963) 41 Ill.App.2d 240, 190 N.E.2d 849 (App.Ct. 1963), cert. denied (1964) 379 *U.S.* 945, 85 *S.Ct.* 444, 13 L. Ed. 2d 545.
58. *Gleitman v. Cosgrove* (1967) 49 NJ 22, 227 A.2d 689 (1967).
59. For academic discussions with this conclusion, see Lisa K. Lockhart; Peter H. Ditto; Joseph H. Danks; Kristen M. Coppola; William D. Smucker (2001) "The Stability of Older Adults' Judgments of Fates Better And Worse Than Death," *Death Studies*, 25(4): 299–317; Susan Macran and Paul Kind (2001) "'Death' and the Valuation of Health-Related Quality of Life," *Medical Care*, 39(3): 217–227.
60. Herbert Harrer (1994) "Aspects of Failed Family Planning in the United States of America and Germany," *The Journal of Legal Medicine*, 15(1): 89–127; Tony Sheldon (2003) "Court Awards Damages to Disabled Child for Having Been Born," *British Medical Journal*, 326(7393): 784, available at http://bmj.bmjjournals.com/cgi/reprint/326/7393/784/b.pdf; Jean Cohen (2001) "Wrongful Life Verdict in Paris," *Reproductive BioMedicine Online*, 2(1): 2, available at http://www.rbmonline.com/4DCGI/Article/2001/021/wp_2001_021.pdf; Nanette van der Laan (2001) "France Debates Right *Not* To Be Born," *The Christian Science Monitor*, December 7, 2001, webpage version available at http://www.csmonitor.com/2001/1207/p1s3-woeu.htm; and Julian Savalescu and Mary Spriggs (2002) "The Perruche Judgment and the 'Right Not To Be Born'," *Journal of Medical Ethics*, 28: 63–64, also available at http://jme.bmjjournals.com/cgi/data/28/1/DC1/21.
61. See the case *Zeev Yarmulovitz and the State of Israel v. Moshe Hovav* (Civil Appeal 518/82, 540/82), in Joel Levi (1987) "Wrongful Life Decision in Israel," *Medicine and Law*, 6: 373–374; *Williams v. State of New York* (1966) 18 NY.2d 481, 276 NY.S.2d 885, 223 NE.2d 343 (App.Ct. 1966); *Zepeda v. Zepeda* (1963), *op. cit.*; and Athena Nga-Chee Liu (1987) "Wrongful Life: Some of the Problems," *Journal of Medical Ethics*, 13(2): 69–73.

in a change of language, namely that wrongful life suits for reasons of illegitimacy, poverty, or race are called 'wrongful *dissatisfied* life suits',[62] whereas wrongful life suits for reasons of disability of the child are called 'wrongful *diminished* life suits'.[63]

In the case of *wrongful birth suits,* we have a similar change of language. Wrongful birth suits today normally relate to the birth of a disabled child, whose parents initiate a lawsuit on their own behalf, typically against a doctor who is accused of not performing proper genetic screening or not adequately counseling prospective parents. The essence of wrongful birth is that the defendant's negligence resulted in the birth of a disabled child whom the mother would have aborted had she received adequate medical information. In practice, wrongful birth suits are genetic or prenatal or preimplantation *malpractice suits.* They seem reasonable if we take into account the general view: that the disabled child is a burden, a damage to the mother. As Botkin says: "Parents of a child with unwanted disability have their interests impinged upon by the efforts, time, emotional burdens, and expenses added by the disability that they would not have otherwise experienced with the birth of a healthy child."[64] In order to cover the case of the birth of an unwanted but non-disabled child, the term *wrongful pregnancy* is used, meaning that a woman becomes pregnant without wanting it. This may happen when a pregnancy is not detected or a sterilization procedure fails.

Wrongful birth and wrongful pregnancy suits discriminate in two ways. One is based on the way the sufficient reasons for compensation are viewed. The parents may receive compensation for the birth of a disabled child reflecting the estimated amount of money needed until the child reaches adult age (or even longer).[65] But if, in a wrongful pregnancy case, a mother sues to be compensated for the cost of upbringing an unwanted but non-disabled child, she will only be compensated for the cost of the failed sterilization procedure but *not* for the expenses this additional child will cost her until the child reaches adult age.[66] The rationale for not giving costs compensation to the mother in case of an unwanted non-disabled child is that having a child is so highly valued that you cannot get reimbursed for it. However, in the case of the disabled child this argument is not used because having a disabled child is truly not a good thing and the mother was harmed.[67] Another way to discriminate is based on

62. Mary Pelias (1986) "Tort of Wrongful Birth and Wrongful Life," *American Journal of Medical Genetics*, 25: 71–80, see p.73.

63. *Ibid.*

64. See Jeffrey R. Botkin (1995) "Fetal Privacy and Confidentiality," *Hastings Center Reports*, 25(3): 32–39, see pp.36 and 37.

65. Harrer, *op. cit.*

66. This is the case in many countries, including the United States. *Ibid.*

67. *Ibid.*

the fact that mothers are not allowed to act on certain characteristics of their child: many countries have laws or law proposals prohibiting sex selection. That means the mother cannot sue if a predictive prenatal test misinterprets the sex of the fetus, as the mother would not be able to act on it anyway, but she can sue if a predictive prenatal test designed to detect certain 'disabilities' in a fetus fails to do so.

Wrongful life suits and wrongful birth suits are a logical consequence of the prevalent views that "it is socially irresponsible knowingly to bring an infant with a serious[68] genetic disorder into the world in an era of prenatal diagnosis"[69] and that "it is unfair to the child to be born with a disability."[70] In this kind of climate, the parents will likely accept that there is the option of preventing the birth of a disabled

68. The term 'serious' or 'severe' is often used to imply that there is a qualitative or ethical difference between targeting a so-called 'severe' or 'serious' and 'non-severe' or 'non-serious' disease, disability, or defect. However, this is a smokescreen. Ultimately, there is no way to distinguish between different so-called disabilities, diseases, defects, disorders, and impairments. Reality tells us that ethics and other policies do not draw a line between 'severe' or 'serious' and 'non-severe' or 'non-serious' but between using the concepts of 'medical reasons' versus 'social reasons'. See the following documents which state that a distinction between disabilities (sM) based on 'severity' cannot be done: Group of Advisers on the Ethical Implications of Biotechnology (GAEIB) (1996) *The Ethical Aspects of Prenatal Diagnosis.* Opinion of the Group of Advisers on the Ethical Implications of Biotechnology to the European Commission, No.6, February 20, 1996 (Brussels: The European Commission), available at http://europa.eu.int/comm/european_group_ethics/gaieb/en/opinion6.pdf; Marcus E. Pembrey (1998) "In the Light of Preimplantation Genetic Diagnosis: Some Ethical Issues in Medical Genetics Revisited," *European Journal of Human Genetics*, 6(1): 4–11, at p.9. See also International Bioethics Committee of the UNESCO (2003), *op. cit.*, p.9, paragraph 65: "An often-debated subject is line-drawing in case of the indications both for PGD and PD. So far all professional organisations in clinical genetics and reproductive technology and all advisory groups on bioethics have argued against lists of diseases that can be defined as severe enough to justify PGD or PD. The number of monogenic diseases alone exceeds 5,000 and nearly each of these has variants of different severity and clinical course. Also the same disease may be perceived differently by different couples depending on their family history, religious and socio-economic background, life situation and future expectations."

69. There was also widespread agreement with the statement that "it is socially irresponsible knowingly to bring an infant with a serious genetic disorder into the world in an era of prenatal diagnosis." Although there is no legal definition of 'serious', more than 50 percent agreed in South Africa, Belgium, Greece, Portugal, Czech Republic, Hungary, Poland, Russia, Israel, Turkey, China, India, Thailand, Brazil, Columbia, Cuba, Mexico, Peru, and Venezuela; as did 26 percent of US geneticists, and 55 percent of US primary care physicians. See Dorothy C. Wertz (1998) "Eugenics Is Alive and Well: A Survey of Genetic Professionals Around the World," *Science in Context*, 11(3–4): 493–501, see p.501.

70. When 2901 genetic professionals from 36 countries were asked in a survey performed between 1994 and 1996 whether they agreed with the statement that "it is unfair to a child to be born with a disability," the majority in 24 countries agreed along with 40 percent in the United States, Canada and Chile; 36 percent in Finland and the United Kingdom; 33 percent in Switzerland and the Netherlands; 29 percent in Argentina; 27 percent in Australia; 25 percent in Sweden; and 18 percent in Japan. Further, 52 percent of US primary care physicians agreed. See Wertz, *op. cit.*, p.501.

child: either through the use of prenatal testing and selective abortion after the fetus was labeled as having a disability (mM) or through pre-implantation diagnostics and the selection of an embryo deemed as being non-defective or non-disabled for implantation into a womb using *in vitro* fertilization. If they (the disabled and the non-disabled alike) are conditioned to think that the occurrence of the disabled child was preventable and that it is bad to live as a disabled, it is clear that many parents would sue on their own behalf and on the behalf of the child.

7. Outlook: New Forms of Genetic and Non-Genetic Discrimination Still to Come

If "it is socially irresponsible knowingly to bring an infant with a serious genetic disorder into the world in an era of prenatal diagnosis"[71] and if "it is unfair to the child to be born with a disability"[72] should it not also follow that it is irresponsible not to provide your child with a cure, as the bioethicist John Harris claims?[73]

Furthermore, would not parents be *obliged* to give their child non-genetic cures? Would not deaf parents be obliged to give their deaf child hearing abilities through cochlear implants (non-genetic cure) independent of whether or not they see 'being deaf' as a variation of being and believe in a 'deaf culture'?[74] Moreover, what would the answer be if parents could add 'advantageous' genes to their children's genetic make-up?[75] For example, if a gene enhancement could transmit immunity to AIDS,

71. See footnote 69.
72. See footnote 70.
73. "I do not believe there is a difference between choosing a preimplantation deaf embryo and refusing a cure to a newborn. Nor . . . an important difference between refusing a cure and deliberately deafening a child." He asks, "But [assume] now a cure for congenital deafness is discovered, it is risk free and there are no side effects. Would the parents in this case be right to withhold this cure for deafness from their child?" John Harris, *op. cit.*, p.97.
74. A first lawsuit in the USA just happened where child welfare took away the kids from their mother because she refused to give them cochlear implants. She won her case but she still has not received her kids back. See Cal Montgomery (2002) "Mom Can Refuse Sons' Cochlear Implants, Says Court," *Ragged Edge Magazine*, Oct. 4, 2002, at http://www.raggededgemagazine.com/drn/10_02.shtml#446.
75. See James J. Hughes (1996) "Embracing Change with All Four Arms: A Post Humanist Defense of Genetic Engineering," *Eubios Journal of Asian and International Bioethics*, 6(4): 94–101, also available at http://www.changesurfer.com/Hlth/Genetech.html. See also Nick Bostrom (2002) "A Transhumanist Perspective on Human Genetic Enhancements," at the author's own website: http://www.nickbostrom.com/ethics/enhancements.html. See further the "Human enhancement" and "Transhumanism" sections of the Institute for Bioethics, Culture and Disability website, at http://www.bioethicsanddisability.org/geneticnongeneticenhancement.htm and http://www.bioethicsanddisability.org/transhumanism.htm.
76. International Bioethics Committee of the UNESCO, *op. cit.*, p.13, paragraph 96.

would not parents be obliged to add this 'advantageous' gene to their children's genetic make-up? It has yet to be seen if routine somatic and germ-line gene therapy ever becomes feasible, but if that happens, would not parents be obliged to give their children (who fit the norm) genetic enhancements (to better the norm) as long as it is seen to be advantageous to the child?

And what do we do in the case of non-genetic enhancements and cures? Let's have a look at artificial legs. The bionic artificial legs of the future, which will take advantage of the advancements in the fields of nanotechnology, cognitive science, artificial intelligence, and information technology, will be more able than human legs. One could see these bionic legs as enhancements, not just cures. Why would parents only choose artificial legs which would bring their 'below the norm' child 'up to the norm' if legs are available which would give the child a competitive edge by having legs with 'above norm' abilities? These dynamics will lead to a 'normative creep' where the bell curve is moved towards the right to higher abilities and 'improvement of the norm'.

It is interesting that a draft report on pre-implantation genetic diagnostic and germ-line genetic intervention by the International Bioethics Committee of the UNESCO seems to be in favor of enhancements as long as they are done for 'medical reasons' when it says: "Without further elaboration, our Committee rejects the idea of testing and/or enhancing any human characteristic other than those of importance in alleviating suffering by disease."[76] Their reasoning for rejecting enhancement for 'social reasons' is the following:

- "With regard to germ-line interventions, the most fundamental argument is that we do not have the right to predetermine characteristics of future generations. The notion of justice between generations, defended by philosophers from completely different backgrounds also demands respect for the living conditions of future individuals who should be free to develop their potentialities without being biologically conditioned by the particular conceptions of 'good' and 'bad' human traits that were dominant at the time they were conceived. Neither PGD nor genetics in general should become an instrument for 'intergenerational tyranny'."[77]
- "Another argument against genetic enhancement of normal human characteristics is that it would profoundly affect our self-perception as 'persons'—that is as autonomous beings. Instead we might consider ourselves as mere 'things' or biological artifacts designed by others."[78]
- "A final objection against testing for normal characteristics, selection, and enhancement is that even if social agreement on the 'ideal' human being is reached,

77. *Ibid.*, p.13, paragraph 97.
78. *Ibid.*, p.13, paragraph 98.

it will inevitably reinforce stigmatisation and discrimination of those who do not fall into the accepted standards of genetically desirable traits. And who is able to define now the ideal human characteristics for the future?"[79]

However, the same arguments cited above can be used for objection against testing for so-called 'sub-normal' 'medical' characteristics, for selection and enhancement. Defining something as 'subnormal' and then allowing enhancement towards the norm "will inevitably reinforce stigmatisation and discrimination of those who do not fall into the accepted standards of genetically desirable traits" and it will "affect the self-perception [of the 'subnormal'] as 'persons'." Furthermore, it could also be said that "PGD and genetics in general . . . become an instrument for 'intergenerational tyranny'" if they are used for fixing or elimination of so-called 'subnormal' characteristics.

Interestingly, the above arguments reflect the arguments of disabled people: namely, that the elimination of their characteristic affects their own self-perception and that it increases stigmatization against disabled people. Furthermore, if "the notion of justice between generations . . . demands respect for the living conditions of future individuals who should be free to develop their potentialities without being biologically conditioned by the particular conceptions of 'good' and 'bad' human traits,"[80] then it means that these 'traits' are up to interpretation within cultural settings and that future generations have the right to develop without being burdened by these cultural prejudices of today's societies. However, if this is true, does that not mean that we cannot determine diseases, defects, and disabilities? Are we not subjectively, based on societal prejudice, deciding what we view as diseases, defects, and disabilities?[81] Are they not the above mentioned 'bad traits'? Being gay was seen not to long ago as a disease, a 'bad trait' in North America and is still seen as a disease, a 'bad trait' in some cultural settings in some societies. The arguments used above are true for all characteristics targeted. The only way to justify the use of therapy and enhancement technologies for disease, defects, and disabilities (the 'bad traits') but not for the 'good traits' lies in the basic fact of labeling something as a disease, defect, disability (that is, 'bad trait'), or 'good trait', which is a political, cultural, and societal decision, not an ethical one.

79. *Ibid.*, p.14, paragraph 99.

80. See footnote 77.

81. Gregor Wolbring (2003b) "The Social Construct of Health and Medicine," *Health Ethics Today*, 13(1): 5–6.

8. Conclusion

This chapter argued that disabled people are not protected from genetic discrimination by the current laws prohibiting genetic discrimination. Genetic discrimination for disabled people appears not only in relation to insurance and employment, but also in the use of prenatal and pre-implantation genetic testing and diagnosis, as well as of genetic and non-genetic therapy and enhancement tools. The justification of genetic discrimination against disabled people is based on labeling disabled people and the characteristic disability as a *medical problem* and therefore interventions are labeled as done for 'medical reasons', thus they are acceptable. Non-disabled people and 'non-disability' characteristics, on the other hand, are seen within a social justice framework and therefore interventions leading to discriminations are not acceptable. The above justification of genetic discrimination based on defining 'disability' within a medical framework is not accepted by disabled people who believe in the medical model, but in the social model or the transhumanist model.

The paper also showed that disabled people, for the most part, do not see themselves as a medical problem but rather as a social justice problem, and therefore on par with other human rights groups. This is a view mostly not accepted in the debates surrounding genetic discrimination and other bioethics issues. Disabled people, alternatively, may see every human being as a disabled person with a medical problem (transhumanist model of disability—a view that so far has been hardly debated in the literature. It was furthermore suggested that new technologies on the horizon, such as genetic and non-genetic enhancements, would change the dynamic of the usage of technology. In the current discourse, there are non-disabled people (mM) who are seen as normal and disabled people (mM) who are seen as subnormal, so we use technology to fix and 'elevate' the subnormal up to the norm. However, this old dynamic might soon be obsolete, as new technologies of enhancement will allow us to fix the subnormal in such a way that their capabilities actually exceed the norm.[82] These new capabilities could be seen as beneficial to human beings and therefore might become the 'new norm' which would mean that all the 'non-disabled' who still fit the old norm could be then be seen as subnormal. Therefore it might be better for society to see the disabled and the non-disabled through a social model, supporting both groups without prejudice and eliminating genetic discrimination for the benefit of all.

82. Enhancement technologies could be justified by saying that the fix was done for medical reasons to eliminate the subnormal state. This argument seems to fit with the International Bioethics Committee of the UNESCO, which states "Without further elaboration, our Committee rejects the idea of testing and/or enhancing any human characteristic other than those of importance in alleviating suffering by disease." International Bioethics Committee of the UNESCO, *op. cit.*, p.13, paragraph 96.

Insurance and Genetic Testing in Germany and the International Context

Jürgen Simon

1. Introduction

Genetic tests were originally used in the field of health care and in biomedical research, but over the past fifteen to twenty years, their application has become much more widespread. Genetic test results are now analyzed in forensics and paternity cases, and found valuable for employment and insurance purposes. The utilization of genetic testing and the resulting genetic information, however, is not without controversies. For example, laws in various countries have adopted general prohibitions against genetic testing for employment or insurance purposes. Some countries extended their existing human rights laws to this new domain; others enacted new genetic privacy and genetic discrimination laws. Prohibiting the use of genetic testing as a condition for health and life insurance, on the other hand, may raise another legal issue to grapple with: the duty to disclose information. This chapter provides a comparison and analysis of the solutions of various legal systems, with a special emphasis on the German legal model.

In the case of the insurance industry, the use of genetic information as a result of genetic testing may be regulated according to the following three models:

1. unlimited permission to conduct genetic tests for insurance purposes, including requiring genetic testing as a precondition for making an insurance contract;
2. access to the results of genetic tests already performed; and
3. general prohibition of requiring genetic test results as a precondition of an insurance contract.

Each of these models has advantages and disadvantages for the insurance companies as well as for the insurance applicant. A consideration of the rights concerned, the legitimate interests, and the economic aspects points to a *general prohibition* as the best model of regulation, if it is connected with the insurance applicant's duty to specify and disclose all relevant medical findings before the contract is made. In developing the best solutions for regulating the use of genetic test results, the fundamental differences between health insurance and life insurance have also to be taken into consideration.

In the case of *health insurance*, a complete disclosure obligation is ethically acceptable only if basic protection is secured by a general mandatory health insurance system (like in the case of the German *Bürgerversicherung*). In any other type of health insurance system it would be ethically unacceptable to require complete disclosure of medical information. A second best solution could be a duty to disclose the results of genetic tests *without* an obligation to undergo testing before the conclusion of an insurance contract.

In the case of *life insurance*, complete disclosure obligation, including undergoing a genetic test, is ethically acceptable only if there is an access to *general* life insurance without a disclosure obligation on an individual basis. Today persons who want to conclude an insurance contract must declare, for example, if their parents have had Huntington Chorea. On the other hand, a complete prohibition of disclosure obligation would also be ethically unacceptable because it would lead to *adverse selection*[1] and unequal handling of people who have a *conventional* (non-genetic) predisposition.

Many countries, including Germany, plan to reduce the range of public compulsory health insurance benefits. This means that additional benefits have to be provided by private insurance companies. But a more privatized insurance market would allow insurance policies on stricter conditions, and might eventually offer only basic coverage if they are not allowed to require genetic test results for their risk calculation. For certain social groups that could mean a reduced insurance coverage due to their genetic status.[2]

On the other hand, as genetic tests become widely accessible (in the form of 'home test kits', for example), the insurance applicants themselves also could make use of them and initiate an insurance contract with the knowledge of a present or future health condition. This would likely increase adverse selection on the life insurance market. As a response, the insurance companies would increase the premiums and a secondary market[3] might develop, producing a vicious circle. On the secondary market individuals with a high risk sell their policies to others with a lower risk. As a consequence, the life insurance market could ultimately break down.[4] This is because

1. The tendency that those people apply for insurance contracts (and thus constitute market demand) who pose a greater risk, with their health status, medical history, or living conditions.

2. Kurt Bayertz, Johann S. Ach, and Rainer Paslack (1999) *Genetische Diagnostik: Zukunftsperspektiven und Regelungsbedarf in den Bereichen innerhalb und außerhalb Humangenetik, Allgemeinmedizin, Arbeitsmedizin und Versicherungen* (Münster: Argos und Berlin: TAB), p.261.

3. Hennen, Petermann, and Sauter, *op. cit.*, p.127.

4. A similar dealing existed years ago when life insurance policies of HIV infected persons were offered on the capital market. If such a secondary market exists, the premiums must rise so as to avoid a deficit. In the worst case, the premiums are so high that the market collapses. See Achim Wambach (1999) "Die ökonomischen Auswirkungen von Gentests auf Versicherungsmärkte," Manuscript, p.10.

high risks—in this case, persons with positive test results—have to be insured by a higher sum, while low risks with a lower sum.[5] Thus, in order to avoid running a deficit, the insurance companies would have to increase the premiums. The consequence of this, in turn, would be that the already insured persons would also have to bear the increased risk. An argument against this scenario is that insuring bearable conditions are principally possible in case of both the health and the life insurance contracts.[6]

In the following, this chapter presents an overview of the present regulations of using genetic test results in Germany, with a special emphasis on the duty to disclose information. The next section reviews the solutions of international legal standards and national laws. The third part of this chapter then turns back to Germany, and formulates proposals for further legislative development.

2. Legal Regulation of Using Genetic Testing in Insurance Contracts in Germany

In Germany, on grounds of the welfare state principle, parliamentary acts regulate the operation and management of the health and pension insurance institutions. The social security system protects the citizens from the consequences of illnesses and loss of employment, provides health care services and a range of benefits.[7] Furthermore, those who are already insured under the state social security schemes may freely initiate insurance contracts with private health or life insurance companies as well. Of course, the realization of such contracts depends solely on the result of individual risk calculation.

5. Wambach, *op.cit.*, p.9.

6. For example in 1999 Swiss RE published a mathematical study concerning Huntington Chorea proving that persons with positive genetic test could be insured up to bearable condition. See further Robert Pokorski (1998) "Genetische Tests: Herausforderung für die Versicherungswirtschaft," *Die Kölnische Rück, Themen 7* (Köln: General Cologne Re), 12, p.30.ff.(33.f.); Irène Dietschi (1999) "Wie versicherbar sind 'schlechte Gene'?" in *Neue Zürcher Zeitung*, December 7, 1999, 285: 15; Jörg Schmidtke (1998) "Gentests in der Lebensversicherung," in *Versicherungsmedizin*, 50(3): 110–111, p.110ff.

7. According to Books V and VI of the Social Code *(Sozialgesetzbuch)*. In Germany, nearly 90 percent of the health insurance policyholders are members of the state health insurance scheme. The situation is similar in the case of pension insurance.

2.1 The Duty to Disclose Information in Insurance Applications

According to Article 16 of the Insurance Contract Act[8], the insurance applicant is obliged to disclose all information he has about the circumstances of his health which are relevant for the insurance contract and the insurer. Obligation means that if the insurance applicant fails to disclose the relevant information, then the insurer might withdraw the contract and the insurance applicant might not have a right to the insurance benefits. Article 16(1) further specifies that all circumstances are relevant that the insurer has explicitly and in a written form asked for. Moreover, all relevant circumstances have to be disclosed, even if the insurer has not explicitly asked for them. The insurer, on the other hand, has only this general right to obtain the relevant information; thus the duty of the insurance applicant to disclose information only exists if the relevant circumstances are adequate, necessary, and proportional for the insurer to assess the risk.[9]

The insurer may require that the insurance applicant undergo a *medical examination* or may look into the existing health documents[10], under the precondition that the insurance applicant indicates the names of his physicians who are then no longer bound to observe the principle of medical confidentiality. The insurance company may also choose to conduct the medical examination by its own doctors before making a decision over a contract. If the risked capital exceeds the amount of 125,000 Euros in the cases of capital, pension, or nursing care insurance, the insurance applicant has no choice but to undergo a medical check-up. During such medical examinations, information about family medical history may also be obtained. If the insurer requires it, the examining physician has to ask the insurance applicant about the past or present occurrence of diabetes, mental illness, or menstruation problems among close family members. Following this practice, insurers might actually include genetic testing among the required medical examinations, as long as the data coming from these tests are adequately relevant from the perspective of risk assessment.

The purpose of this general risk assessment is to make risk more calculable for the insurer, to guarantee the fulfillment of the insurer's obligation, and reduce the danger of adverse selection.[11]

8. *Gesetz über den Versicherungsvertrag (VVG).* Insurance Contracts Act of 1908, as amended. For the official text, see http://bundesrecht.juris.de/bundesrecht/vvg/gesamt.pdf.

9. Berberich, *op. cit.*, p.156ff.

10. Wolfgang Schulz-Weidner (1993) *Der versicherungsrechtliche Rahmen für eine Verwertung von Genomanalysen: die Bedeutung präklinischer kausalitäts-, Risiko- und Eigenverschuldungs-bewertungen* (Baden-Baden: Nomos), p.218ff.

11. Taupitz, *Genetische Diagnostik und Versicherungsrecht, op. cit.*, p.10ff.

2.2 Genetic Testing and the Duty to Disclose Information

Now we turn to the question of requiring and using *genetic tests* in the risk assessment made by the insurance company, and the current legal regulations in Germany concerning this issue.[12]

First of all, the use of genetic testing and test results must correspond to the German Constitution, and must not violate any constitutional rights. Constitutional rights, however, are not affected in all cases. It depends on whether a private or a state insurance scheme is involved; whether new genetic tests can be required or the results of an earlier test may be obtained; and whether the potential policyholder, the insurer, or a third person initiated the test.

The *right not to know* is the legal basis for the protection from unauthorized release of genetic test results. It is part of the personal rights, stated in Article 2(2) of the German Constitution, and based on the principle of human dignity, according to Article 1(1) of the German Constitution. Thus, if the individual is informed about his genetic make-up and condition against his will, then it violates basic personal rights. The right not to know protects the individual from any disturbing knowledge of future conditions, and preserves his freedom to choose his own life-style.

It is in the fundamental interest of the insurance applicant that the result of his genetic testing is kept secret. This right is part of the *right to 'informational self-determination'* derived from the general personal rights, as stated by the German Federal Constitutional Court in its famous census case decision in 1983 *(Volkszählungsurteil)*. The Court held that the individual has the right to decide on the use of his own personal data. This right is also binding in legal relations between an insurance applicant and the insurance companies, in that it protects the individual from having personal information, including that coming from genetic test results, revealed. The requirement to make the individual undergo genetic testing as a prerequisite for the conclusion of a contract could demote that individual to a mere object of economic interests—especially because without the free and informed consent of all parties a contract cannot be concluded. If the individual agrees to undergo genetic testing and accepts that the results will be revealed under social and economic pressure from the contracting partner, then the right to self-determination and the right not to know are violated. In this case, the formal 'consent' of the individual does not mean relinquishing the right not to know, as the individual himself did not initiate it.

12. Taupitz, "Genetische Tests," *op. cit.*, pp.72ff, pp.82ff.; Jürgen Simon (1993) "Rechtliche und rechtspolitische Aspekte der gegenwärtigen und zukünftig erwartbaren Nutzung genanalytischer Methoden am Menschen. Gutachten im Auftrag des Büros für Technikfolgenabschätzung beim Deutschen Bundestag," in *Bericht des Ausschusses für Forschung, Technologie und Technikfolgenabschätzung zur Technikfolgenabschätzung Genomanalyse*, BT-Drucks, 12/7094, p.13–83, see p.77ff.

Ultimately, it is the duty of the legislator to safeguard the free decision of the individual to know his genetic disposition by regulations in private and insurance law. This interpretation of free will and protection from intrusions into the private sphere creates more restrictive conditions for requiring genetic testing and using test results than it is in the case of regular medical examinations and diagnoses. But this added restriction on the use of genetic tests is justified on grounds of the particular sensitivity of the genetic information involved.

From the perspective of constitutionality, it is necessary that the refusal to undergo genetic testing or to disclose the results of previous genetic tests does not lead to legal disadvantages. Furthermore, the individual still has the right to consent to undergoing genetic testing and disclosing the results for preventive or therapeutic reasons.

We must see that, in fact, genetic testing does not fundamentally differ from other medical examinations[13] in that the threat of discrimination or the misuse of sensitive information is the same.[14] But the right to informational self-determination of the insurance applicant, concretely realized in the right not to know his genetic make-up, could be violated. In Germany, therefore, the insurer is not allowed to require genetic testing from the insurance applicant, although it may require the disclosure of previous genetic test results of earlier medical examinations.[15]

3. The International and Supranational Situation

3.1 National Laws

In the United States, life insurance companies are basically allowed to require genetic test results from the insurance applicants, as a precondition to conclude an insurance contract.[16] The laws in different federal states may impose restrictive con-

13. Jürgen Simon (1993) "Risikoregulierung und Rechtspolitik im Bereich Genomanalyse," in Reinhard Damm and Dieter Hart (eds.) *Rechtliche Regulierung von Gesundheitsrisiken* (Baden-Baden: Nomos), pp.19–37, see pp.24, 31.; Taupitz, *Genetische Diagnostik und Versicherungsrecht, op. cit.*, p.27ff.
14. Taupitz, *op. cit.*, p.29ff.
15. Jürgen Simon, (2001) *Gendiagnostik und Versicherung. Die internationale Lage im Vergleich* (Baden-Baden: Nomos), p.137.
16. Alexander M. Capron (1999) "Genetics and Insurance: Accessing and Using Private Information," in Ellen Frankel Paul, Fred D. Miller, Jr, and Jeffrey Paul (eds.) *The Right to Privacy*, Volume 17, Part 2 (Cambridge: Cambridge University Press), pp.235–275, see p.269; Mark A. Hall (1999) "Legal Rules and Industry Norms: The Impact of Laws Restricting Health Insurers' Use of Genetic Information," *Jurimetrics: The Journal of Law, Science, and Technology*, 40(1): 93–122, see p.98.

ditions to the use of genetic data in life insurance contracts, but overall prohibition does not exist in any state.[17]

The situation is different in the case of health insurance.[18] There are significant differences between the regulations of individual federal states,[19] although in most cases it is prohibited to use genetic test results in health insurance contracts.[20] The laws of the federal states do not prescribe a duty to disclose the results of previous genetic tests, and it would also contravene the Health Insurance Portability and Accountability Act of 1996 (HIPAA). In most of the federal states, it is also not allowed to use the results of genetic tests obtained before contracting, as the insurance company is not permitted to consider genetic information (in contrast to the life insurance contracts).[21]

The legal-political evaluation of these laws is extremely difficult because the regulations vary from one federal state to another. In the legal system of the United States, the existing laws on private health insurance prohibit genetic discrimination, although the term 'genetic information' is interpreted differently from one state to another.[22] However, this prohibition does not exist in the case of private life insurance. Private life insurance, as an instrument for retirement arrangements, plays a less important role in the United States than in Germany, and therefore it is not extensively regulated. The insurance applicant of health insurance contracts is thoroughly protected by law, insofar as in nearly all federal states a prohibition of genetic testing related to health insurance contracts exists.

17. Simon, *op. cit.*, p.61ff.
18. Simon, *op. cit.*, p.52.
19. Council for Responsible Genetics (2001) *A Comparison of Enacted State Genetic Discrimination Legislation* (Cambridge, MA: CRG). See the webpage http://www.gene-watch.org/programs/privacy/GDcomparisonchart.pdf.
20. Deborah A. Stone (1996) "The Implications of the Human Genome Project for Access to Health Insurance," in Thomas H. Murray, Mark A. Rothstein, and Robert F. Murray, Jr. (eds.) *The Human Genome Project and the Future of Health Care* (Bloomington: Indiana University Press), pp.133–157, see p.151ff.
21. Simon, *op. cit.*, p.56ff.; Christopher M. Keefer (1999) "Bridging the Gap Between Life Insurer and Consumer in the Genetic Testing Era: The RF Proposal," *Indiana Law Journal*, 74(4): 1375–1397; Trudo Lemmens (2000) "Selective Justice, Genetic Discrimination, and Insurance: Should We Single Out Genes in Our Laws?" *McGill Law Journal*, 45(2): 347–412.
22. Simon, *op. cit.*, p.55ff.

In the United Kingdom[23], the Netherlands[24], France[25], Switzerland[26], or Austria[27], laws do not regulate the use of genetic tests and test results as a precondition for a conclusion of either a life or a health insurance contract. In the United Kingdom and the Netherlands, insurance companies distinguish between the contracts according to the sum insured: if a certain limit is exceeded, there is at least the possibility to require the insurance applicant to disclose the results of previous genetic tests. In Austria, Switzerland, and France the disclosure of test results is neither a right nor a duty.

In every European country there is a regulation that requires the applying person to disclose information about all facts relevant for the insured risk (for example diseases). If the applicant fails to provide this information, the insurer is allowed to terminate the contract. But only the United Kingdom[28] derives from this duty a disclosure obligation in the case of life insurance contracts. In the Netherlands[29] there is a disclosure obligation if the information refers to diseases that are already recognizable. In this legal interpretation genetic information does not have a predictive but a diagnostic character, just like the results of conventional medical examination methods.

Concerning procedural law, there are two extremes. On the one hand, there is Austria with a complete and detailed legal regulation; on the other hand, there is the United Kingdom without a legal basis but a multitude of reports and codes of practice. Whereas Switzerland approaches the Austrian situation in its relevant bill[30], there is a very strict prohibition of any kind of genetic testing in French law.

Despite these large formal differences, all investigated countries show efforts to provide an optimal protection of sensitive genetic information in health insurance, following the principles of the welfare state.

There are a number of other countries that have either adopted laws regulating the use of genetic tests for insurance contracts or considered the issue in other legal norms. The unlimited use of information gained from genetic testing is allowed in

23. Simon, *op. cit.*, p.36ff.; Berberich, *op. cit.*, p.1190ff.

24. Simon, *op. cit.*, p.46ff.; Berberich, *op. cit.*, p.1193.

25. National Consultative Ethics Committee for the Health and Life Sciences (CCNE) (1995) *Opinion and Recommendations on "Genetics and Medicine: From Prediction to Prevention,"* Report, No.46, October 30, 1995 (Paris: CCNE). Also available at http://www.ccne-ethique.fr/english/avis/a_046.htm.

26. Simon, *op. cit.*, p.29ff.

27. *Ibid.*, p.27ff.

28. *Ibid.*, p.39ff; Thomas H. Murray (1992) "Genetics and the Moral Mission of Health Insurance," *Hastings Center Report*, 22(6): 12–17.

29. Simon, *Gendiagnostik und Versicherung*, p.47.

30. Bundesamt für Justiz, Begleitbericht zum Vorentwurf für ein Bundesgesetz über genetische Untersuchungen beim Menschen, www.ofj.admin.ch/d/index.htm, Stand 07.10.2000.

Canada, Spain,[31] and Japan. In Canada, the insurance applicant has a duty to disclose genetic information, and genetic test results can be used for concluding a contract.[32] In Japan the use of genetic testing is allowed, as there is no explicit law to regulate the issue. But because of the particular features of the Japanese health insurance system, genetic testing can be applied for life insurance contracts only. Negative results known by the insurance applicant of former genetic investigations have to be revealed.[33] In Denmark there is a duty to disclose genetic information if it concerns the current or former health condition of the applicant, which means that information about the future health condition cannot be considered in deciding the insurance contracts.[34] In Sweden, an agreement between the state and the national insurance association regulates the use of genetic testing, and the insurance applicant has to agree with the use of genetic information related to him or her.[35]

3.2 International Legal Norms

The harmonization of legal rules on an international and supranational level is essential if one considers the prospect of multinational insurance companies competing with each other on domestic markets, with the possibility of getting around national regulations by contracting in other countries.[36] In the European Union, legislative procedures on the national level have to respect European regulations and conventions, even if they are not yet ratified.[37]

3.2.1 The 1989 Resolution of the European Parliament

In 1989, the European Parliament adopted a resolution on the ethical and legal problems of genome analysis, which also dealt with the issue of how insurance companies might handle genetic data.[38] In Section 19 of this resolution, the European Parliament states that "insurance companies have no right to demand that genetic testing be carried out before or after the conclusion of an insurance contract nor to

31. Simon, *op. cit.*, p.72ff.
32. Simon, *op. cit.*, p.77.
33. Simon, *op. cit.*, p.76.
34. Article 3(a) of the Danish Insurance Contract Act.
35. *Avtal mellan staten Försakringsförbund avseende genetiska undersökningar, Regierungsschriften*, 1998/99: 136, p.33.
36. Schöffski, *op.cit.*, p.188.
37. *Ibid.*
38. Resolution of the European Parliament on the Ethical and Legal Problems of Genetic Engineering, Doc. A2-372/88, *Official Journal of the European Communities, C 096*, 17.4.1989, 165–171.

demand to be informed of the results of any such tests which have already been carried out and that genetic analysis should not be made a requirement for the conclusion of an insurance contract." Section 20 further asserts that "the insurer has no right to be notified by the insurance applicant of all the genetic data known to the latter." It follows from these statements that genetic analysis shall not be a prerequisite for the conclusion of an insurance contract. As the notion of 'genetic data' is wider than the notion of genetic testing results, a larger protection has been intended.[39] But it is not obvious if and how these principles should be applied in case of health and life insurance.[40] Although the European Parliament Resolution is a mere expression of opinion and does not have a binding force, it already indicates a restrictive tendency for later formulations of international legal norms.

3.2.2 The 1992 Recommendation of the Committee of Ministers of the European Council

The Committee of Ministers of the European Council adopted a Recommendation on genetic testing and screening for health care purposes on February 10, 1992. According to Principle 7 of this Recommendation, insurance companies "should not have the right to require genetic testing or to inquire about results of previously performed tests, as a pre-condition for the conclusion or modification of an insurance contract."[41] All governments represented in the Committee of Ministers of the European Council, except that of the Netherlands, consented to this recommendation.

3.2.3 The 1997 WHO Statement

Following an expert meeting organized by the World Health Organization at the end of 1997, the WHO proposed a guideline on ethical issues in medical genetics.[42] The core principle stated in this guideline proposal was that the results of genetic test-

39. Berberich, *op.cit.*, p.364.
40. Schöffski, *op.cit.*, p.189.
41. Recommendation No. R (92) 3 of the Committee of Ministers to Member States on Genetic Testing and Screening for Health Care Purposes, February 10, 1992, available at the webpage http://cm.coe.int/ta/rec/1992/92r3.htm; see also Berberich *op. cit.*, p.364; Spranger, *op. cit.*, p.818.
42. Proposed International Guidelines on Ethical Issues in Medical Genetics and Genetic Services, Report of a WHO Meeting on Ethical Issues in Medical Genetics, Document Ref. WHO/HGN/GL/ETH/98.1, December 15–16, 1997, Section 6, Table 4, see http://www.who.int/ncd/hgn/hgnethic.htm; Tade Matthias Spranger (2000) "Prädiktive genetische Tests und genetische Diskriminierung im Versicherungswesen," *Versicherungsrecht*, 51: 815–821, see p.818.

ing cannot be used without the consent of the individual affected. This document, however, is not legally not binding. Moreover, certain fields of potential application of genetic information are not exactly specified in this proposal, such as the possible use by third parties of genetic data acquired by insurance companies; the circumstances in which the insurance applicant is obliged to reveal genetic data connected to him or her; or the specific limits the companies can exert on the insurance applicant to reveal such data. According to this WHO proposal, there "shall be no compulsory genetic testing of adult individuals or populations. Every test shall be offered in such a way that individuals and families are free to refuse or accept according to their wishes and moral beliefs . . . Genetic data should not be given out to insurance companies, employers, schools or governments, other than after the full informed consent of the person tested."

3.2.4 The Convention on Human Rights and Biomedicine

The Convention on Human Rights and Biomedicine[43] protects the dignity and identity of all human beings (Article 11). According to Article 2, the interests and welfare of the human being shall prevail over the sole interest of society or science. Predictive genetic testing shall only be made for health purposes or for scientific research linked to health purposes. Otherwise genetic test results could influence the individual private sphere in an undesirable way.[44]

At first glance this formulation could be interpreted to mean that genetic testing for insurance purposes should be generally forbidden.[45] But this interpretation is not cogent. With respect to the Explanatory Report to the Convention on Human Rights and Biomedicine, containing explanations of the general rules of the convention without a legally binding character, predictive genetic testing is only forbidden if

43. Convention for the Protection of Human Rights and Dignity of the Human Being with regard to the Application of Biology and Medicine: Convention on Human Rights and Biomedicine (ETS No. 164). Signed in Oviedo, April 4, 1997. For the full text, see the Appendix of this volume, and the official webpage http://conventions.coe.int/treaty/en/treaties/html/164.htm.

44. Explanatory Report to the Convention for the Protection of Human Rights and Dignity of the Human Being with regard to the Application of Biology and Medicine: Convention on Human Rights and Biomedicine (ETS No. 164), authorized on December 17, 1996. For the full text, see http://conventions.coe.int/treaty/en/reports/html/164.htm. See also Birgit Lammersmann (1999) "Medizinische Eingriffe an einwilligungsunfähigen Personen: Die Position der Biomedizin-Konvention des Europarats im Spannungsfeld zwischen Forschungsbedarf der Medizin und Selbstbestimmungsrecht des Patienten," *Recht und Psychiatrie*, 17: 157–164; Spranger, *op. cit.*, p.819.

45. Theresia Degener (1998) "Chronologie der Bioethik-Konvention und ihrer Streitpunkte," *Kritische Vierteljahresschrift für Gesetzgebung und Rechtswissenschaft*, 1: 7–33, see p.14.

it is not made for health purposes.[46] This article of the Convention does not include a *general* prohibition of genetic testing.[47] Furthermore, although it can be read in this Report that the use of genetic testing for other than health purposes, for example, for an employment or an insurance contract, would be an unacceptable intrusion of the individual's right to privacy, a corresponding rule in the Convention does not exist. According to the Report of the German Government concerning the negotiation conditions of the second draft of the Convention "there was no agreement if an employer or insurer should have access to genetic data."[48] Therefore the steering committee decided to do without such regulation.[49] Thus, it is up to the national legislature and the involved parties of the procedure to make a decision,[50] especially in view of the possibility that a Member State might express its reservation according to Article 36. Nevertheless, it is a clear tendency in the Convention that the conclusion or modification of a contract should not depend on conducting predictive genetic tests.[51]

Risk assessment and questions concerning diseases that already existed in a family are not excluded by the Convention.[52] Furthermore, it seems to be possible to present negative genetic testing results so as to get a more favorable premium calculation.[53] In the opinion of Berberich[54] and Spranger[55], a prohibition on including genetic testing within a risk assessment (cf. Article 12) would actually not cause serious disadvantages.

46. Explanatory Report to the Convention on Human Rights and Biomedicine, *op. cit.*, Article 85 states: "Article 12 prohibits the carrying out of predictive tests for reasons other than health or health-related research, even with the assent of the person concerned. Therefore, it is forbidden to do predictive genetic testing as part of pre-employment medical examinations, whenever it does not serve a health purpose of the individual. This means that in particular circumstances, when the working environment could have prejudicial consequences on the health of an individual because of a genetic predisposition, predictive genetic testing may be offered without prejudice to the aim of improving working conditions. The test should be clearly used in the interest of the individual's health. The right not to know should also be respected."
47. Marie-Therese Tinnefeld (2000) "Menschenwürde, Biomedizin und Datenschutz," *Zeitschrift für Rechtspolitik,* 1: 10–13.
48. Degener, *op. cit.*, 14.
49. *Ibid.*, with reference to BT–Drucks, 13/5435, 2.
50. For another opinion, see Spranger, *op. cit.*, p.820.
51. Spranger, *op. cit.*, p.819.
52. Berberich, *op. cit.*, p.366.
53. Spranger, *op. cit.*, p.820; Berberich, *op. cit.*, p.366; for another opinion, see Ludger Honnefelder (1997) "Das Menschenrechtsübereinkommen zur Biomedizin des Europarats: zur zweiten und endgültigen Fassung des Dokuments," *Jahrbuch für Wissenschaft und Ethik. Band 2: Beiträge, Berichte, Dokumentation* (Berlin and New York: Walter de Gruyter), pp.305–330, see p.308ff.
54. Berberich, *op. cit.*, p.366.
55. Spranger, *op. cit.*, p.820.

The Convention entered into force as the necessary number of States has expressed their consent to be bound by the Convention.[56] It is a widespread view that the Convention was elaborated to create a minimum standard.[57] But Germany fears that only this minimum will be accepted everywhere. According to Article 27, none of the provisions of this Convention shall be interpreted as limiting or otherwise affecting the possibility for a party to grant a wider measure of protection with regard to the application of biology and medicine than is stipulated in this Convention. If Germany signs the Convention, the above mentioned rules have to be accepted in the area of insurance, unless a reservation is made (according to Article 36). This gives the German legislature a certain legal option.

3.2.5 Summary

Following the 1989 Resolution of the European Parliament, it is a central theme of all statements, decisions, and recommendations of international and supranational organizations, that the insurer should not have the right to require genetic test results before and/or after contracting. In respect to Germany's and other countries' future, the European inclusion of the Convention for the Protection of Human Rights and Dignity of the Human Being with regard to the Application of Biology and Medicine[58] is most important. Contrary to the German fears of its 'low standards',[59] it prohibits genetic tests if they do not serve the health of the individual person.

56. Status as of August 2003: Fifteen signatures not followed by ratification, sixteen ratifications or accessions, see http://conventions.coe.int/Treaty/EN/searchsig.asp?NT=164&CM=8.

57. Taupitz, *op. cit.*, 542; Köhler, *op. cit.*, 8.

58. Convention for the Protection of Human Rights and Dignity of the Human Being with regard to the Application of Biology and Medicine: Convention on Human Rights and Biomedicine (ETS No. 164). Signed in Oviedo, April 4, 1997. For the full text, see the Appendix of this volume, and the official webpage http://conventions.coe.int/treaty/en/treaties/html/164.htm.

59. Taupitz (1998), p.542f.; Michael Köhler (2000) "Europäische Bioethikkonvention: Beitritt unter Vorbehalt," *Zeitschrift für Rechtspolitik*, 1: 8–10, see p.8. Citing 'low' or 'minimum' standards refers to the German fear that rights guaranteed by the German legal system are not entirely protected by this Convention.

4. Directions of Further Legislation in Germany

4.1 Health Insurance

According to the current distribution of risk in Germany and other countries, the applicant is obliged to disclose all relevant information about his former and current state of health. Therefore he has to release his physician from medical discretion, which raises the issue of personal rights and the right to self-determination.

At present, 99.87 percent of the German population is insured. Thus, risk assessment is just a question of changing from public (compulsory) to private health insurance or within the private insurance sector. In this situation one cannot speak of a violation of personality rights through the dissemination of data. The right to self-determination is not affected either, even if a third party (e.g. relating family diseases) is included. However, health insurance officials usually do not request a physical examination from the applicant before the conclusion of a contract in order to gain new medical information. What is important is only the present distribution of risks. To maintain this situation, the insurance companies should also not be entitled to request a genetic test in the future.

On the other hand the applicant should not be entitled to refuse the inclusion of results of gene tests in the 'health check' (*bona fide* principle), if those results are to be handled like other medical findings, that is, if they do not have any novel and special quality. Results of genetic tests should be handled basically in the same manner as other medical examination results. They can indicate a future health condition just like other medical diagnoses.

This parity of genetic results and medical findings does not mean that genetic tests cannot take an exceptional position due to their significance. Genetic information does not change with age, clinical state, or tissue type. Among the other distinguishing features, one might mention that genetic data are stable; they can be derived from a very small sample; they affect not only the tested individual but his or her family; there is a coherence between genetic variations and ethnic affiliation; and so on. Genetic tests do not create new problems, but they might put existing problems in a different light. But this cannot justify attributing a new quality to genetic results. Therefore genetic test results have to be indicated in the same way as other medical findings, and the German legislature has to take this into consideration.

In certain countries the law requires the insurance applicant to provide *comprehensive* information about his or her state of health. This means that besides reporting current diseases, the insurance applicant has to specify the prognostic results of examinations as well. This includes hereditary diseases that may develop in the future but can be prognosticated now on the basis of present findings. If the results of genetic tests are to be treated like medical findings, it is consistent to treat the legal aspects of disclosure obligations relating to known health risks the same way as it has been the case with medical information. This means that knowledge about risks revealed by existing genetic tests has to be disclosed before the signing of a contract, just as it has been the case with the knowledge of medical risks.

This solution has the advantage of being medically and legally based on the current distribution of risks. Moreover it also allows for disregarding the questionable distinction between present and prognosticated medical information.

4.2 Life Insurance

The situation concerning life insurance is different. The insurer usually obtains information about the applicant's current state of health and possible future health developments in order to calculate risk. But, in addition, if the applicant for a life insurance contract wants to be insured at a higher sum, he or she has to undergo a medical check-up. To maintain the current distribution of risks it would be consistent to include in this obligation the use of genetic tests. This view applies the idea of the 'parity of information'. Thus, it appears appropriate to treat the insurance applicant who conceals a known disease at the signing of a contract in exactly the same way as one who conceals a corresponding genetic disposition.

In summary, it seems prudent to proceed with a gradated solution. Primary care should be guaranteed without a compulsory genetic test. In this manner, the right to self-determination in the form of the right not to know is taken into sufficient consideration. Beyond the level of primary care, compulsory genetic tests may be used as a precondition for an insurance contract, in a way similar to what is applied in the United Kingdom.

5. Conclusion

At the moment, genetic testing and the resulting genetic information are not used thoroughly in the fields of health and life insurance. From the perspective of the insurance applicant, however, this is a crucial issue, thus the possible legal solutions are intensively discussed throughout the world. There is one common interpretation in the propositions of all national and international regulations and supranational recommendations, namely that the necessity to protect an insurance applicant is smaller in the case of life insurance contracts than in the case of health insurance contracts. While applying for health insurance, the applicant should be globally protected against the use of genetic testing, but only on the condition that the applicant himself or herself does not conceal information on the results of previous genetic tests.

IV.

Genetic Databanks

The Constitutionality of the Icelandic Act on a Health Sector Database

Ragnar Aðalsteinsson

1. Introduction

In December 1998, the Icelandic Parliament (Alþingi) passed an act of law at the initiative of deCode Genetics Inc., a US for profit company, and Íslensk erfðagreining ehf, its subsidiary in Iceland. According to the Health Sector Database Act, the government was authorized to give an exclusive license for twelve years to a company in order to build a database containing existing medical records in Iceland. It was allowed without the informed consent of the patients, but with a possibility for them to opt out before a certain date.[1] The medical data is supposed to be non-identifiable. DeCode was granted the exclusive license in January 2000. The licensee is to negotiate agreements with health institutes and individual physicians in private practice about the conditions for transfer of data from their medical records to the Health Sector Database (HSD). DeCode has already concluded agreements with several hospitals in Iceland, but not yet with the largest hospital in the country. The regulatory agencies have not yet accepted the security arrangement set up by deCode and the transfer of health data has consequently not yet been started.[2]

2. Legislation and Political Debates

In 1997 the Icelandic subsidiary of deCode Genetics, a US for profit company, sent by telefax draft legislation on health sector databases in Iceland to the Ministry of Health. On the basis of the de Code draft the Minister of Health had a bill drafted

1. *Lög um gagnagrunn á heilbrigðissviði.* Act No. 139 of 1998 on a Health Sector Database. Passed by the Alþingi (Parliament) on December 17, 1998. See the Appendix to this volume. For the official English text, see the webpage http://government.is/interpro/htr/htr.nsf/pages/gagngr-log-ensk.

2. At the time of preparing this contribution for print in April 2003, the regulatory authorities in Iceland had not yet accepted the security arrangements proposed by deCode and consequently no transfer of health data into the HSD is in sight.

which was introduced to the Alþingi (the Icelandic Parliament) in April 1998. Information concerning the communications and negotiations between deCode and the Icelandic Government prior to the introduction of the bill in Parliament is still not available. The government bill was prepared in secrecy without consulting professionals outside government or other interested parties. The members of parliament were also unaware of the bill being drafted and of the negotiations between the government and deCode. In February 1998 deCode signed an agreement with the Swiss company Roche Pharmaceuticals in the presence of the Icelandic Prime Minister, Mr. Oddsson. His presence was generally understood to indicate that deCode enjoyed the enthusiastic support of the government.

It was the intention of the government in April 1998 to have the bill on the HSD passed by Alþingi within a few weeks and not later than before the summer recess in May. The government seems to have believed that the bill would not meet with any real opposition within and outside Alþingi. The wishes of the Government did not come true. The bill was subjected to heavy criticism both within and outside Alþingi. The government had to accept that the bill could not be passed in the spring as planned due to the unforeseen opposition. A new bill was drafted during the summer of 1998 and sent to various interested parties for comments and consultation. A final bill was drafted by the Ministry of Health in the autumn of 1998 and introduced to Alþingi in October 1998. The original draft from the spring had been amended to a considerable extent and a new provision added on the possibility for patients to opt out of the Health Sector Database before a certain deadline defined in the Act. During the second reading of the bill in parliament, a provision on a genealogical database was added and during the third reading of the bill, it was amended to cover the third database, a genetic database, and the possibility of having all three databases interconnected. The last amendment to the bill was made only two or three days before the final vote on the bill in late December 1998. It was passed with 37 votes against 20.

There was no public debate in Iceland prior to the bill being introduced to Alþingi simply because nobody outside a very limited circle of politicians and civil servants had any knowledge of the intention of the government to propose a bill to this effect. The notes to the first and the second bill as presented to Alþingi did not contain objective arguments for and against the proposed HSD. All the arguments in the notes were *for* the project—as if it did not have any negative aspects.

Even though the name of deCode as the prospective exclusive licensee for 12 years did not appear in the text of the bill itself, there was never any doubt that deCode was the pre-selected licensee. During this period, the government and deCode jointly held several town meetings where the planned project was unilaterally explained and praised. The final bill was sent to interested parties for comments by the relevant parliamentary committee and the newspapers published hundreds of articles, the majority of which were opposed to the bill on the HSD. The same applies to the comments received by the parliamentary committee from interested parties. Most of them were opposed to the project for several reasons, including the lack of informed consent; possible interference with privacy; discrimination as regards scientific freedom; the exclusive right to be granted to the licensee to the restraint of competition; erosion of the confidential relationship between doctors and patients; etc.

Unfortunately people in Iceland were not prepared for an intensive and in-depth public debate on the ethical and legal issues raised by the bill on the HSD. The Ministry of Health had recently issued a policy statement where it was said that there was no intention to establish a centralized database in Iceland containing the health data kept by the different health institutes and individually practicing physicians. Alþingi had in 1997 passed a fairly modern Act on the Rights of Patients providing for informed consent and medical data protection.[3] Most people were taken by surprise. The ethical questions raised by biotechnology and genetics combined with the new information technology had not been debated. Objective information was not available and the debate on the HSD project was immediately polarized and politicized.

So even though the debate was intensive, it was not very educating. The government parties emphasized the economic side of the project; it would bring back a lot of Icelandic scientists working outside the country; it would be the start of biotechnology industry in Iceland; the medical records were a national resource like the fish stocks in the sea; and it was the role of Alþingi to decide on the utilization of this resource as any other resource not subject to private property legislation. The government said that as the data in the database would not be personally identifiable, informed consent was not required. Anyhow those in doubt could opt out. The opponents on the other hand argued that privacy of the participants would be invaded as the data would be personally identifiable. They argued that since the health data in the database would not be disconnected or unlinked from the identifiers such as names and ID numbers, but only decoded, a key to the database could either be said to exist or could at least always be made. They pointed out that since new health data was constantly to be added to the data already in the database and since the genealogical and genetic data on the individual patients was to be linked to the data on the same persons in the HSD, the risks were overwhelming. The opponents also said that individual data subjects, the patients, had according to the bill no right to know what was registered and therefore no right to correct data incorrectly registered. The opponents considered this to be in violation of the most basic principles of data protection.

3. Act No. 74 of 1997 on the Rights of Patients.

In my opinion the process leading up to the passing of the HSD Act can hardly be said to have been democratic in the sense that the government did not inform the public and the Alþingi about all the aspects of the project, including the ethical issues. Further, the debate focused on the health sector database as an isolated databank. The possibility of it being connected to two other databases, genealogical and genetic databases, did not come into light until the very last moment in the parliamentary process and was never debated before the passing of the bill. The proposed health sector database ended up being much more by including in the Act and thereby legitimizing the genealogical database and the genetic database.

3. The Role of the Medical Profession and the Issue of Consent

It is noteworthy that according to the HSD Act, the boards of the health institutes are to represent the institutes actually in negotiation with deCode and the only role given to the medical staff of the institutes is an advisory role. The boards are politically appointed and their members do not necessarily belong to the medical professions. And as it is the case, most of them do not. The health data has been given by the patients to the physicians in confidence and on the implied condition that the physicians would not pass the data on unless it were for the direct benefit of the patient and only to those who needed to know. According to the Act, the control of the medical records is removed from the medical staff and given to the administrative boards as regards the conditions for transfer of medical data to the HSD. The government's justification for transferring the control to the boards was that medical records are not subject to property rights and are a national resource that would be without any value residing in the dispersed medical databases of the various hospitals and individual physicians. As the government was not able to fund the building of such a databank, the government said it had to license a private for-profit company to do the building and provide the funds against an exclusive right for 12 years each time.

In the first draft of the bill all medical records on paper of persons living or deceased were to be digitalized and transferred into the databank. In the final text this was amended and every living person now has the opportunity to opt out by informing the Director General of Public Health of his/her wish. The Director is to make a coded register of the opt-outs to be used by those that carry out the entry of data into the HSD. In the words of the government those who do not opt out are presumed to have given their consent to participate in the database as regards their medical records. There are no specific provisions in the HSD Act concerning those who are not able to make decisions of their own, such as children less than 18 years of age and persons that are incompetent for other reasons such as mental health reasons. Therefore only the parents of children and the legally appointed representatives of incompetent persons can opt out on their behalf under the general provisions of the legislation. If parents have not opted out their children under 18 years they can themselves opt out when they reach the age of maturity, but only as regards their future data. What has already been registered will remain in the databank under the law.

Let me digress a little at this point to tell you that last year deCode and the Medical Association of Iceland issued a joint statement to the effect that deCode would in the future erase, at the patient's request, all data on that patient already in the database. This statement, however, has no legal force and it is difficult to foresee how it will function in real life. On the other hand, through this joint statement, deCode has more or less managed to silence the criticism of the database coming from the Medical Association.

But what about medical data on patients already dead? It is not disputed that the medical data on persons who have opted out cannot be transferred to the database after their death. On the other hand the government's standpoint concerning persons deceased before the HSD Act came into force is that their descendants or spouses have no rights to opt out on their behalf. The court case that is already before the courts deals with the issue of medical data of deceased persons and I will come to this later.

The concept of informed consent is well known in connection with specific scientific research where the participants are informed of the aims, the methods, and the risks. Additionally, they are told that they can withdraw at any time from the research project and that all personal data will be destroyed at the project's conclusion. The patient is supposed to understand what is involved and to consent freely after having been so informed. Such an informed consent cannot be applied in the case of the HSD. The objectives of the research are vague and the data can be used for whatever purpose decided by the licensee, subject to approval of the regulatory bodies. Participation in the genetic database is, on the other hand, based on what is said to be freely given informed consent. DeCode has already succeeded in collecting tens of thousands of blood samples from patients.

Let me also digress at this point to tell you that a National Bioethics Committee complying with international standards had been set up in Iceland before this issue arose. When the Minister of Health found the committee was not reacting positively enough to requests from deCode, the Minister dissolved the committee and amended the rules so as to have government ministers appoint the committee members, instead of having them named by expert bodies like the Law Faculty of the University of Iceland and the Institute of Ethics of the same university.

4. Identifiability of Data

I now come to the point that will probably be the most important in the case already before the Icelandic courts and that, of course, is identifiability of the data. It has been maintained that research on data from medical records that cannot be linked to individual patients is an exception to the general rule on informed consent. The standpoint of the government and deCode is that this exception applies as the data can not be linked to identified or identifiable individuals as the one-way coding provided for in the Act renders the data non-personally identifiable. The critics of the database maintain on the other hand that the exception does not apply at all. They say that a key to the database either exists or can easily be made. They maintain that a table can be built which is a key and they also point out that keys can be built from comparisons of patterns of family trees as well as by putting generally available information into context. Therefore they say it is right and reasonable to obtain *a priori* consent of patients for the transfer of their health data to the database.[4] In this context they refer to the definition in the data protection Directive 95/46/EC, stating that personal data means:

> any information relating to an identified or identifiable natural person ('data subject'); an identifiable person is one who can be identified, directly or indirectly, in particular by reference to an identification number or to one or more factors specific to his physical, physiological, mental, economic, cultural or social identity.[5]

The directive was incorporated into Icelandic law in 2000, as Iceland is a member of the European Economic Area.[6]

I do not want to go into detail about the technicalities of identifiability here, but there are also more general considerations that should be borne in mind, such as the small size of the population of Iceland (280,000 persons). Further, there is a

4. Einar Árnason (2002) "Personal Identifiability in the Icelandic Health Sector Database," *The Journal of Information, Law and Technology (JILT)* 7(2). An electronic publication, see http://elj.warwick.ac.uk/jilt/02-2/arnason.html.

5. Article 2(a) of the Directive 95/46/EC of the European Parliament and of the Council of 24 October 1995 on the Protection of Individuals with Regard to the Processing of Personal Data and on the Free Movement of Such Data. For the official text, see *Official Journal of the European Communities, L 281, 23.11.1995*, pp.31–50, at http://europa.eu.int/comm/internal_market/privacy/docs/95-46-ce/dir1995-46_part1_en.pdf.

6. Act No. 77 of 2000 on the Protection and Processing of Personal Data, as amended by Acts No. 90 of 2001, No. 30 of 2002, and No. 81 of 2002.

difference between a Health Sector Database as a separate entity and a HSD linked to a genealogical and genetic database. The latter, in my opinion, inevitably increases the risks of misuse. On January 1, 2001 a new Act on Biobanks came into force in Iceland and it was passed almost without any debate in Parliament.[7] This Act also uses the concept of presumed consent by a patient that has had a biological sample taken from her or him for a clinical test. If the patient has not expressed an unwillingness, the sample can be permanently preserved in a biobank for use in scientific research if approved by the Data Protection Authority and an ethics committee, provided that important interests are at stake and that the potential benefits outweigh any potential inconvenience to the donor or other parties.

The question has also been asked whether it is of importance that approximately 90 percent of the medical data of a whole nation is transferred to a for-profit corporation that is one hundred percent foreign owned. The company is, of course, established and operated to return profits to the shareholders and investors. The profit motive must influence the policy decisions of the board and management of the company, especially under financial difficulties. Will that increase the risks? I think this is a reasonable question that must be addressed. It can be added that in the spring of 2002, Alþingi passed an act authorizing the Minister of Finance to guarantee a loan to deCode Genetics up to the amount of 200 million USD.[8] If the state takes on this guarantee, then the state will be very much involved in the interests of the company, potentially influencing government decisions in the future.

5. Testing the Constitutionality of the HSD Act

Finally let me tell you about the case already brought before the Reykjavík District Court. The plaintiff is a young girl of 17 years who lost her father approximately ten years ago. She requested the Director General of Public Health to opt out the data in the medical records of her father. The Director refused the request in an administrative decision and the reasons given were that even though the medical data regarding deceased persons was not mentioned in the text of the HSD Act itself, the legislative

7. *Lög um lífsýnasöfn.* Act No. 110 of 2000 on Biobanks, passed by the Alþingi (Parliament) on May 13, 2000. See the Appendix to this volume, and for the official English text, the webpage http://government.is/interpro/htr/htr.nsf/pages/Act-biobanks.

8. Act No. 87 of 2002 on the Authorization for the Minister of Finance on Behalf of the Treasury to Guarantee Bonds in Order to Finance New Activities of Íslensk erfðagreining ehf (the Icelandic subsidiary of deCode Genetics Inc.).

will to include medical data of dead persons appeared in the notes to the bill. It was also argued that the privacy of persons already dead did not enjoy the protection of the law. The claim of the plaintiff in the case against the Director General of Public Health is to have the decision invalidated and her demand for the opt-out of her father's data thereby granted. The reason for selecting this case as the first one is that there are certain problems of standing in cases like this. Namely, that the government can maintain that as everyone can elect to opt out of the HSD, no individual can have a legal interest in having the courts passing a declarative judgment on the constitutionality of the HSD Act.

This plaintiff, however, has such a legal interest. First of all, she has maintained that the privacy of deceased persons enjoys the protection of the law in the same way as their reputation does. Second, medical records contain information not only on the patients themselves but possibly also on their descendants and other family members. This especially applies in cases where medical records contain genetic data on a patient from which deductions can be made on the genotype of his or her children. The District Court of Reykjavík at an early stage of the proceedings dismissed the case at the request of the state on grounds of lack of legal interest by the daughter, but the Supreme Court on appeal overturned the decision and sent the case back to the District Court. In the premises of the Supreme Court's decision, it is accepted that the daughter can have an individual legal interest in having a substantive judgment passed on her claim, as the data in her father's medical records can reflect on her. She has herself opted out of the HSD. Written evidence from computer experts will be presented to the District Court concerning the identifiability of data in the database. This is significant, as I think the case will turn on whether the plaintiff will be able to convince the court of the identification risks through the presentation of such expert evidence.

The main argument of the government is that all the data in the HSD will be non-identifiable and consequently there will be no invasion of privacy or violation of other human rights protected in the Constitution.[9] If the state's arguments prevail it

9. On March 6, 2003 the Reykjavík District Court passed judgment in the case and found for the state. The court accepted that the plaintiff had an individual legal interest in having the court make a substantive decision on her claim. In principle, the court based its decision on the arguments presented by the state, that is, that all the security arrangements provided for in the Act, in administrative regulations and in the license considered as a whole made the health data in the HSD secure against external interferences by third parties. Thus, these data are claimed to be non-identifiable "in the legal meaning" of that term as stated by the court without defining its understanding of the term. The court, on the other hand, did not take any position on the risks of internal unlawful interferences with the health data, even though it is widely acknowledged that such attacks are quite common in practice.

means that under the Icelandic Constitution or Iceland's international obligations, citizens cannot rely on the traditional confidential relationship of medical doctors and patients. The interest of scientific research by privately owned for-profit corporations will prevail over the privacy interests of individuals.[10]

10. Article 10 of the Universal Declaration on the Human Genome and Human Rights (UNESCO, 1997) states: "No research or research applications concerning the human genome, in particular in the fields of biology, genetics and medicine, should prevail over respect for human rights, fundamental freedoms and human dignity of individuals or, where applicable, of groups of people." For the full text of the UNESCO Declaration, see the Appendix to this volume.

The Estonian Genome Project

Ants Nõmper and Krista Kruuv

1. Introduction

In September 2002, Estonia launched the Estonian Genome Project. Preparations for the project started already in 1999 and were monitored closely by the foreign press.[1] The first part of the project was successfully completed at the beginning of 2003. The principal goal of this pilot phase was testing, in three counties, the technical and organizational readiness for carrying out the main project. The second stage started in March 2003, with getting the populations of more and more counties involved in data collection. In addition to family physicians, specialists have also started to collect blood samples and medical history records. By the end of 2003, the whole population of Estonia will be covered by the project. For the year of 2003, the principal aim is to collect the descriptions of health status and tissue samples of 10,000 gene donors.

One of the cornerstones of the Estonian Genome Project is the Human Genes Research Act[2], which entered into force at the beginning of 2001. This Act represents one of the first attempts to codify law in the genetic era, and it has already served as a model for other countries intending to launch their own genome projects. The purpose of this article is to give an overview of the Estonian Genome Project and the Human Genes Research Act, as well as to compare them to some extent with the genome project of deCode in Iceland.[3]

1. For example, the magazines *Science, Nature, Nature Biotechnology*, and *Der Spiegel* and the daily newspapers *Financial Times, Frankfurter Allgemeine Zeitung, The Guardian*, and *The New York Times*. See media watch http://www.genomics.ee/index.php?lang=eng&show=16. Last accessed on July 20, 2003.

2. *Inimgeeniuuringute seadus* (Human Genes Research Act), passed by the *Riigikogu* on December 13, 2000 (as RT I 2000, 104, 685), entered into force on January 8, 2001. RT is *Riigi Teataja* (State Gazette). See http://www.legaltext.ee/text/en/X50010.htm for the official English translation.

3. For a comprehensive bibliography discussing the Icelandic genome project and the "deCode controversy," see http://www.raunvis.hi.is/~sksi/kit.html. Last accessed on July 20, 2003.

2. Overview of the Estonian Genome Project[4]

The goal of the Estonian Genome Project is to establish a database (the Gene Bank) consisting of phenotype and genotype data of the majority of the Estonian population. The Gene Bank will make it possible to carry out genetic and health studies in order to identify genes that cause and influence common diseases and to develop new medicines based on these discoveries.

The chief processor of the Gene Bank is the non-profit foundation Eesti Geenivaramu (EGV), which was established by the Republic of Estonia. The EGV is a foundation under private law, not a state office. It is governed by the Supervisory Board, which consists of nine members appointed by the *Riigikogu* (the Parliament of Estonia), the Estonian Government and the Estonian Academy of Sciences. According to the Human Genes Research Act (hereinafter HGRA), the objectives of the EGV are to

- promote the development of genetic research;
- collect information on the health of the Estonian population and genetic information concerning the Estonian population;
- use the results of genetic research to improve public health.

In order to achieve its objectives, the competence of the chief processor of the Gene Bank includes the establishment and maintenance of the Gene Bank. Under Article 3(1) of the HGRA, the EGV, as chief processor, has the right to

- organize the taking of tissue samples;
- prepare, code, decode, store, destroy, and issue descriptions of state of health and genealogies;
- perform genetic research; and
- collect, store, destroy, and issue genetic data.[5]

The costs of implementing the Estonian Genome Project are estimated to be about 150 million USD. As a non-profit foundation with seed capital of 70,000 USD,

4. For a more detailed comparison of the International Human Genome Project and other national genome projects, see Sasha Beljin and Hermann Fenger (2001) "Nationale Humangenomprojekte," in Stefan F. Winter, Hermann Fenger, Hans-Ludwig Schreiber (eds.) *Genmedizin und Recht. Rahmen-bedingungen und Regelungen für Forschung, Entwicklung, Klinik, Verwaltung* (München: C.H. Beck), pp.269–275. See also Gisli Palsson and Kristin E. Hardardottir (2002) "For Whom the Cell Tolls. Debates about Biomedicine," in *Current Anthropology*, 43(2): 271–301, especially pp.272–273; and Norwegian Ministry of Health (2002) *The Norwegian Biobank Report* (NOU 2001:19) (Oslo: Ministry of Health).
5. See the Appendix, p.345.

the EGV is not able to carry out the project without significant additional investment. Taking into account the amount of investment needed, the current economic situation in Estonia, and the public resources available for scientific research and the health care system, it became obvious that the size of the investment needed would have to come mainly from the private sector and most likely from foreign countries.

Therefore, in order to finance the project, the EGV has established a public limited company, EGeen. Another company, EGeen International, was incorporated in the United States to market the Estonian Genome Project, to target potential investors, and to mediate the negotiations between the EGV and the investors. In July 2003, the EGV exchanged its 100 percent share in EGeen for 2.5 percent share in EGeen International.

In order to attract investment into EGeen and to make EGeen lucrative for domestic and foreign investors, the EGV and EGeen signed a Gene Bank Agreement on September 19, 2001. The Gene Bank Agreement grants to EGeen the right to commercialize the data stored in the Gene Bank. The EGV has the right to conduct genetic research for scientific purposes and to transfer Gene Bank data free of charge to gene donors, physicians of gene donors, state agencies responsible for the organization of health care in Estonia, and to research institutions. For its right, EGeen is obligated to pay certain fees and charges to the EGV and to reimburse the foundation for all costs associated with the creation and maintenance of the Gene Bank. In fact, EGeen International will pay these charges and costs, as EGeen, like other spin-off biotech companies, does not make a profit yet. Supported mainly by that agreement and the HGRA, EGeen was able to raise enough money to finance its activities and to pay fees to the EGV, which enabled the foundation to launch the pilot phase of the project.

3. The Estonian and the Icelandic Genome Projects

In this chapter, we do not intend to provide a detailed comparison between the Estonian Genome Project and the Icelandic one. Nevertheless, it is useful to present a general overview of the economic aspects of both projects. Clearly, there are several other important factors that determine the sums paid by investors and therefore the following is just an illustration. The idea for such comparison arose in connection with an article written by Henry T. Greely, in which he argues that Iceland seems to have "got a bad deal."[6] The following describes briefly what kind of payments EGeen

6. Henry T. Greely (2000) "Iceland's Plan for Genomics Research: Facts and Implications," *Jurimetrics: The Journal of Law, Science, and Technology*, 40(2): 153–191, p.187.

in Estonia and the subsidiary of deCode Genetics (Íslensk erfðagreining ehf) in Iceland have to perform within the framework of each genome project.

Table 1. The Estonian and the Icelandic Genome Projects

	Iceland[7]	**Estonia**[8]
Annual payment	900,000 EUR, indexed	300,000 EUR, indexed
Annual profit payment	6 percent calculated on EBIT[9], max. 900,000 EUR	0.5 percent calculated on EBIT, unlimited
Compensation of costs related to the establishment of the Gene Bank	1,400,000 to 22,000,000 EUR, i.e., less than 1,000 EUR per person	65,000,000 to 100,000,000 EUR, i.e., less than 1,000 EUR per person
Payments to doctors	Applicable to doctors handing over the data Total sum unknown	Applicable to doctors collecting the data and taking blood samples, Circa 13 EUR per gene donor, in total cca. 13,000,000 EUR
Patent rights	No transfer of patent rights as in-kind contribution	All the patents are going to be in equal co-ownership of EGeen and EGV
Intellectual property rights	No payments on commercialization of intellectual property rights	Payment equal to 3 percent of the turnover from transfer of created intellectual property rights
Shareholdings in companies	No shareholding in the company commercializing the data	Guaranteed shareholding in the company commercializing the data; 2.5 percent anti-dilution clause[10]

7. Sources: Explanatory report to Icelandic Act on a Health Sector Database, see the webpage http://government.is/interpro/htr/htr.nsf/pages/gagngr-log-ensk; License Agreement between the Ministry of Health and Social Security and Íslensk erfðagreining ehf, see the webpage http://brunnur.stjr.is/interpro/htr/htr.nsf/Files/Aggreement/$file/AGREEMENT-english.pdf, both accessed on August 1, 2002.
8. Source of information: various Estonian newspapers.
9. EBIT stands for earnings before interests and taxes.
10. The Ethics Committee of the Human Genome Organization recommends a one to three percent profit-sharing clause. HUGO Ethics Committee, Statement on Benefit-Sharing, April 9, 2000. See the webpage http://www.hugo-international.org/hugo/benefit.html, accessed on July 24, 2003.

There are no good criteria to measure the advantageousness of the economic conditions of the projects. Also, there are no good arguments to value the projects in order to argue that the payments should be higher or lower. Therefore, the biggest difference must not be seen in annual payments but merely in the fact that the EGV acquires some patent rights and payments related to commercialization of intellectual property rights and, more importantly, shareholding in the company commercializing the data (that is, in EGeen International). The monetary value of this shareholding will be known after the initial public offering.

4. Human Genes Research Act

4.1 Background to the Human Genes Research Act

From the very start of the Estonian Genome Project, the adoption of an act to regulate human gene research and the operation of a gene bank has been considered indispensable. The initiators of the Genome Project set this objective out of their own free will as neither Estonian law nor recommendations by international organizations require the adoption of such an act.[11] There exist some stipulations in the Convention on Human Rights and Biomedicine of the Council of Europe calling for adoption of necessary legislation to protect the principles laid down in the Convention (for instance the Articles 23 and 25).[12] However, taking into account the broad and soft wording of the Convention,[13] it becomes obvious that the Convention does not require the adoption of a human genome research act. Moreover, the Estonian Human Genes Research Act was adopted more than a year before Estonia ratified the Convention. Therefore, the adoption of the HGRA with only a few substantial changes (for example, regulation of genetic testing was omitted from the draft act) reflects a strong sense of responsibility by the initiators of the Estonian Genome Project and the high standards of Estonian legal culture.

11. For example, it has been argued that there is no need for a separate Act to regulate the protection of genetic data as such data does not constitute qualitatively new data. See Martin Koppernock (1997) *Das Grundrecht auf bioethische Selbstbestimmung. Zur Rekonstruktion des allgemeinen Persönlichkeitsrechts* (Baden-Baden: Nomos), p.86.

12. Convention for the Protection of Human Rights and Dignity of the Human Being with regard to the Application of Biology and Medicine: Convention on Human Rights and Biomedicine (ETS No. 164). Signed in Oviedo, April 4, 1997. For the full text, see the Appendix to this volume.

13. Cf. Hans-Ludwig Schreiber (2000) *Das europäische Menschenrechtsübereinkommen zur Bioethik: Ein Schritt auf dem Wege der Einheit des Biorechts*. Aulavortrag September 14, 1999 (Tartu: Tartu Ülikool, The University of Tartu Press), p.14.

The HGRA, as a law specific to Estonia, was drafted by Estonian geneticists, lawyers, physicians, ethicists, data protectors, officials, and politicians. This Act is one of the pioneers in legislating genomic law around the world. A lack of predecessors to follow, however, does not mean that the working group only relied on its own knowledge or that it did not pay attention to opinions published in law journals, the legislation of other countries, or foreign expert advice. In some respects, nevertheless, the HGRA is likely to set an example for the legislature of other countries. Its principles and structure have already been a model for legislators in Latvia, where in June 2003 the Latvian Parliament passed the amended Act on Human Genome Research.

4.2 Main Aspects of the Human Genes Research Act

The HGRA consists of seven chapters. The first chapter is dedicated to general provisions; definitions used throughout the Act; composition and functions of the EGV and other authorized processors of the Gene Bank; and to a declaration to limit the scope of the Act only to genetic research, thus not covering genetic testing.

The second chapter contains a list of gene donor's rights and regulates the consenting procedure. Under HGRA, every gene donor has among other rights the following rights:

- to remain anonymous after coding;
- to permit disclosure of his or her identity;
- to disclose the fact of being or not being a gene donor and the circumstances thereof, unless otherwise prescribed by law;
- not to know their genetic data;
- to access their data for free except genealogies stored in the Gene Bank and to receive genetic counseling upon accessing their data;
- to submit additional information on themselves to the chief processor and to prohibit the supplementation, renewal and verification of descriptions of their state of health stored in the Gene Bank.

The HGRA and the Estonian Genome Project are based on the principle of expressed informed consent. In order to become a gene donor, a person has to go through a counseling process and then sign a written gene donor consent form. This consent form has been adopted by the Ministry of Social Affairs of the Republic of Estonia and covers the following main aspects, as also stated in Article 12(4) of the Act:

- granting the consent is voluntary;
- the gene donor is not entitled to request a fee for providing a tissue sample, preparation and study of a description of his or her state of health or genealogy, or use of the research results;

- data on hereditary characteristics and genetic risks obtained as a result of genetic research may be unpleasant for the gene donor;
- the gene donor has the right not to know his or her genetic data;
- the gene donor has the right to know his or her genetic data, except genealogy;
- the gene donor has the right to apply to the chief processor for the destruction of data which enables decoding or, upon unlawful disclosure of his or her identity, for the destruction of the tissue sample, description of DNA, and description of his or her state of health;
- the gene donor has the right to withdraw his or her consent until his or her tissue sample or the description of his or her state of health is coded.

Besides listing the rights of the gene donor and the possible consequences of donating tissue samples and health descriptions, many other important details should also be disclosed to the donor, such as the description of the operation of the EGV, the aims and financial background of the project, the possible further uses of data, ownership of data and samples, and coding and decoding procedures. The main purpose of the consent form is to provide limited but essential information, which is necessary for a person to make an informed 'to do or not to do' decision. Additional information is available in the form of the so-called 'gene donor information kit' which contains leaflets, contact information, the text of the HGRA, etc. Even though Russian is not an official state language of Estonia, all the documents are translated into Russian.

Despite the availability of written information one should not forget the importance of discussions between the gene donor and a general practitioner. Therefore, the procedure of obtaining consent from a gene donor constantly underlines the need to discuss the issues with the possible gene donor.

The following chapter of the HGRA regulates processing data in the Gene Bank. The medical procedure of taking a tissue sample is performed by a health care professional. This person also collects health data of the gene donor and submits the data and blood sample to the EGV in a secured form, keeping the identity of the gene donor confidential. The EGV carries out further coding of information and is entitled to process the data. Tissue samples, descriptions of state of health, other personal data, and genealogies are transferred into the ownership of the EGV from the moment when the tissue sample or personal data is provided or from the moment that the state of health or genealogy is prepared. Tissue samples and uncoded information owned by the EGV, as well as the written consent of gene donors are not transferable. The Act regulates the storage of data, the use of this data by authorized processors, the cases in which data shall be destroyed, as well as the ways these actions are to be recorded.

A separate chapter of the HGRA deals with data protection issues. As a basic and unexceptional rule, all data in the Gene Bank shall be processed in compliance with the highest standard of data protection. No data shall be issued from the Gene Bank in uncoded form, except in cases expressly set forth by the HGRA. For the gene donors, the possibility of data leakage poses the greatest risk, because it allows decoding and therefore breaches confidentiality. The HGRA fights this possibility in several ways. First of all, data enabling identification of gene donors shall not be available through the external computer network of the Gene Bank. Only a few members of the EGV staff shall have access to such data and these persons are closely monitored.

Second, the coding procedure itself must be approved by the Estonian data protection supervision authority and meet at least the following requirements. Each tissue sample, description of DNA, description of state of health, and genealogy shall receive a unique code consisting of at least sixteen random characters immediately after receipt in the Gene Bank. This code shall replace all data concerning a tissue sample, description of DNA, and description of state of health which enables the reverse identification of the gene donor, including the name, personal identification code, date of birth, and residence. The same code shall be indicated on the written consent of the gene donor and the written consent shall be the only possible key for decoding. In the interests of greater security, the EGV shall carry out an additional coding.

In order to prevent indirect identification of a gene donor, the chief processor may not issue data from the Gene Bank even in coded form if no less than five persons can be the source of that information, that is, who meet the criteria of issued data.

One of the distinguishing features of the Estonian Gene Bank is that the information in the bank can be decoded. For instance, in contradiction to the name of the company behind the Icelandic project, deCode, the methodology applied in working with the Icelandic database does not allow decoding. However, Henry T. Greely has quite convincingly argued that there must exist a way in Iceland to trace back to a particular gene since otherwise the updating of the databank would be impossible.[14]

Article 24 of the HGRA allows decoding in the following cases and for the following purposes:

- in order to destroy a tissue sample, a description of DNA or a description of state of health or data which enables decoding;

14. Henry T. Greely (2000) "Iceland's Plan for Genomics Research: Facts and Implications," *Jurimetrics: The Journal of Law, Science, and Technology*, 40(2): 153–191, p.184.

- in order to enable access to data on a gene donor stored in the Gene Bank, except genealogies, at the written request of the gene donor;
- in order to renew, supplement or verify a description of the state of health of a gene donor without contacting the gene donor unless the gene donor has prohibited the supplementation, renewal or verification of the description of his/her state of health;
- in order to identify a gene donor on the proposal of the chief processor and with the consent of the Ethics Committee, to contact the gene donor and to renew, supplement or verify a description of his/her state of health with his/her written consent;
- in order to identify a gene donor and, with the gene donor's written consent, to take a new DNA sample if a tissue sample has been destroyed or does not contain sufficient DNA;
- in order to identify a gene donor within the Gene Bank and amend his/her genealogy if the results of DNA research contradict the previously known genealogy or provide new information concerning the genealogy;
- in order to issue a description of the state of health of a gene donor to the doctor of the gene donor at the request of the doctor of the gene donor and with the consent of the gene donor or, in urgent cases where a gene donor is unable to grant consent and has no legal representative or guardian or if the legal representative or guardian of the gene donor is unavailable, without the consent of the gene donor but in his/her interests and according to his/her presumed intentions.

The fifth chapter of the HGRA contains very strict rules regarding discrimination. The prohibition of discrimination entails the constitutional principle that all persons are born equal. The HGRA sets forth a general prohibition to restrict the rights and opportunities of a person or to confer advantages on a person on the basis of the structure of the person's DNA and the genetic risks resulting therefrom. Since in some areas the improper wish to discriminate on a genetic basis is greater, there are also separate legal norms prohibiting any discrimination in employment and insurance relations and providing for criminal sanctions.

However, one should note that in the Estonian health insurance practice genetic difference is not considered a risk. The Estonian health insurance system is based on the principle of solidarity and is financed by taxpayers. Everyone having a job or not able to work due to age or physical disability are covered by the health insurance scheme irrespective of their genetic make-up.

4.3 Withdrawal of Consent

One of the issues criticized in Estonia in connection with the Estonian Genome Project and the HGRA is the regulation of withdrawal of consent and its implications. Pursuant to Article 12(4)(7) of the HGRA, a gene donor has the right to withdraw his or her consent until a tissue sample or the description of state of health is coded. In such cases, the gathered information and blood sample shall be destroyed. After the coding is done, a gene donor has the right to apply, at any time, to the chief processor for the destruction of data that enables decoding. However, destruction of data that enables decoding (i.e. anonymization) does not also mean the destruction of biological material or other data. A gene donor has the right to request destruction of biological material and other data available in the Gene Bank if the identity of a gene donor is unlawfully disclosed. It has been argued that such a limitation of gene donor rights is in contradiction with the Articles 5 and 16 of the Convention on Human Rights and Biomedicine, which state that the person concerned may freely withdraw consent at any time.

As stated in the first section of the Article 5 of the Convention on Human Rights and Biomedicine, the whole article regulates intervention in the health sphere. In the framework of the Project, intervention in the health sphere occurs during the drawing of tissue samples from the gene donor. As the tissue sample drawing is always performed prior to coding the tissue sample, the regulation of the HGRA shall not restrict the right of the gene donor to withdraw his or her consent. Article 16 of the Convention on Human Rights and Biomedicine, regulating the scientific research, has been drafted taking account Article 3 of the Recommendation R (90) of the Committee of the Ministers of the Council of Europe.

The specified recommendation regulates the implementation of medical research, where medical intervention into a human body is continuous. Thus, it is obvious that the Convention on Human Rights and Biomedicine is aiming to ensure unlimited right to withdraw consent if the person is still under medical intervention. The goal of the Convention is to protect a person against intervention into his or her body. The supplementary risks that may occur together with intervention into the body are significantly bigger than risks that may occur while studying the already segregated tissues. Therefore it is not correct to extend the regulation of the Convention automatically to the study of the tissues already segregated from the body. The corollary is that the HGRA does not violate the requirements of the Convention by giving to a gene donor only a limited right to withdraw his or her consent.

At the same time, the study of the tissues already segregated from the body does contain some risks for the person. These risks, however, are not so much connected with the physical aspect of self-determination, as with the informational aspect of self-determination. Taking into consideration such transformation of legal values to be protected, the HGRA provides for the transforming of the right to withdraw the consent (measure of the protection of physical aspect of self-determination) into the requirement of disabling decoding of the existing data (measure of the protection of informative aspect of self-determination). After the data that allows decoding is destroyed, the health data and the tissue samples of a gene donor, stored the Genome Bank, are anonymous. Thus, the regulation of personal data protection shall not extend to the tests performed with them. Therefore we are of the opinion that the regulation of the HGRA regarding withdrawal of consent is in accordance with the Convention on Human Rights and Biomedicine.

5. Conclusion

After the successful pilot stage, the Estonian Genome Project has entered its main phase of data collection. The initial plan of the project has turned out to be adequate, the cooperation of different parties has been fruitful and the technical systems have proved to be reliable. During the main phase, the Eesti Geenivaramu exercises control over the project according to the Gene Bank Agreement and the Human Genes Research Act, rather than through its shareholding in EGeen or EGeen International, Inc.

As data collection expands to cover the population of the whole country, data processing and storage capacities are also increased. According to the development plan for the 2003–2007 period, the Gene Bank will start its operation. Information technology solutions will enable the Gene Bank to issue personalized data to the gene donors, to produce gene maps of the donors, to carry out scientific research on this data, and to start marketing the genome database and the products of the Gene Bank.

Together with the expansion of the database, various public relations activities are planned with the purpose of informing the Estonian population about the development of the project and giving supplementary training to the project parties. In order to guarantee the high quality and security of data processing, required also by the ISO, the chief processor cooperates with a number of international partners, including the participation in the work of the International Consortium of Public Population Genomics Projects (P3G). Moreover, the comparison with the Icelandic genome project of deCode shows that if the Estonian Gene Bank is successful, the EGV and the Estonian society will probably benefit to a greater extent than the Icelanders.

The Human Genes Research Act has been in force for almost three years, without the need to make amendments to it. This proves that the leap into the unknown, which was made by the Estonian legislature at the end of 2000 by adopting the HGRA, has been a successful one. The main principles of the HGRA have already found their way into the legislation of other countries, particularly Latvia. Estonia is looking forward with great interest to seeing whether the gene-banking regulations on the European and global levels will require amendments to the HGRA.

The Latvian Genome Project

Valdis Pīrāgs and Elmārs Grēns

1. Genome Projects in Service of Fighting Against Diseases

The European public health care systems are currently facing enormous problems due to a dramatic increase of common multifactorial diseases related to population aging and various environmental factors. Cardiovascular diseases, cancer, autoimmune diseases, and metabolic disorders, including obesity and diabetes mellitus, have become the main causes of morbidity and mortality in most countries. It is, however, difficult to find the exact diagnosis and proper treatment of these diseases in concrete cases, as the different genetic background of the affected individuals causes different responses to 'standard' treatments, and even the same diseases might have different clinical features.

From the viewpoint of philogenesis, the selection of our genes occurred under prehistoric conditions of limited food supply, high physical activity, and a clean natural environment. The common diseases among modern humans are caused by the consequences of lifestyle changes: over-nutrition, physical inactivity, environmental pollution, and the influence of all these factors on the slowly changing human genome. The advances in deciphering the genome of different organisms allow for the implementation of principally new strategies of prevention and treatment of multifactorial diseases within the next ten to twenty years, based on the discovery of the links between genetic predisposition, environmental and behavioral factors, as well as pharmaceutical agents. Genetic research is postulated to be one of the priorities of the forthcoming Sixth Framework Program of the European Commission.

Many research groups throughout world are trying to identify the major genes involved in development of metabolic, cardiovascular, and malignant diseases. Genome databases will become, in the near future, a powerful tool in this search, and they will eventually lead to the discovery of the complex molecular pathogenesis of these multifactorial diseases. The Latvian Genome Project[1] may play a particularly important role in this, as it is going to have one of the most extensive databases, with

1. The official Latvian name is *Vienota Latvijas iedzivotaju genoma datu baze*, or Unified Genome Database of the Latvian Population.

large cohorts of exactly defined phenotypes of the most common diseases, like cancer, diabetes, and coronary heart disease.

2. The Aim and Current Status of the Project

The aim of the Latvian Genome Project is to create a unified national network of genetic information and data processing, to collect a representative amount of genetic material for determining the genotype of the Latvian population, and to compare genomic data with clinical information and family medical history data. As a direct outcome of the project, the participating individuals will be able to know the likelihood of developing certain diseases due to their genetic features. By the application of individualized therapy based on the genetic characteristics of the patient, these risks might be decreased, even eliminated, in the future.

In the preparation phase of the Genome Project, a joint program of several Latvian research institutes was launched on January 1, 2001. This research program, titled "Genomic studies of the Latvian population, their application for diagnosis and prevention of human diseases"[2] aims to investigate the genetic causes of monogenic diseases (cystic fibrosis, hemophilia B, and fragile X syndrome) and multifactorial diseases (breast cancer, colorectal cancer, familial melanoma, coronary heart disease, obesity, and diabetes mellitus). It also has the aim of studying the ethnogenesis of Latvians. The research program is funded by the Latvian Council of Science, and involves molecular biologists and geneticists from the Biomedical Research and Study Center (BMC) at the University of Latvia, the Latvian Medical Academy, as well as the State Center of Medical Genetics. The diagnostic facilities of specialized clinical centers at the Paul Stradins Clinical Hospital, the Latvian Center of Cardiology, the Latvian Center of Oncology, and the State Center of Hematology are also mobilized in the research program.

The Latvian Unified Genome Database will be even more fruitfully utilized if its findings are compared to those of other large population-based studies: the Estonian Genome Project, the UK Population Biomedical Collection (UK Biobank), the Botnia (Finland) Study on the genetics of diabetes, and the European Prospective Investigation into Cancer (EPIC), and others. The Latvian Genome Project might become a 'pilot project' for other European countries and our experience should help other countries to deal with the ethical, legal, and social problems in the genetic field.

2. *Latvijas populācijas genofonda pētījumi un to izmantošana cilvēka patoloģijas diagnostikā un profilaksē*, funded by the research grant of the Latvian Council of Science, No. 01.0023.

3. The Legal Basis of the Project

The legal basis of large-scale genetic research in Latvia was recently established by the Human Genome Research Act[3], which has three main objectives:

1. to regulate the creation and the operation of the Unified Genome Database of the Latvian Population and the various genetic research projects connected with the database;
2. to ensure the voluntary nature of gene donation and the confidentiality concerning the identity of gene donors; and
3. to protect gene donors from misuse of their data and from discrimination based on the interpretation of their DNA.

The Unified Genome Database is going to be established after the Human Genome Research Act enters into force on January 1, 2004. Then the Latvian government will appoint and authorize the Main Processor of the Genome Database and found the Main Processor Board and State Population Genome Register. The Main Processor will be a state-owned research institution with international experience in genetic research. The principal objectives of the Main Processor will be:

1. to obtain tissue samples and prepare health descriptions and genealogies;
2. to code, store, and destroy tissue samples, health descriptions and genealogies; and
3. to organize and conduct genetic research and store the research findings.

The Main Processor will have the right to delegate the rights of processing, except for coding and decoding, to one or more Authorized Processors (such as hospitals, research institutions, and private companies).

The Main Processor Board will be a supervising authority appointed by the government. In order to represent the plurality of interests in society, the parliament, the government, Central Medical Ethics Committee, and various NGO's (the Latvian Academy of Science, the Latvian Medical Society, the Latvian Association of Lawyers) will delegate members to the Board.

The State Population Genome Register will be established under the supervision of the Ministry of Welfare with the purpose of creating the database of personal data on gene donors, and of preserving the genetic data and written consents of the gene donors received from the Main Processor. Thus, the Main Processor will store coded

3. *Cilvēka genoma izpētes likums* (The Human Genome Research Act) was adopted by the *Saeima* (the Latvian Parliament) on June 13, 2002, and was amended on June 12, 2003. See the Appendix for the first English language translation of these documents in a unified form.

tissue samples, DNA and health descriptions, while the State Population Genome Register will store personal information and codes.

4. Informed Consent and the Rights of the Gene Donors

Participation in the Latvian Genome Project will be voluntary and without any direct financial benefit for the gene donors. Prior to the decision on donating genes to the Genome Database, the gene donor receives a written information sheet discussing the following questions:

- "Why is this study being done?" (information about the aims and content of the genome project);
- "What is involved in the study?" (information about obtaining a tissue sample, questionnaire, medical records);
- "What are the risks of the study?" (information about the potential informational risks);
- "What are my rights as a participant?" (information about the right to withdraw the donor's consent, the right to apply for the destruction of tissue samples or data which enables decoding);
- "Are there any costs and payments involved?" (information about the fact that there will be no cost for participating and no personal financial benefits for gene donors);
- "How will information about me be kept private?" (information about the data protection system).

After learning the above information, the individual might decide to give his or her consent to become a gene donor. The consent form shall be prepared in two copies and signed by the gene donor and the person appointed by the Main Processor. The form contains an agreement

- that the gene donor will provide a tissue sample for the Genome Database;
- that the Genome Database will obtain the health description and genealogy of the gene donor; and
- that the tissue sample, DNA description, health description, and genealogy will be used only for genetic research, public health research, and statistical purposes.

One copy shall be stored in the State Population Genome Register; another copy shall be given to the gene donor. The format of informed consent and the procedure for the completion and preservation of the consent shall be established by the Cabinet of Ministers. Consents shall be preserved for 75 years after the last change is performed.

The gene donors have the right to access their data stored in the Genome Database and the right to genetic counseling. They have the right to submit additional information on themselves to the Main Processor, as well as the right to prohibit the supplementation, renewal, and verification of health descriptions stored in the Genome Database. The gene donors also have the right to withdraw, at any moment, their consent to be a gene donor for the Latvian Genome Database. In that case the tissue sample, description of DNA, health description, and genealogical data shall be destroyed.

5. Issues of Data Protection

In order to ensure the highest standard of data protection, the Main Processor gives each tissue sample, DNA description, health description, and genealogy a unique code immediately after the receipt of these data in the Genome Database. The Main Processor replaces all data with a code, which enables the reverse identification of the gene donor, including the name, personal code, and residence.

The code shall be indicated on the written informed consent form signed by the gene donor. Specific persons shall be appointed to perform coding and to issue coded tissue samples, DNA descriptions or health descriptions to the Authorized Processors. The Main Processor may issue tissue samples, DNA descriptions or health descriptions only in a coded form. Within three days after the coding is done, the Main Processor shall deliver the written and signed consent form together with the code indicated thereon to the State Population Genome Register and the latter shall have the only possible key for decoding.

The State Population Genome Register is permitted to decode data only in the following cases:

- to destroy a tissue sample, a DNA description, a health description or genealogy;
- to enable access to data on a gene donor stored in the Genome Database (except genealogies) at the request of the gene donor;
- to renew, supplement, or verify the health description of a gene donor unless the gene donor has prohibited this; and
- to issue a health description of a gene donor to the donor's physician at the request of the physician and with the consent of the gene donor.

The State Population Genome Register shall appoint the specific persons who will perform decoding and have access to the written consent forms signed by the gene donors.

The Main Processor shall store coded tissue samples, DNA descriptions, and health descriptions within the territory of the Republic of Latvia. The Central Medical Ethics Committee may, at the request of the Main Processor, grant permission for a limited number of tissue samples to be stored abroad, if appropriate research methods are not available in Latvia. The Main Processor shall provide an effective control over the tissue samples and ensure that the tissue samples cannot be used in a manner prohibited by Latvian law.

The State Data Inspection Agency shall exercise the supervision over the collection, coding and decoding, and processing of tissue samples, descriptions of DNA, health descriptions and genealogical data. It shall also approve the method of generating the codes. The Central Medical Ethics Committee shall oversee the ethical issues during the creation of the Genome Database and processing of the data according to the generally recognized medical and scientific research ethical standards and international conventions.

6. Future Perspectives

The Project is planned for a period of 10 years and has been divided into three stages:

1. In *Phase 1* (2001–2003), a national network of genome research groups is created. The phenotypes and genotypes associated with certain monogenic diseases are analyzed in the framework of a pilot project, which uses a database for patients from two hospitals: the Paul Stradins Clinical University Hospital and the Latvian Center of Oncology.
2. In *Phase 2* (2004–2006), the Unified Genome Database will be created for major multifactoral diseases in Latvia. The new single gene polymorphisms (SNP) related to common multifactorial diseases are going to be identified on the basis of data collected from patients treated in the largest hospitals in Latvia.
3. In *Phase 3* (2007–2009), the full-scale Genome Database of the Latvian Population will be built. Data stored will be made available for research on the genetic background of common diseases, for follow-up studies of disease development among the gene donors, as well as for research on how environmental factors and medications interact with the genetic make-up of the donors.

After the Human Genome Research Act enters into force on January 1, 2004, the legal environment for genetic data processing and data protection will be clear and transparent. It will also be possible to establish a nationwide infrastructure for large-scale genomic research on patients with cancer, cardiovascular diseases and diabetes mellitus, as the necessary funding for population-based genome studies in Latvia will be finally available.

Population Genome Project in Latvia: Exploring the Articulation of Agency[1]

Aivita Putniņa

1. Introduction

This paper explores the ways human population genetic research is articulated in Latvia. Assumptions about the nature of genes and their ownership determine the way the Population Genome Project is defined and realized. I propose that instead of discussing broader ideas, one should investigate "the question of ways and means" underlying these ideas.[2] Thus, one can observe that the fundamental discourses on the Population Genome Project have been framed in terms of the development of *science* and *nation*, and only lately has attention turned towards the rights of donors. I argue that the original framing not only curtails the donor's rights, but also limits the scope for public debate, thus leaving the public without the necessary discursive tools to argue. Deconstructing the framing of the discourse makes it possible to demolish the closed boundaries of the biotechnological field and offers a common language for negotiations.

2. The Genome Project

The first activities of the Latvian population genome project started in 2000, initiated by a molecular biologist professor, Elmārs Grēns. These efforts resulted in a long-term state-funded genetic research program, the passed and recently amended Human Genome Research Act[3], and a program to establish the Unified Genome

1. An early version of this article appeared in *Soziale Technik* (2002) 3: 9–11. The research was funded by the Open Society Institute, International Fellowship Program.
2. Bruno Latour (1988) *The Pasteurization of France* (Cambridge, MA: Harvard University Press), p.47.
3. *Cilvēka genoma izpētes likums* (The Human Genome Research Act) was adopted by the *Saeima* (the Latvian Parliament) on June 13, 2002, and was amended on June 12, 2003. See the Appendix for the first English language translation of these documents in a unified form.

Database of the Latvian Population (the Latvian Genome Project *per se*).[4] The Unified Genome Database Program, however, has not yet received any funding, as a result of government change and budget shortage.

Human genetic research in Latvia concentrates on exploring specific monogenic and multifactoral diseases.[5] Since no regulations were applied to genetic research until 2002, foreign research companies could receive research approvals relatively easily. Several ethics committees, giving such approvals, were established at various hospitals or medical schools. The local benefit of such international research was receiving some hospital equipment or medicines in exchange for collaboration. Representatives of the local research community, however, voiced their criticism and ethical concerns when these international research projects tested children, or a group of people who had worked in Chernobyl after the nuclear accident.

Population genetic research, in contrast, has been more academically driven. However, these research projects explored only a small part of the DNA, and aimed at finding correlations between genetic markers and social variables within the population. For example, a long-term project lead by Professor Artūrs Sočnevs explored the incidence of human leukocyte antigen (HLA) alleles amongst the Latvian native population, in order to establish a relationship between genetic data and the donors' place of residence and language use. Another direction of population genetic research, led by Professor Raisa Deņisova, studied the 'ethnogenesis' of Latvians.[6] The media caught the preliminary research results naming them 'the finding of Latvian gene'.

This paper concentrates on the discursive aspects of the Latvian Genome Project, and identifies three basic discourses that justify its launching and operation. The first one relates to the particular socio-historic context of Latvia. Political concerns about independence, positioning Latvia on the global map, and nationalism are reflected in the arguments for the genome project. The second discourse situates biomedical research in the broader socio-political environment in a way to uphold the boundaries between the science and practice of biotechnology and the general public. The genome project is articulated in exclusively biotechnological and biomedical terms, which also results in the exclusion and neglect of the donor's point of view. The third

4. For a short description of the Unified Genome Database of the Latvian Population *(Vienotas populācijas genoma datubāzes)*, see http://forum.europa.eu.int/irc/rtd/cogene/info/data/pub/Latvian Genome Project.htm.

5. Valdis Pīrāgs (2002) "The Latvian Genome Project." Presented at the International Workshop *Law in the Genetic Era*, at the Central European University, Budapest, June 7–9, 2002. See also the contribution to this volume.

6. See, for example, Raisa Deņisova (2001) "The Most Ancient Population of Latvia," on the webpage http://vip.latnet.lv/hss/denisova.htm.

discursive orientation concerning genetic research defines standards and protocols for public health policy, rights and obligations, and frames the expectations of the participants.

3. Methodology of Research

The research for this chapter originated as a part of a broader policy analysis project in public health. It consists of document and media analysis, supported by interviews with the main participants of the project. Unfortunately, it was impossible to interview politicians during the first stage of drafting the Human Genome Research Act and to participate in the preparation of the Act. The Act was passed in a hurry (within only half a year), and without much publicity: there were no public debates, and the *Saeima* (Parliament) website was initially silent about it. Eventually, six scientists and doctors participating in the project and/or conducting human genome research were interviewed. Identities of the respondents remain anonymous, as some issues they discussed were controversial.

After the passing of the Human Genome Research Act, several actors affected initiated various amendments to the law. In the second phase of drafting the amendment, a working group of the Parliament of Latvia and the Ministry of Health discussed the possible changes to the law. I managed to participate in the working group meetings and to make observations. I also conducted interviews with journalists and lay people, but these did not become the main focus of research.

4. The Genome Project and the Ideology of the Nation-State

The discursive context of the Latvian Genome Project is strongly influenced by the ideology of the nation-state. First, it is linked to general understandings of the ownership of science by the state, and the responsibility linked to that ownership. Second, genes in particular are viewed in terms of state ownership. The Human Genome Research Act stipulates that the state is the owner of the intellectual property produced by genetic research.

The research agenda of genetics is also closely linked with the state ideology. Research incentives themselves contain a great deal of fascination about the future of the Latvian nation-state. Valdis Pīrāgs, a member of the Latvian Genome Project team, opened his presentation at an international seminar on the legal aspects of genetic research by suggesting that the genome project was important because it shows the 'civilized' face of the Latvian nation and it sheds light on the origin of the nation.

The Project Draft of the Unified Genome Database strongly allies genetic research with the progress and development of nations:

> In case we were to refuse carrying out the formation of a genetic database or were to delay its creation, medicine and the health care system in Latvia would inevitably lag behind the standards and the level of developed countries.[7]

State ideology is also articulated through the motive of resisting colonization, stressing the independence of Latvian State. The research also leads to the objectification of the Latvian nation through the socio-political history inscribed in the genes.

4.1 De-Colonization Motive

Latvia had been a German, Polish, Swedish, and Russian colony successively for a period of some 800 years. This colonial experience is articulated in research, too. The loss of the genes as research resources to foreigners would lead to a loss of Latvian sovereignty and create certain dependence on foreign agents. Thus, the Latvian Genome Project is also linked to the aims to eliminate the former colonial experience among the Latvians and to establish the Latvians' own control over their genetic legacy in terms of research. The de-colonization motive based on former historical experience is also present in the discourse on the development of biomedical science and research. As genetic research is pictured as an instrument for colonization, (re-)gaining control over it is crucial.

This was one of the main reasons why Latvian scientists objected to the first attempts of launching the project in cooperation with foreign investors. The first initiatives came from a foreign diagnostics company. The Latvian side would collect the samples, work on them using the company's software. The plans for cooperation came to an end when the foreign company insisted on storing the data themselves: "I saw how it was beginning to look. They can export our [genetic] data and sell it. And then that all [data] might disappear."[8]

The same fear of losing control over data was expressed in the working group of the Parliament, which drafted the Amendment to the Human Genome Research Act. The Association of Geneticists proposed that only the residents of the Republic of Latvia and nationally based companies should carry out the genetic research. Globalization of research posed serious threats to keeping the research at a national

7. Elmārs Grēns, Valdis Pīrāgs, Juris Šteinbergs (2002) *Project Draft of the Unified Genome Database of the Latvian Population* (Rīga: University of Latvia), p.7.
8. Interview with professor Elmārs Grēns, 2002.

level. This was also reflected in the debate about the legal difficulty to ensure such national ownership. The Parliament members were worried about the exporting of tissue material, proposing that the transfer should be enacted at the government level.

4.2 The State Monopoly over Research

The first version of the Human Genome Research Act established a state monopoly over all genetic research. However, certain misunderstandings about the difference between various kinds of genetic research caused problems. According to the 2002 Act, all researchers working in Latvia were obliged to submit their database to the state-managed Unified Genome Database. Individual research depended on various permits required by the Genome Database Committee. The committee, however, did not represent the interests of the patients or donors. Several scientists interviewed indicated that the restrictions posed by the Act would disturb genetic research outside the domain of the database.

The confusion grew when the Minister of Health, Āris Auders introduced his own amendments to the Act. These amendments intended to cover not only individual genetic research but also genetic testing by the regulations of the Act. The amendments also clearly stressed that the Main Processor is a state scientific institution, delegated by the government and controlled by the state. The working group, however, agreed that all research other than that connected with the Population Genome Database should be taken out of the scope of this Act, allowing the special government orders to regulate genetic testing and individual human genetic research projects.

4.3 Constructing Latvians

Anthropologists[9] argue that genetic research constructs cultural group identities. The Human Genome Research Act identifies the exploration of ethnogenesis as one of the major research directions of population genome research. One of the subprojects within the Latvian Genome Project deals with DNA polymorphisms of the Latvian basic population, thus producing "deep and maximally objective data on ethnic composition, migration, and regional differences in the Baltic and Northern Europe."[10]

The project group announced the first discovery of a 'Baltic gene' in 2000: the polymorphic marker 'Langsteiner Wiener B' (LWb), which was most expressed

9. See, for example, Margaret Lock (2001) "The Alienation of Body Tissue and the Biopolitics of Immortalized Cell Lines," *Body and Society*, 7(2–3): 63–91.
10. Project Draft of the Unified Population Genome Database, *op. cit.*, pp.20 and 33.

among Latvians (5.9 percent, compared to 5.7 percent among Lithuanians, 4 percent among Estonians, 3 percent among Finns and 2.2 percent among Northwestern Russians).[11] Although it appeared to be just a small fragment without much significance and its "dense concentration" was not statistically valid, the finding of "the genetic marker of Balts"[12] was warmly embraced by the Latvian media. Thus Latvians became 'objectified' both scientifically and socially.

These scientific efforts were based on the politically charged category of 'basic population' in Latvia. The research results also supported the political agenda to locate Latvia within the area of Northern Europe and not Russia. However, as a result of the migration policy under the Soviet regime, Latvians comprise only 57 percent of the population. In this context, evoking the category of 'basic population' was intended to justify the history and moral rights of Latvians over their territory, as well as to serve as a criteria for the self-identification of being Latvian for several generations.

The issue is potentially quite sensitive in respect to the large Russian minority whose interest was eventually excluded from the state-funded program of Latvian ethnogenesis. However, the issue did not cause any political debate. The attitude of the Russian-speaking media towards the historically grounded pedigree and long suffering of Latvians so far has been rather ironic.[13] The scientists themselves also did not see any political implications, seeing it as a purely academic task with a potentially practical significance. Keeping the politicized context of the genetic research in mind, strained ethnic relations might emerge as one of the side effects when the Genome Database Project is carried further.

5. Developing the Science of Biotechnology

When Romualds Ražuks, Member of Parliament and medical doctor, first introduced the Human Genome Research Act to the *Saeima* in the Autumn of 2001, he stated two important reasons for creating such law. First, that it is the most advanced branch of biomedicine; and second, that our northern neighbors, the Estonians also have such a project.

11. Gunita Nagle (2000) "The Gene Unifying the Baltic Nations Discovered," *Diena*, December 28, 2000, p.3.

12. *LETA*, December 28, 2000.

13. Sergejs Kruks and Ilze Šulmane (2001) "Etniskie stereotipi Latvijas presē (Ethnic Stereotypes in Latvian Media)," in *Mēdiju analīze II*. (Rīga: LU SZF, Faculty of Social Sciences, University of Latvia).

As Jean Comaroff[14] writes, medicine assumes a privileged role in the constitution of experience in any society. Medical knowledge serves as a basis to produce other kinds of knowledge. Medical knowledge is power, and medical discourse allocates it to the doctor creating asymmetrical power relations between the doctor and the patient. Framing the genetic research as a matter of medical progress limits the discussion to professionals and silences other interest groups. Thus, genes and genetic research have become new grounds for re-establishing the authority of the medical profession, which had weakened by the end of the twentieth century.

David Armstrong[15] argues that the medical and human sciences only recently started to focus on the subject. Subjectivity has rather been 'liberated', given its own moral rights, than 'created'. The patient emerged as "the product of a series of smaller discourses (on compliance, communication, etc.) which, though intertwined with one another, have contributed separate elements to the final perception of the patient: a 'subject' imbued with personal meanings, constructs, feelings, subjectivity, etc."[16] Once this subject was reinvented, it was provided with a history of oppression and a universal status.

This liberation also entered the public health discourse and gave rise to the concepts of equal health care opportunities, patient's choice, and informed decision-making. Genetic data and its interpretation offer a new challenge to patient's choice, opening the question on knowledge possession and the rights to possess and interpret this data.

At one extreme stands a vision of the patient–scientist relationship offered by a population geneticist, who stressed the academic significance of his research:

> We don't have time to speak with the donor when we take the blood sample. Actually, the blood donation brigade takes the samples, we then join them. We receive the tube with the knot from the individual blood sacks. We have a short time to approach and ask [the donor] about the language [he or she uses]. In case the donor comes from mixed families or [his or her] mother comes from Byelorussia, we do not take blood. We don't ask anything personal, that does not interest us. Later, back in Rīga, we take the marked

14. Jean Comaroff (1982) "Medicine: Symbol and Ideology," in Peter Wright and Andrew Treacher (eds.) *The Problem of Medical Knowledge: Examining the Social Construction of Medicine.* (Edinburgh: University of Edinburgh Press), pp.49–68, see p.63.

15. David Armstrong (1982) "The Doctor-Patient Relationship: 1930–80," in Peter Wright and Andrew Treacher (eds.) *The Problem of Medical Knowledge: Examining the Social Construction of Medicine* (Edinburgh: University of Edinburgh Press), pp.109–122.

16. *Ibid.*, p.119.

> samples. . . . I know that clinical scientists need long pages of ethic approvals. We do [the research] anonymously.[17]

As the scientist explained later, they did not ask for the consent of the donors, since the purpose of the research was purely scientific and explored only the spread of alleles in a given territory. They felt this legitimized their use of the samples without the knowledge of the donors.

Clinically applied research is much different. Current clinical genetic research in Latvia focuses on particular diseases and allows establishing partnerships between the researchers and patients. Donation works as a reciprocal relationship: exchange of information between the patient and the doctor. Donors submit their tissue samples, as they expect to benefit from the exploration of their disease. This relation might become threatened in the case of Population Genome Project. As one of the scientists involved in the project expressed himself,

> I did not like the idea [of the Genome Database] at the beginning. What worries me is that molecular biologists are not medical professionals or geneticists. What is the benefit the women [in case of breast cancer research] receive? She has already got it [breast cancer]. The information carries significance for the family. The risks for breast cancer [are important]. How many of the relatives are screened? The preventive part is absent in such a project.[18]

The researchers with a clinical research agenda stressed that the Population Genome Project should deal with the same research directions, focusing on applied research. The database itself was seen as a rather distant target. In the course of the drafting the Amendment Act, evoking the social importance of scientific activities also became a shield, used to protect the project from accusations of commercialization. As a result, the phrase "medical products and methods" was replaced by "diagnostic and treatment methods" in the amended text of the Act.[19]

The benefits of genetic research for a broader population are seen in its ability to predict and control the health behavior of the population. On the one hand, research might lead to more individualized treatments, decreased number of tests, and more effective therapies and drug treatment. On the other hand, the research results might indicate individual health risks and point to directions of healthier lifestyle to minimize these risks. In both cases, however, the biomedical control over people's lives would increase and options for the individual's choice would narrow.

17. Interview with a population geneticist, Rīga, 2002.
18. Interview with a geneticist, Rīga, 2002.
19. See the Appendix for Article 8 in the two versions of the Human Genome Research Act.

The science-dominated approach gradually changed when social consequences were discussed amongst the scientists. Social implications were also included in the Project Draft. Similarly to the Estonian Human Gene Research Act, the new version of the Act allowed donors not to know one's genetic data. The amendments propose different degrees of donor involvement, giving them partial control over the donated information.

However, genetic data was seen as primary to the medical and social information offered by the donor. The Act allows modifications in the medical history of the donor when the information submitted by the donor contradicts the genetic research data. Thus, biomedical and genetic truth becomes prioritized over social knowledge.

6. The Public Policy Arena

6.1 The Main Players

Only two active agents—politicians and scientists—emerged as key figures in the debate on the Latvian Population Genome Project. The tension between the two is clearly seen in evaluating the control mechanisms described in the original Act. Many scientists involved in genetic research claimed that the Act limited the freedom of science. The scientists participating in the working group, however, did not object to the draft Act since their priority was to create a legal framework for further promotion of the project. They believed any discussions concerning the Act would slow down the project, and amendments were expected anyway after the project received financing. Their passivity in drafting the original Act therefore was a clever move as it eventually led to asserting their capacity for agency and ability to set the rules.

The Act was passed rather rapidly and without much public involvement. When the Latvian Association of Anthropologists wanted to participate in the debating the Act, the Parliament Commission suggested that it would be possible only after its adoption by the *Saeima*. As broader public discussions were absent at the time of passing the Act, it was followed by the media only sporadically. Moreover, journalists found the framing of the law too difficult to analyze and interpret for the wider public.

The situation changed when the new government and parliament withdrew the Act. All the sides interested were invited to the working group established by the Ministry of Health and the *Saeima*. The contents of the Act changed significantly; the amendments enhanced donor's rights, and listed public representatives in the Board of Main Processor. It also withdrew the reference to a state monopoly over genetic research.

6.2 The Lack of Public Debate

The Latvian population on the whole was absent from the initial discussions of the project. This absence was very much linked to the way politicians and scientists imagined the public. The MP Ražuks put 'the population' in the project in a quite loose manner:

> The main aim of the research is to ensure the voluntary participation of *donors, that is people* [my emphasis] offering the data, to guarantee safety and *computer anonymity* [*sic*, my emphasis], to protect the donor from the unlawful use of the genetic information, because data would be stored in computers. That means to guarantee the human rights and to regulate the collection, processing and use of genetic information, ensuring state control in this area.[20]

This creates "people" as a synonym for "donors", whose "human rights" should be guaranteed. Anonymity is not seen as something inherent to donors but rather as a technical tool. Therefore the main debate in the parliament concentrated on the safe storage of data, rather than on the private feelings of prospective donors. The first draft of the Human Genome Research Act completely ignored the norms of informed consent. The only means of control over the donated material and processed information allocated to prospective donors was through claiming it back when data proved to be misused. This raised some objections in the Latvian media articulated by an anthropologist and the Central Medical Ethics Committee.[21]

The original Act was significantly improved before the final reading, as a chapter on informed consent and enlarging the rights of the donors was added. Researchers have to obtain informed consent by means of asking the donor to fill out a written information form concerning donor's rights, obligations and the potential medical risks involved in donating tissues. Rights of legally incapable persons were established. Donors also gained rights to withdraw their donated material and information related to it at any time. They could also demand a limited scope of information on their donated material. As noted earlier, these improvements of the draft Act enhanced

20. Parliament debate on October 25, 2001, see http://www.saeima.lv/steno/2001/st_2510/st2510.html.

21. See Aija Lulle (2001) "Iebilst pret genoma izpētes likumprojektu (The Central Medical Ethics Committee Objects the Human Genome Act Project)," *Neatkarīgā Rīta Avīze*, August 2, 2001; Aivita Putniņa (2001) "Kam vajadzīgi mūsu gēni? (Who Needs Our Genes?)," *Neatkarīgā Rīta Avīze*, July 24, 2001. p.2; and Aivita Putniņa (2002) "Jaunajā pasaulē vecajās čībās (Wearing Old Slippers in the Bright New World)," *Neatkarīgā Rīta Avīze*, June 13, 2002.

the control of donors over the donated materials, allowing conditional participation in the project. The Amendment of 2003 also prohibits the use of tissue material obtained from a dead person, and defined the circumstances in which persons without a legally recognized capacity to act (now including children) could become donors, and thus participate in the Genome Database Project.

Although the wording of the project draft was sympathetic to the public and its ultimate aim was the improvement of public health and quality of life, the image of the public still remained ambivalent in it:

> A project of this scale should be based on the support of the major part of the population. Therefore a careful educational campaign and free access to information is needed at any stage of the project. At the same time the project should be protected against the rapid changes in public opinion or political situation.[22]

Framing the issue of genetic research within the discourses of national politics and biomedicine makes the utility and public benefit of the project self-evident, and simultaneously empowers the two groups of professionals: the politicians and biomedical scientists. The lay public is not expert in either one of these fields. Widening the scope of debate was further made difficult by the neglect of the government and the parliament to publish information on the drafting of the Act and the Genome Project itself on their webpages. Most debates in the *Saeima* (which was the only official and open source of information) were dedicated to the regulating functions of genetic research. Even the print and electronic media experienced difficulties in reporting on the project, perceiving it as a very particularistic issue. Journalists sensed the social risks beyond the project but did not feel competent enough to voice them.

The field of biomedical science was also not interested in informing the public about the Genome Project: it was a way to reassert the medical authority that doctors and scientists felt was weakening. As a result, now the patients are weakened in their competence to make informed decisions on genetic data. This might be another reason the Genome Project was so quickly and silently launched. As several geneticists believed, the donors would not be prepared to accept and interpret data 'as it was' and would not be capable of evaluating the significance of the project. Seeing it as a purely scientific endeavor helped them to avoid facing dilemmas about public participation.

22. Elmārs Grēns, Valdis Pīrāgs, Juris Šteinbergs, *op. cit.*, p.23.

7. Conclusion: Strategies for Public Participation

Enlarging donor rights in the Amendment of the Human Genome Research Act, however, would not ensure the exercise of donor choice since the overall context of choice is not present. First, patients generally still have a rather low authority in medical decision-making. Second, consequences and social risks of the donation are not adequately considered from the perspective of the donor.

A strong advantage of Latvian genome research lies in its current approach which focuses on specific applications. This way, biomedical research directly contributes to the diagnostic interests of the donors. Participation in each of the current research projects is based on close treatment relationships between the doctor and the patient in the hospital. However, this relationship might change when the focus of research becomes more general. The donors could be empowered when the informed consent procedure makes informed decision-making possible. This requires not only the information itself, but also a favorable context for such decision-making: equal and trustful doctor-client relationships, and the presence of alternative choices in receiving information from genetic testing and screening.

However, the dual discourses of state ideology and biomedicine do not leave much space for public debate. They both make 'the public' invisible. The Icelandic Population Genome Project was made 'public' when concerns over legitimacy arose. This allowed for opening questions on property rights, decision-making and social consequences of the project. The Estonian Genome Project found the solution in launching an information campaign, which created a free of charge hotline, and led to the production of various booklets, TV, radio, and Internet materials. General practitioners were trained specifically as the main collectors of genetic material. At the beginning of the campaign, only one third of the Estonian population supported the project and was willing to offer their tissue material.[23] No such opinion poll has been conducted in Latvia yet. Interviews with lay persons suggest that the attitude towards gene donation is diverse. For some, it is a potentially dangerous revelation of most personal information; for others, it is like an ordinary blood test.

Transparency is a crucial aspect of the project. Even though the Latvian Genome Project working group resisted the temptation to trust the collection of genetic material to a foreign private company, as was done in Iceland, state control over the project alone does not ensure the realization of public interest. Moreover, the use

23. Krista Kruuv (2002) "The Estonian Genome Project." Presented at the International Workshop *Law in the Genetic Era*, at the Central European University, Budapest, June 7–9, 2002. See also the contribution to this volume.

of state budgetary means for the operation of the project and the conceptualization of genetic material and information as public property demands responsibility and accountability in decision-making. A stronger emphasis on the public utility of the project would help opening a space for public involvement.

The discussions should also focus on the political risks of interpreting the data. The current ethnogenesis project has already accented social differences amongst ethnic groups in Latvia. Besides ethnic segregation, other forms of discrimination might also occur in the course of handling genetic data.

The project presupposes a silent and supportive public. Playing a passive role, the public becomes a tacit, uncritical supporter of the project. More active public involvement, however, can be achieved only by forming efficient partnerships between all the participants in genetic research and shifting the discussion content from the biotechnological field to its politics. The problem of the Latvian Genome Project lies not as much in the genetic data *per se* as in the application of the data. This can be solved by opening the project for public debate and reformulating it along a different discursive line.

The UK Biobank

Alan Doyle, Frances Rawle, and Peter Greenaway[1]

1. Introduction

The UK Biobank[2] is a major infrastructure resource project designed to build upon the large and successful investment that there has been in human sequencing over the last 10 years. With the overall goal of making genetic information relevant in the context of patient outcomes and in healthcare, it is proposed to study the multifactorial diseases of adult life in 500,000 volunteers aged 45–69, selected at random from the UK population. Included in this study will be important lifestyle information; certain questionnaire-derived baseline measures; as well as a collection of DNA, plasma, and serum. Follow-up will be achieved by tracking through healthcare records over an extended period including the use of existing disease registers. The database and biological samples will be made accessible to both the academic and commercial research communities under a carefully planned ethical and legal framework on an anonymized basis. As volunteers will be recruited on an entirely opt-in basis, there has been and will continue to be considerable open debate and consultation on all of the issues surrounding the project.

The Funders of the project, the Wellcome Trust (a leading medical research charity) and two UK government institutions, the Medical Research Council (MRC) and the Department of Health are anxious to maintain a positive climate amongst all interested parties and stakeholders for the project to be successful. The allocation of earmarked resources to the MRC for this initiative from recent government spending reviews indicates support for the Biobank initiative from the Office of Science and Technology and the Department of Trade and Industry. The Department of Health has been an active partner in the development of the initiative to date, and the Secretary of State for Health underlined the importance of the UK Biobank to the National Health Service (NHS).

1. The authors would like to acknowledge the invaluable assistance of Prof. Tom Meade, Chairman of both the Protocol Development Committee and the earlier Expert Working Group, and Dr. Emily Banks, who drafted the protocol.

2. "Several issues were debated, including the types of information that the project should gather, the diseases that might be investigated, the role of GPs in the study, whether the volunteers should be adults or children, and the size of study needed to obtain statistically significant results." See the official website of the UK Biobank, at http://www.ukbiobank.ac.uk/status.htm.

Outlined below are the steps that have been taken to plan this major undertaking and ensure that it remains an inclusive process and takes into account the views of the potential participants in the study as well as the user groups and those professionals likely to be affected by it.

2. Initial Concepts and Protocol Development Process

In May 1999 an initial workshop was held to bring together the biomedical scientists involved in most of the existing major long-term longitudinal studies within the United Kingdom. The intention was to determine whether any of these existing studies had the potential to form the basis of a much more large-scale cohort as envisaged to create the UK Biobank. The outcome of this meeting was the agreement to establish a joint working group with representation from the relevant scientific disciplines to explore and elaborate upon some of the basic parameters for a major study to fulfill the expectations of the research community. At this stage it was decided that none of the existing cohorts was likely to match the necessary requirements and a cohort would need to be established *de novo*. Besides the problem of the size and scale of the existing studies, another issue was the prevailing consent arrangements.

In June 1999, the Wellcome Trust and the MRC committed funds in principle to the UK Biobank (then known as the UK Population Biomedical Collection). An expert working group, chaired by Professor Tom Meade, Director of the MRC Epidemiology and Medical Care Unit, Wolfson Institute, London, was established to discuss the project's feasibility and the possible ways the Collection might be organized.

The working party reported to the Wellcome Trust and MRC in April 2000 and recommended the establishment of a major adult cohort and also a childhood cohort; the latter was deemed to be of lower priority at this stage. The funders established an Action Team to explore the concept. As it followed, a further workshop with over 150 participants from the scientific community met in April 2001 to discuss the measures that would be prioritized in such a major study. This led in turn to the establishment of a Protocol Development Committee[3] under the Chairmanship of Prof. Meade.

3. *Membership in the Protocol Development Committee:* Prof. Tom Meade (Chairman) (MRC), Dr. Emily Banks (Oxford), Prof. Paul Burton (Leicester), Prof. Lon Cardon (Oxford), Prof. Nick Day (Cambridge), Prof. Anna Dominiczak (Glasgow), Mr. Mark Duman (Manchester, lay member), Dr. Rhyddian Hapgood (Sheffield, GP member), Prof. Mark McCarthy (London), Prof. David Porteous (Edinburgh), Prof. Martin Prince (London), Prof. Ian Purves (Newcastle), Dr. Anne Richardson (London, lay member), Prof. Alan Silman (Manchester), Prof. David Strachan (London), and Dr. Ron Zimmern (Cambridge). For a description of the protocol development as well as the report from the Protocol Development Workshop held on April 17, 2001, see the UK Biobank webpage "Protocol Development" at http: //www.wellcome.ac.uk/en/1/biovenpopprt.html.

This group was aided by an experienced protocol writer, Dr. Emily Banks, from the University of Oxford, and a first draft protocol was completed in November 2001. This protocol was subjected to international peer review in an unusual iterative process that enabled further refinement of the protocol in light of the referees' comments. Following internal review meetings at the Wellcome Trust and MRC in early 2002, a final funding decision was taken by their respective ruling bodies in April 2002 with an initial sum of 45 million GBP (MRC 20 million GBP; Wellcome Trust 20 million GBP; Department of Health 5 million GBP) allocated to the study. It is anticipated that further refinement of the protocol will enable more accurate budgeting as the project develops and the final level of funding required is likely to be higher than this sum (the current estimate is 63 million GBP).

2.1 The Protocol

The cohort will comprise 500,000 individuals aged 45–69 at the time of recruitment. The prospective numbers of deaths and illness episodes for a given range of disease are presented in Table 1.

Table 1. The UK Biobank cohort size: Predicted events

	In 5 years	In 10 years
	Deaths	
Heart	2,200	4,300
Stroke	540	1,100
Lung	290	590
	Cancer Cases	
Breast	3,010	6,270
Large bowel	2,140	5,410
Prostate	1,090	3,290
Lung	980	2,540
Ovary		1,120

SOURCE: Tom Meade, personal communication.

It is anticipated that the study design will reveal enough cases to allow the study of factors involved in the development of a range of cancers (bladder, breast, ovarian, lymphoid, prostate, colorectal) and also of cardiovascular diseases, diabetes, osteoporosis, arthritis, and neurological disorders (including dementia and Parkinson's disease).

Baseline data is expected to include:

- socio-economic data: age, sex, postal code, education, socio-economic status;
- lifestyle: exercise, diet, alcohol, smoking;
- current health: diagnosis and treatments (volunteers should not be pre-screened to remove those currently experiencing ill-health);
- personal and family history of disease;
- environmental data: occupation, sibling order (some connecting with siblings and spouses may be advantageous), family size, toxicological exposure;
- other factors: body mass index, history, birth weight, infection history.

The following measures will be taken by a research nurse during a proposed 45-minute clinical evaluation (this will also be part of the consent procedure; see below):

- anthropometric measures: waist, hip, weight; body mass index, leg length;
- physiological measures: BP, ECG;
- lifestyle factors: exercise, diet, biochemical measures.

2.2 Recruitment Procedure

Recruitment will occur through local general practices (GPs) selected for research capabilities. However, they will need to form a representative and broad spectrum from across the UK population. Recruitment will be arranged administratively through a network of centers that will operate on a regional basis. It is anticipated that a publicity campaign through local and national media will precede the arrival of a brochure and letter from the GP inviting the patient to participate. At the same time a consent form and questionnaire will be provided and, in addition, first language contact will be a requirement in the recruitment of ethnic minorities.

Those who accept the invitation will need to arrange an appointment with a research nurse based at the local GP surgery. Extra consideration will have to be given regarding participation from individuals from socially deprived areas who may not be convinced of the value of participation in the study. Piloting of the procedures will be necessary to validate the methodology before embarking on full recruitment. The recruitment interview will include signing the patient consent and collecting a blood sample, which will be taken as a long-term source of DNA, plasma, and serum for subsequent analysis.

2.3 Follow-Up

Participants will be flagged as participants in the UK Biobank through the Office of National Statistics, providing routine follow-up data regarding cause-specific mortality and cancer incidence. Data regarding incident morbidity will be obtained through regular follow-up via hospitalization data and general practice records, with confirmation of diagnoses using standard criteria. Every two years a subset of around 2,000 participants will be resurveyed to allow correction for regression dilution and the entire cohort will be resurveyed by postal questionnaire every five years to update exposure data and to ascertain self-reported incident morbidity. While follow-up is already feasible using existing hospital and general practice resources, it will also make use of NHS information systems where appropriate and the study design will be adapted as new information resources become available.

2.4 Anticipated Results of the Study

Conducting a long-term study on this scale (500,000) it is necessary to ensure that in the context of complex gene–environment interactions there are sufficient examples of spontaneous disease occurrences to prove beyond doubt there is a link between cause and effect. This is especially the case when dealing with multiple factors (genetic and environmental) that are behind the development of major diseases of adult life.[4]

The main means of assessing the combined effects of genotype and exposure on the risk of disease will be through a series of case-control studies nested within the cohort. These will provide information regarding the risk of disease associated with exposures of interest, for individuals with and without a particular genotype, and will also give information regarding the effect of genotype on the risk of disease in people with and without particular exposures. Evidence of formal statistical interaction between genotype and exposure will also be examined. Studies based on the determination of continuous phenotypic outcomes are also envisioned within the cohort

4. For conditions yielding around 5,000 or more incident cases (e.g., diabetes mellitus, ischaemic heart disease mortality, myocardial infarction, colorectal cancer, breast cancer), the study will have the power to detect minimum relative risks of disease associated with exposure (within the genotypes of interest) of 1.5 and an interaction ratio of around 1.4, for exposures and genotypes present in around 20 to 80 percent of participants (at 95 percent power and 0.1 percent significance). For the same exposure, genotype, power, and significance values and outcomes with around 1,000–2,000 events (e.g., rheumatoid arthritis, Parkinson's disease, hip fracture, ovarian cancer, bladder cancer), the study should be in a position to detect relative risks of disease associated with a particular genotype and/or exposure occurring of around 1.8–2.0 and an accompanying interaction ratio of 1.7–2.0.

subject to approval by the Scientific Management Committee[5], which will oversee the overall scientific objectives of the study.

3. Management of the Project and Governance Issues

The funders aim is to create a management structure that provides a long-term legal status for an organization charged with the delivery of the project. This will enable the project to gain an identity separate from the funders and build a business plan eventually based upon a diverse portfolio of users providing income. The simplest model to achieve these aims is through the establishment of a charitable company limited by guarantee, which is operated by a Board of Trustees with representation from the funders. This legal entity would be responsible for the operation of a coordinating center (the so called 'Hub') with defined tasks in the execution of the project and guided by a Chief Executive Officer who will hold the responsibility for the successful conduct of the study.

3.1 The Role of the Hub

In addition to coordinating the activities of the spokes (see Figure 1), the Hub organization will be responsible for database management and quality assurance, computing and financial management and will be the formal custodian of the data and the biological samples. It will have responsibility for auditing the scientific activities of the spokes and other users of the data, and will also be responsible for the financial management of the study. It will manage access to the data and samples in accordance with policies agreed by an oversight body.

Figure 1. The UK Biobank—overall management structure

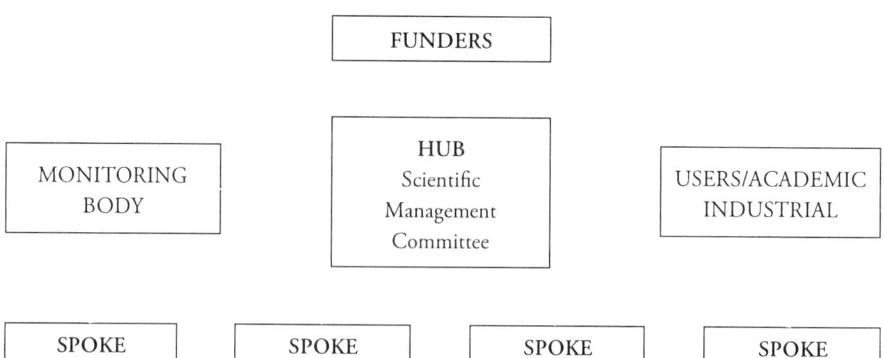

5. See Section 3.4 below.

3.2 Spokes

The spokes are likely to be based in university academic departments with expertise in epidemiology and/or bioinformatics and experience of running large-scale studies. They will have day-to-day responsibility for organizing the recruitment of participants within their local area, which will include setting up clinics, managing the nurses and organizing the collection of samples and data. They will also manage relations with the participating general practices.

3.3 Oversight Body

An oversight body will be responsible for development, application, and monitoring of protocols on access to data and samples, and for ensuring that the hub has appropriate measures in place to protect the confidentiality of study participants. It will be accountable to the participants and to the public, and have the power to veto uses of the data or samples that it considers against the interests of the participants or likely to damage the reputation of the study. In particular, if, as intended, license fees are charged for access to data as a means of cost recovery, it will be important to ensure that decisions on granting access are not inappropriately influenced by financial considerations.

Following legal advice, it is currently proposed that the oversight body will be the Joint Committee of the Medical Research Council and the Wellcome Trust. There will be an open advertisement for membership on the committee, with selection according to UK government approved 'Nolan' principles, and a fixed membership term.

3.4 Scientific Management Committee

A Scientific Management Committee will be established to advise the Board of the Hub Company on all scientific issues related to the UK Biobank. The Committee will include members representing each of the spokes, several independent members designated by the funders and chosen for their scientific expertise. There will be an independent chair, who will be an internationally recognized scientist with appropriate expertise, appointed by the funders. The committee will be responsible for finalizing the design of the study and the study instruments; agreeing to modifications as it progresses; setting performance targets; monitoring the progress of recruitment and follow-up; approving add-on studies requiring additional data collection; and advising on the prioritization of requests for the use of biological samples. It would also be responsible for monitoring scientific outputs.

4. Information Technology Issues

4.1 Overall Information Technology Strategy

The UK Biobank is certain to present large-scale information technology challenges. The scale of these will reflect the enormous amount of data that will be collected at the baseline and during the follow-up, conceivably for very many years, on the 500,000 participants. At baseline, the data will be collected from well-defined sources: the participants' questionnaires and the analysis of the blood samples and the DNA. From a managerial and technical point of view, this is relatively straightforward; and the UK Biobank can draw upon the experience of other large-scale epidemiological studies.

The major informatics challenge relates to long-term follow-up, when data will be collected from a variety of sources, extracted from many different existing NHS systems and datasets, standardized and finally stored on what could become one of the largest medical databases in the UK. Appropriately-sized equipment and reliable networking will be required to achieve this, and a methodology will need to be devised to maintain confidentiality and to ensure anonymization.

4.2 Two Main Foci of the IT Strategy

In planning this project the Information Technology Strategy has two main facets:

1. Examining and evaluating the context in which follow-up data will be obtained, and identifying means of transferring this data to the UK Biobank as seamlessly as possible. This largely (although not exclusively) relates to the interface between the UK Biobank and the NHS, and can be defined as:
    - data sources: where it is (primary care, secondary care, non-NHS sources, and so on); what form it is in; and what quality standards apply.
    - data extraction and database design: interfacing with the data source; retrieving this data; standardizing data format; and defining the structure of the recipient UK Biobank database.
    - framework issues: meeting the ethical and legal requirements, notably on confidentiality; managing the collaboration with the providers of data, notably GPs; and dovetailing with the NHS information strategy.

2. Defining the overall IT architecture. This means identifying the needs for:
    - overall shape of the system (informatics relationship between the hub, spokes, and the outside world; possibility of distributed functions);

- servers and other computer hardware;
- size of data requirements and data storage;
- networking, including possibility of Grid-based solutions;
- security;
- disaster recovery;
- IT staffing requirements;
- presentation of data to users; and
- long-term plans to keep data usable across several generations of systems and software.

5. Consultation with Healthcare Professionals

5.1 Consultation with General Practice Professionals

A consultation workshop was held with primary healthcare professionals from the Trent Focus Collaborative Research Network in the north of England in the spring of 2000. The aim was to identify potential barriers to participation in the project.[6]

Participants had experience of research in primary care, but this mostly involved people who already had a disease diagnosis, so recruitment of healthy participants to a cohort study was less certain. They were generally positive about the project and thought that their patients would be too. However, it was thought unlikely that general practices would be able to provide physical space for recruitment clinics, and practices would need resources to cover the extra workload (for example, referrals for abnormalities found at the time of recruitment). There would also be a significant training requirement for practice staff about the project and about genetics in general, to enable them to answer patients' questions prompted by the study. Participants thought the hardest groups to recruit would be healthy men, people in full time employment, and members of ethnic minority communities, and highlighted the need for some clinics to be held outside normal working hours.

6. For more, see the UK Biobank webpage "Public Perceptions of the Collection of Human Biological Samples: Attitudes of GP's and Practice Nurses," at http://www.wellcome.ac.uk/en/1/biovenpopcolatt.html.

5.2 Consultation with NHS Managers

Senior Managers in the National Health Service were invited to consultation meetings, held regionally in the spring of 2001. Key points from this exercise included:

- There is a need to manage expectations, especially among scientists, as this project will be very useful but cannot be all things to all people.
- The NHS plan and overall strategy in the UK will help the project, as it will require GPs to improve their record keeping for the purposes of clinical governance.
- In a time of change in the NHS, academic clinical genetics networks could be used as a point of continuity and information about genetic databases.

6. Public Consultation and Communication Issues

6.1 Overview

As already emphasized, public understanding and support for the UK Biobank over the long term is essential for the initiative to succeed. There are two key aspects to public acceptability. First, the project depends on the ability to recruit a very large number of volunteers who will receive no personal direct benefit from being participants. The concerns of volunteers, their motivation and all the factors that might affect their willingness to participate as well as their information needs must be understood and addressed.

Second, the project must be understood and accepted by the public at large as a laudable and worthwhile national endeavor that will bring long term benefits for healthcare. The project must be able to generate a high level of public trust and a positive profile, in order to withstand criticism from pressure groups that are often indiscriminately opposed to genetic research. Indeed, as a positive benefit, the sheer scale of the project can be seen as providing a significant opportunity for public engagement and thus improving the level of understanding of genetic research and its potential benefits for health in the wider community.

6.2 Public Consultation, Phase I

Two phases of public consultation have taken place on the UK Biobank so far. The first was conducted in spring of 2000, at an early stage in the project development.[7]

7. For more, see the UK Biobank website "Public Perceptions of the Collection of Human Biological Samples: Key Findings," at http://www.wellcome.ac.uk/en/1/pinbiobnkbiopub.html.

The MRC and the Wellcome Trust jointly commissioned research consultants Cragg Ross Dawson to carry out qualitative research to explore public attitudes towards the use of human biological samples in research and their linkage to medical databases.

The aim of conducting the study was to inform policy making, as to what safeguards would be most acceptable to the public when establishing a large population cohort, and also to gauge the immediate response from the public to the proposed 0.5 million cohort study. The research included focus group discussions with members of the general public, including people from ethnic minority groups. There were also a small number of in-depth interviews with GPs and practice nurses, individuals with experience of disease or disability, religious and community leaders, and representatives from organizations with a special interest in the subject. Key findings included:

- There was widespread acceptance of the use of human biological samples for medical research, provided informed consent is obtained.
- The use of such samples for genetic research was less readily accepted, largely because this area of medical research was less well understood.
- The initial response to the proposed cohort was favorable but unconsidered. Issues highlighted for further discussion prompted concerns, but additional information tended to restore a positive view. People with direct experience of illness and their relatives were generally more supportive.
- There was a majority view that the proposed large cohort should be in public ownership, overseen by an independent body with a known public figure at its head, including representation from minority communities. There should be mechanisms by which benefits from the research should be retained by the public or the NHS.
- Consent is a crucial issue, and adequate information on the project needs to be given at the outset, including information on what the resource will be used for.
- Feedback on results needs to be handled carefully, and volunteers need to understand what feedback they can expect and how it will be given.
- Mechanisms for ensuring anonymity are important, and volunteers will need to be reassured as to how information stored about them will be safeguarded and kept confidential.
- Commercial access was clearly a concern; there was some suspicion of pharmaceutical industry motivation and a view emerged that the industry should not profit from freely donated samples. Arrangements for commercial use must be carefully thought through and explained to participants when consent is sought. However, there was recognition that the pharmaceutical industry is part of the infrastructure of healthcare provision.

6.3 Public Consultation, Phase II

A second phase of public consultation took place in January 2002, following the completion of the draft UK Biobank protocol.[8] The research consultants People, Science, Policy Ltd. were commissioned to look at public attitudes towards the ethical framework and management issues surrounding the project, and explore factors that would affect recruitment. The consultation also encompassed the five main questions raised in a critical report from the pressure group GeneWatch UK, published in December 2001.[9]

Three similar but separate events were held in different locations in the United Kingdom, involving a total of 59 people in the age range of potential UK Biobank participants. At each location, there was an introductory evening session of 1.5 hours, which was followed by a four-hour interactive daytime session with officers of the funding agencies four days later. Thus, participants had the opportunity to further consider the issues raised and were encouraged to discuss the project with colleagues, family, and friends. Many used the opportunity to do so. As a result, the exercise worked well in generating thoughtful debate on the issues and concerns. There was active participation and, indeed, useful ideas to support the development of the project were also generated.

The participants identified five main issues: the burden of taking part in the project; the question of confidentiality in the access to samples and data; the possible future use of data; the question of control or governance; and value for money. Key points that arose included:

- Participants were very supportive, but there was some evidence that people can overstate altruistic motives.
- Most potential barriers to participation were practical, such as how far to travel to attend a clinic and how much time it would take.
- Concerns were raised about differing response rates from people from poorer backgrounds. There were concerns also about the potential impact of the study on the availability of GP appointments.
- Complete confidentiality for individually identifiable data was clearly important—some participants did not want their GPs to have access to their lifestyle data.

8. For more, see the UK Biobank webpage "Ethical Aspects and Public Consultation: A Question of Trust," at http://www.wellcome.ac.uk/en/1/biovenpopethtwo.html.

9. Kristina Staley (2001) *Giving Your Genes to Biobank UK: Questions to Ask*, A Report for GeneWatch UK, December 2001 (Buxton: GeneWatch UK), available at http://www.genewatch.org/humangen/Publications/Reports/BioRport.pdf.

- Many respondents wanted access to their own records and some assumed that a benefit of participation would be access to health information that they would not otherwise get.
- The idea of access by commercial organizations raised some concerns but on reflection most people realized it was the only way that new therapies and medicines would be developed. Fees for access were seen as appropriate.
- People were interested to know how the data would be used and which diseases would be prioritized for study. Moderators raised the potential for studies of genes related to behavior, but this did not elicit much interest or concern.
- Participants recommended some form of oversight body and developed two main models; a stakeholder panel headed by a well-known person or a lay member group with no vested interest in the UK Biobank, supported by professional staff.
- Many assumed that much more is known about the impact of genes on disease than is really the case, and on this basis questioned the need for the study. The majority thought the project would be a good investment that could save NHS money and reduce the cost of healthcare in the long term.

7. Consultation Mechanisms on Ethical Issues

A long-term prospective genetic study of this nature gives rise to important ethical and legal issues. These center around four key areas: obtaining informed consent; feedback of research data to individuals; confidentiality and data protection; and commercial access issues. These have been a focus of stakeholder consultations throughout the development of the project and the outcome of these consultations was taken into account in drafting the protocol. In January 2002, ethics and legal experts reviewed the protocol, and the issues highlighted have been addressed further in the next stages of project development.

7.1 Informed Consent

Consent for participation in the UK Biobank will be broad in nature but not 'blanket' consent. Participants will be given information about the purpose and nature of the study and examples of the type of research project that might be undertaken. However, they will not be informed about the details of the genetic variants or biochemical markers to be tested, nor about the specific analyses to be conducted. They will be given ongoing access to information about studies being undertaken using the resource (probably in a newsletter) and would have the right to withdraw at any time. The likelihood that commercial companies could have access to the samples and data will be explicitly mentioned, and it will be made clear that participants would have no entitlement to a share of possible financial benefits. Consent will also be obtained to access the participant's NHS medical records on an ongoing basis, and there will be an explanation of the measures in place to protect confidentiality. At an April 2002 workshop on ethical issues it was suggested that the consent procedures should be validated to ensure that people agreeing to participate really understood what they were agreeing to. It was also suggested that research should be carried out at an early stage to establish what conditions or types of research people would expect to fall outside the scope of a broad consent.

Much discussion and comment to date has centered around the extent to which properly 'informed' consent can be obtained when it is not possible to specify at the outset exactly what will be done with the data and the samples. And indeed, there is no agreement on whether the 'individual informed consent' model is applicable to the UK Biobank. The legal validity of 'generic' consent has also been questioned, and it has been pointed out that there must be mechanisms to take into account changes in societal views and ethical thinking over the long time-scale of the project. It has been suggested that some areas of research, such as the genetics of intelligence, should be proscribed at the outset, and that participants should be presented with lists of research areas and allowed to indicate any for which they would not wish their sample to be used. However, this latter proposal could lead to significant practical difficulties. It has been suggested that providing regular updates to participants on research being undertaken within the cohort, together with the provision for people to withdraw at any time, would be an important factor in ensuring that the consent remained valid.

7.2 Feedback to Participants

Consent to participate in the project would be on the basis that individual research results, for example, results of genotypic or biochemical testing, would not be fed back to participants. Instead, they would be informed of any findings of possible clinical significance (for example, high blood pressure) arising at the recruitment interview. Feedback of general research findings would be available, and ongoing communication about the research being done and its results is seen as very important in establishing the relationship between the UK Biobank and its participants. This position on feedback relates to the fact that research findings are unlikely to have any significant clinical or predictive value for individuals and that criteria for validation of test results in a research setting are different from those required for clinical diagnosis.

Some concern has been expressed that insurance companies or employers might put the participants under pressure to reveal or seek access to their own data, which could lead to discrimination. The British insurance industry has currently agreed to a five-year moratorium on the use of genetic testing in insurance, and has stated that participation in genetic research will not affect insurance eligibility. However, it will not be until a later stage of the project when enough genotype data has been accumulated that research might start to demonstrate certain genetic variants that the issue may really arise. It is clear that insurers or employers should not have access to individual data from genetic research databases, and this must be prevented in the UK Biobank. The UK Human Genetics Commission also concurs with this view.[10]

7.3 Confidentiality and Data Protection

There has also been discussion of confidentiality and data protection, and of conditions under which access to the samples and data should be granted. There are issues as to whether patients have a right to access their own data under the Data Protection Act of 1998. The legality of obtaining and storing information on family members (related to recording family history of disease and identifying people in the cohort who are siblings) without their consent has also been questioned. Protection to ensure that insurance companies or employers cannot have access to individual data is clearly essential.

In April 2002 a one-day Ethics Consultation Workshop was held. The consultation brought together a selection of ethicists, medical lawyers, public policy advisers, social scientists, medical professionals, biomedical scientists, and representatives of patient and other consumer and pressure groups. Discussion focused on issues and questions that have been raised in the course of protocol development, public consultations, and by expert reviewers.[11]

The main consequence of the meeting is the establishment of an interim Advisory Committee on ethics and governance within the UK Biobank so that expert opinion can be consulted in the next phase of project development prior to the establishment of a permanent oversight body.

10. Human Genetics Commission (2002) *Inside Information: Balancing Interests in the Use of Personal Genetic Data* (London: HGC), see also http://www.hgc.gov.uk/insideinformation/introduction.pdf.

11. For more, see the UK Biobank webpage "Ethical Aspects and Public Consultation: An Ethics Workshop," at http://www.wellcome.ac.uk/en/1/biovenpopethwks.html.

Bona fide researchers applying to use the UK Biobank data will be required to enter into a license agreement, which will set out conditions of access and specify the research that can be done, before access is granted. This will provide a means of ensuring that the work undertaken falls within the scope of the consent given by the study participants. New analyses of the samples will require approval by an appropriate NHS ethics committee, as will any projects that require re-contacting of participants and collection of new data. Any studies requiring collection of additional information directly from study participants will have to be approved by the Scientific Management Committee and re-contact will be via a member of the UK Biobank team.

Commercialization will clearly be a controversial area and it will be important that arrangements are transparent to participants at the time of recruitment. The sponsors are committed to providing access to the resource for scientists in both the academic and commercial sectors, seeing this as essential to obtain maximum benefit from the investment. Consultation with companies in the pharmaceutical and biotechnology sector indicate that there is a significant interest, at least in principle, in using the resource, provided that access conditions are right. Discussions about how intellectual property and license arrangements are to be handled are still ongoing, but it has been agreed that the principal means of ensuring that commercial companies contribute appropriately to the costs of the resource will be via license fees. It is also clear that any access that could be construed as selling the samples would be unacceptable to the public. The principle that benefits should flow back to the public, if research using the UK Biobank leads to a profitable new product for a commercial company, is accepted by the sponsors, but, as the HGC observes in their report, it may be difficult to put this into practice.

8. Governance and Ethical Framework for the UK Biobank

The consultations described have indicated that public ownership and oversight by an independent body would be important factors in ensuring that the UK Biobank project commands public trust. An independent oversight body will be set up specifically for the project, which will have formal responsibility for safeguarding the interests of the participants and the public. It will develop policies on access to the data and samples to ensure that they are used in accordance with the consent obtained and the overall objectives of the study, and will ensure that appropriate measures are in place to protect confidentiality. As part of its remit to protect participants' interests, the oversight body will have to be satisfied with the way various ethical issues, such as consent and feedback of information to participants, are handled, but it will not have ultimate responsibility for the ethical approval of the study protocol. This would rest with one of the NHS ethics committees set up to handle multi-centered studies.

The funding partners are still investigating the most appropriate legal status that will allow the oversight body the powers required to exercise its functions and the necessary level of independence. It is clear that the search for applicants for membership of the body must be advertised, those selected should include lay members and selection must be according to the UK standards required for public bodies.

An Interim Advisory Group,[12] established by the funders in January 2003 and chaired by Dr. William Lowrance (Geneva), developed an Ethics and Governance Framework that will be debated before adoption by the Board of Trustees of the Hub Company. This describes the principles on which the UK Biobank will base its operation in terms of all the social and ethical issues relevant to the project.

9. Concluding Remarks: Recent Appointments and Anticipated Schedule of Project Development

The UK Biobank has appointed a Chief Executive Officer, Professor John Newton (a Public Health Epidemiologist) to direct the establishment of the project based upon the selected hub location of the University of Manchester. Six spokes have been selected based upon consortia located at universities in Cardiff (Welsh Consortium), Glasgow (Scottish Consortium), Imperial College (London Consortium), Oxford, Leicester (Fosse Way Consortium), Manchester (Manchester–Wessex Consortium). Thus over twenty academic institutions are to be involved in the project. The Scientific Committee[13] has been created under the Chairmanship of Professor John Bell, University of Oxford. This body will develop the final version of the protocol, undertake piloting in 2004, and launch the full-scale project in early 2005.

12. *Membership of the Interim Advisory Group for UK Biobank:* Dr. William Lowrance (Geneva), Prof. Baroness Onora O'Neil (Cambridge), Prof. Graeme Laurie (Edinburgh), Prof. Jean McHale (Leicester), Prof. Chris Mathew (London), Prof. Alistair Campbell (Bristol), Prof. Erica Haimes (Newcastle), Mrs. Hellen Miller (Independent), and Mrs. Madelaine Wang (Independent).

13. *Membership of the Scientific Committee:* Prof. John Bell (Oxford), Dr. Bernard Keavney (Newcastle), Prof. Raj Bhopal (Edinburgh), Prof. Mike Pringle (Nottingham), Dr. Richard Durbin (Hinxton, Cambridge), Prof. John Todd (Cambridge), Prof. Paul Elliot (Imperial College Faculty of Medicine), Prof. Alan Silman (Manchester), Dr. Jill Pell (Glasgow), Prof. Valerie Beral (Oxford), Prof. Paul Burton (Leicester), Prof. Stephen Palmer (Cardiff), Ms. Madeline Want (Independent), Prof. John Newton (UK Biobank), Prof. Dame Jill Macleod Clark (Southampton), Prof. Hilary Graham (Lancaster), Prof. Catherine Peckham (London). Observers are: Prof. Bill Ollier (Manchester), Dr. Alan Doyle (The Wellcome Trust), Dr. George Sarna (MRC), and Dr. Anthony Horn (Department of Health).

Appendix

Legal and Ethical Norms in the Field of Genetics[1]
A Selection of International and National Legal Documents, Reports, and Policy Recommendations

1. International Organizations

1.1 United Nations (UNESCO, WHO, UNEP)

- Universal Declaration of Human Rights (UN, December 1948), available at http://www.unhchr.ch/udhr/lang/eng.htm.
- International Covenant on Civil and Political Rights (UN, December 1966), available at http://www.unhchr.ch/html/menu3/b/a_ccpr.htm.
- Universal Declaration on the Human Genome and Human Rights (UNESCO, November 1997), available at http://www.unesco.org/shs/human_rights/hrbc.htm.
- International Declaration on Human Genetic Data (UNESCO, October 2003), draft available at http://unesdoc.unesco.org/images/0013/001312/131204e.pdf.
- Convention on Biological Diversity (UNEP, 1992), available at http://www.biodiv.org/doc/legal/cbd-en.pdf.
- Resolution WHA50.37 on Cloning in Human Reproduction (WHO, May 1997), available at http://policy.who.int/cgi-bin/om_isapi.dll?advquery=WHA 50.37&infobase=wha&softpage=Document42.
- Resolution WHA51.10 on Ethical, Scientific and Social Implications of Cloning in Human Health (WHO, May 1998), available at http://www.who.int/gb/EB_WHA/PDF/WHA51/ear10.pdf

2. Regional International Organizations

2.1 European Union

2.1.1 The European Parliament and the Council of the European Union

- Charter of Fundamental Rights of the European Union (December 2000), *Official Journal C 364, 18.12.2000,* pp.0001–0022.

1. In the preparation and revision of the list of ethical and legal documents I received substantial help from Enikő Demény and Ildikó Takács—*Judit Sándor.*

2.1.1.1 Regulations

- Regulation (EC) No. 258/97 of the European Parliament and of the Council of 27 January 1997 Concerning Novel Foods and Novel Food Ingredients, *Official Journal L 043, 14.02.1997,* pp.0001– 0006.
- Regulation (EC) No. 45/2001 of the European Parliament and of the Council of 18 December 2000 on the Protection of Individuals with Regard to the Processing of Personal Data by the Community Institutions and Bodies and on the Free Movement of Such Data, *Official Journal L 008, 12.01.2001,* pp.0001–0022, available at http://europa.eu.int/eur-lex/pri/en/oj/dat/2001/l_008/l_0082 0010112en00010022.pdf.

2.1.1.2 Directives

- Directive 95/46/EC of the European Parliament and of the Council of 24 October 1995 on the protection of individuals with regard to the processing of personal data and on the free movement of such data, *Official Journal L 281, 23.11.1995,* pp.0031–0050, available at http://europa.eu.int/comm/internal_market/privacy/docs/95-46-ce/dir1995-46_part1_en.pdf and http://europa.eu.int/comm/internal_market/privacy/docs/95-46-ce/dir1995-46_part2_en.pdf.
- Directive 98/44/EC of the European Parliament and of the Council of 6 July 1998 on the Legal Protection of Biotechnological Inventions, *Official Journal L 213, 30.07.1998,* pp.0013–0021, available at http://europa.eu.int/eur-lex/pri/en/oj/dat/1998/l_213/l_21319980730en00130021.pdf.
- Directive 98/79/EC of the European Parliament and of the Council of 27 October 1998 on *In Vitro* Diagnostic Medical Devices, *Official Journal L 331, 07.12.1998,* pp.0003–0037, available at http://europa.eu.int/eur-lex/pri/en/oj/dat/1998/l_331/l_33119981207en00010037.pdf.
- Directive 2001/18/EC of the European Parliament and of the Council of 12 March 2001 on the Deliberate Release into the Environment of Genetically Modified Organisms and Repealing Council Directive 90/220/EEC, *Official Journal L 106, 17.04.2001,* pp.0001–0039, available at http://europa.eu.int/eur-lex/pri/en/oj/dat/2001/l_106/l_10620010417en00010038.pdf.
- Directive 2001/20/EC of the European Parliament and of the Council of 4 April 2001 on the Approximation of the Laws, Regulations and Administrative Provisions of the Member States Relating to the Implementation of Good Clinical Practice in the Conduct of Clinical Trials on Medicinal Products for Human Use, *Official Journal L 121, 01.05.2001,* pp.0034–0044, available at

http://europa.eu.int/eur-lex/pri/en/oj/dat/2001/l_121/l_12120010501en0034 0044.pdf.

- Directive 2002/98/EC of the European Parliament and of the Council of 27 January 2003 Setting Standards of Quality and Safety for the Collection, Testing, Processing, Storage and Distribution of Human Blood and Blood Components and Amending Directive 2001/83/EC, *Official Journal L 033, 08.02.2003,* pp.0030–0040, available at http://europa.eu.int/eur-lex/pri/en/oj/dat/2003/l_033/l_03320030208en00300040.pdf.
- Council Directive 86/609/EEC of 24 November 1986 on the Approximation of Laws, Regulations and Administrative Provisions of the Member States Regarding the Protection of Animals Used for Experimental and Other Scientific Purposes, *Official Journal L 358, 18.12.1986,* pp.0001–0028.
- Council Directive 90/219/EEC of 23 April 1990 on the Contained Use of Genetically Modified Micro-Organisms, *Official Journal L 117, 08.05.1990,* pp.0001–0014, available at http://europa.eu.int/eur-lex/en/consleg/pdf/1990/en_1990L0219_do_001.pdf.
- Council Directive 90/220/EEC of 23 April 1990 on the Deliberate Release into the Environment of Genetically Modified Organisms, *Official Journal L 117, 08.05.1990,* pp.0015–0027, no longer in force.
- Council Directive 98/81/EC of 26 October 1998 amending Directive 90/219/EEC on the Contained Use of Genetically Modified Micro-Organisms, *Official Journal L 330, 05.12.1998,* pp.0013–0031, available at http://europa.eu.int/eur-lex/pri/en/oj/dat/1998/l_330/l_33019981205en00130031.pdf.

2.1.1.3 Resolutions

- European Parliament Resolution on Human Cloning of 7 September 2000, available at http://www.europarl.eu.int/comparl/tempcom/genetics/links/b5_0710_en.pdf

2.1.2 European Commission

2.1.2.1 Regulations

- Commission Regulation (EC) No. 831/2002 of 17 May 2002 Implementing Council Regulation (EC) No. 322/97 on Community Statistics, Concerning Access to Confidential Data for Scientific Purposes, *Official Journal L 133, 18.05.2002,* pp.0007–0009, available at http://europa.eu.int/eur-lex/pri/en/oj/dat/2002/l_133/l_13320020518en00070009.pdf.

2.1.3 European Group on Ethics in Science and New Technologies (EGE, GAEIB, European Commission)

2.1.3.1 Opinions

- Opinion No.3. Opinion on Ethical Questions Arising from the Commission Proposal for a Council Directive for Legal Protection of Biotechnological Inventions, 30 September 1993, available at http://europa.eu.int/comm/european_group_ethics/gaieb/en/opinion3.pdf.
- Opinion No.4. The Ethical Implications of Gene Therapy, 13 December 1994, available at http://europa.eu.int/comm/european_group_ethics/gaieb/en/opinion4.pdf.
- Opinion No.6. Ethical Aspects of Prenatal Diagnosis, 20 February 1996, available at http://europa.eu.int/comm/european_group_ethics/gaieb/en/opinion6.pdf.
- Opinion No.7. Ethical Aspects of Genetic Modification of Animals, 21 May 1996, available at http://europa.eu.int/comm/european_group_ethics/gaieb/en/opinion7.pdf.
- Opinion No.8. Ethical Aspects of Patenting Inventions Involving Elements of Human Origin, 25 September 1996, available at http://europa.eu.int/comm/european_group_ethics/gaieb/en/opinion8.pdf.
- Opinion No.9. Ethical Aspects of Cloning Techniques, 28 May 1997, available at http://europa.eu.int/comm/european_group_ethics/gaieb/en/opinion9.pdf.
- Opinion No.11. Ethical Aspects of Human Tissue Banking, 21 July 1998, available at http://europa.eu.int/comm/european_group_ethics/docs/avis11_en.pdf.
- Opinion No.12. Ethical Aspects of Research Involving the Use of Human Embryo in the Context of the Fifth Framework Program, 23 November 1998, available at http://europa.eu.int/comm/european_group_ethics/docs/avis12_en.pdf.
- Opinion No.15. Ethical Aspects of Human Stem Cell Research and Use, 11 November 2000, available at http://europa.eu.int/comm/european_group_ethics/docs/avis15_en.pdf.
- Opinion No.16. Ethical Aspects of Patenting Inventions Involving Human Stem Cells, 7 May 2002, available at http://europa.eu.int/comm/european_group_ethics/docs/avis16_en.pdf.
- Opinion No.17. Ethical Aspects of Clinical Research in Developing Countries, 4 February 2003, available at http://europa.eu.int/comm/european_group_ethics/docs/avis17_en.pdf.
- Opinion No.18. Ethical Aspects of Genetic Testing in the Workplace, 28 July 2003, available at http://europa.eu.int/comm/european_group_ethics/docs/avis18en.pdf.
- Opinion No.19. Ethical Aspects of Cord Blood Banks (under elaboration).

- For more, see the website http://europa.eu.int/comm/european_group_ethics/index_en.htm.

2.2 *Council of Europe*

2.2.1 Treaties and Conventions

- Convention for the Protection of Human Rights and Fundamental Freedoms, as amended by Protocol No. 11 (ETS–005, ETS–155), signed in Rome on 4 November 1950, available at http://conventions.coe.int/treaty/en/Treaties/Html/005.htm.
- European Social Charter (Revised, ETS–163), signed in Strasbourg on 3 May 1996, available at http://conventions.coe.int/treaty/en/Treaties/Html/163.htm.
- Convention for the Protection of Human Rights and Dignity of the Human Being with regard to the Application of Biology and Medicine: Convention on Human Rights and Biomedicine (ETS–164, also referred to as 'Oviedo Convention' or 'Bioethics Convention'), signed in Oviedo on 4 April 1997, available at http://conventions.coe.int/treaty/en/Treaties/Html/164.htm.
- Additional Protocol to the Convention for the Protection of Human Rights and Dignity of the Human Being with regard to the Application of Biology and Medicine, on the Prohibition of Cloning Human Beings (ETS–168), signed in Paris on 12 January 1998, available at http://conventions.coe.int/treaty/en/Treaties/Html/168.htm.
- Additional Protocol to the Convention on Human Rights and Biomedicine, on Transplantation of Organs and Tissues of Human Origin (ETS–186), signed in Strasbourg on 24 January 2002, available at http://conventions.coe.int/treaty/en/Treaties/Html/186.htm.
- For more, see the website http://conventions.coe.int/treaty/en/cadrelistetraites.htm.

2.2.2 Parliamentary Assembly of the Council of Europe

- Recommendation 1160 (1991) on the Preparation for a Convention on Bioethics, available at http://assembly.coe.int/documents/adoptedtext/ta91/erec1160.htm.
- Recommendation 1240 (1994) on the Protection and Patentability of Materials of Human Origin, available at http://assembly.coe.int/documents/adoptedtext/ta94/erec1240.htm.
- Recommendation 1425 (1999) on Biotechnology and Intellectual Property, available at http://assembly.coe.int/documents/adoptedtext/ta99/erec1425.htm.

- Recommendation 1468 (2000) on Biotechnologies, available at http://assembly.coe.int/documents/adoptedtext/ta00/erec1468.htm.
- Recommendation 1512 (2001) on the Protection of the Human Genome by the Council of Europe, available at http://assembly.coe.int/documents/adoptedtext/ta01/erec1512.htm.

2.2.3 Committee of Ministers of the Council of Europe

- Recommendations of the Committee of Ministers to Member States:
- Recommendation R (90) 3 Concerning Medical Research on Human Beings, available at http://cm.coe.int/ta/rec/1990/90r3.htm.
- Recommendation R (90) 8 on the Impact of New Technologies on Health Services, Particularly Health Care, available at http://cm.coe.int/ta/rec/1990/90r8.htm.
- Recommendation R (90) 13 on Prenatal Genetic Screening, Prenatal Genetic Diagnosis, and Associated Genetic Counselling, available at http://cm.coe.int/ta/rec/1990/90r13.htm.
- Recommendation R (92) 1 on the Use of Analysis of Deoxyribonucleic Acid (DNA) within the Framework of the Criminal Justice System, available at http://cm.coe.int/ta/rec/1992/92r1.htm.
- Recommendation R (92) 3 on Genetic Testing and Screening for Health Care Purposes, available at http://cm.coe.int/ta/rec/1992/92r3.htm, with an Explanatory Memorandum, available at http://cm.coe.int/ta/rec/1992/ExpRec(92)3.htm.
- Recommendation R (94) 1 on Human Tissue Banks, available at http://cm.coe.int/ta/rec/1994/94r1.htm.
- Recommendation R (94) 11 on Screening as a Tool of Preventive Medicine, available at http://cm.coe.int/ta/rec/1994/94r11.htm.
- Recommendation R (97) 5 on the Protection of Medical Data, available at http: //www.cm.coe.int/ta/rec/1997/97r5.htm.
- Recommendation R (97) 15 on Xenotransplantation, available at http://www.cm.coe.int/ta/rec/1997/97r15.htm.
- Recommendation R (98) 2 on the Provision of Haematopoietic Progenitor Cells, available at http://cm.coe.int/ta/rec/1998/98r2.htm.
- Recommendation Rec (2001) 5 on the Management of Organ Transplant Waiting Lists and Waiting Times, available at http://cm.coe.int/ta/rec/2001/2001r5.htm.

3. National Bioethics Committee Opinions

3.1 National Consultative Committee of Ethics for Health and Life Sciences (Comité Consultatif National d'Ethique, CCNE, France)

- No.1. Opinion on Sampling of Dead Human Embryonic and Fetal Tissue for Therapeutic, Diagnostic, and Scientific Purposes (May 1984), available in English at http://www.ccne-ethique.fr/english/avis/a_001.htm.
- No.8. Opinion on Research and Use of *In Vitro* Human Embryos for Scientific and Medical Purposes (December 1986), available in English at http://www.ccne-ethique.fr/english/avis/a_008.htm.
- No.9. Opinion on Problems Arising Because of the Development of Methods Using Human Cells and Their Derivatives (February 1987), available in English at http://www.ccne-ethique.fr/english/avis/a_009.htm.
- No.16. Opinion on the Transplantation of Nerve Cells in the Treatment of Parkinson's Disease (October 1989), available in English at http://www.ccne-ethique.fr/english/avis/a_016.htm.
- No.17. Opinion Concerning the Dissemination of DNA Analysis Identification Techniques (Genetic Fingerprinting) (December 1989), available in English at http://www.ccne-ethique.fr/english/avis/a_017.htm.
- No.18. Update on Studies Undertaken by the Committee Regarding Gamete and Embryo Donation (December 1989), available in English at http://www.ccne-ethique.fr/english/avis/a_018.htm.
- No.19. Opinion on Embryo Research Aiming to Achieve Pre-Transfer Genetic Diagnosis for Which a Moratorium Was Declared in 1986 (July 1990), available in English at http://www.ccne-ethique.fr/english/avis/a_019.htm.
- No.20. Opinion on the Present Organization of Gamete Donation and Its Consequences (July 1990), available in English at http://www.ccne-ethique.fr/english/avis/a_020.htm.
- No.21. Opinion that the Human Body Should Not Be Used for Commercial Purposes (December 1990), available in English at http://www.ccne-ethique.fr/english/avis/a_021.htm.
- No.22. Opinion on Gene Therapy (December 1990), available in English at http://www.ccne-ethique.fr/english/avis/a_022.htm.
- No.24. Opinion on Embryonic and Fetal Reduction (June 1991), available in English at http://www.ccne-ethique.fr/english/avis/a_024.htm.
- No.25. Opinion Regarding the Application of Genetic Testing to Individual Studies, Family Studies, and Population Studies (Problems Related to DNA

'Banks', Cell 'Banks', and Computerization) (June 1991), available in English at http://www.ccne-ethique.fr/english/avis/a_025.htm.

- No.27. Opinion that the Human Genome Should Not Be Used for Commercial Purposes (December 1991), available in English at http://www.ccne-ethique.fr/english/avis/a_027.htm.
- No.28. Opinion on Blood Transfusion with Reference to Not Making Commercial Use of the Human Body (December 1991), available in English at http://www.ccne-ethique.fr/english/avis/a_028.htm.
- No.30. Ethical Issues Raised by the Mandatory Genetic Testing for Female Participants in the Albertville Games (January 1992), available in English at http://www.ccne-ethique.fr/english/avis/a_030.htm.
- No.33. Opinion Concerning the Identification of Patients Suffering from Glaucoma in France and on Chromosomal Location of the Causative Gene or Genes (January 1993), available in English at http://www.ccne-ethique.fr/english/avis/a_033.htm.
- No.36. Opinion on the Use of Somatic Gene Therapy Procedures (June 1993), available in English at http://www.ccne-ethique.fr/english/avis/a_036.htm.
- No.40. Opinion on the Transfer of Embryos after the Decease of a Husband or Partner (December 1993), available in English at http://www.ccne-ethique.fr/english/avis/a_040.htm.
- No.42. Opinion on the Evolution of Practices Concerning Medically Assisted Procreation (March 1994), available in English at http://www.ccne-ethique.fr/english/avis/a_042.htm.
- No.45. Ethical Questions Arising from the Transmission of Scientific Information Concerning Research in Biology and Medicine (May 1995), available in English at http://www.ccne-ethique.fr/english/avis/a_045.htm.
- No.46. Opinion and Recommendations on "Genetics and Medicine: From Prediction to Prevention" (October 1995), available in English at http://www.ccne-ethique.fr/english/avis/a_046.htm.
- No.52. Opinion on the Creation of Human Embryonic Organ and Tissue Collections and Their Use for Scientific Purposes (March 1997), available in English at http://www.ccne-ethique.fr/english/avis/a_052.htm.
- No.53. Opinion on the Establishment of Collections of Human Embryo Cells and Their Use for Therapeutic or Scientific Purposes (March 1997), available in English at http://www.ccne-ethique.fr/english/avis/a_053.htm.
- No.54. Reply to the President of the French Republic on the Subject of Reproductive Cloning (April 1997), available in English at http://www.ccne-ethique.fr/english/avis/a_054.htm.

- No.58. Informed Consent of and Information to Persons Accepting Care or Research Procedures (June 1998), available in English at http://www.ccne-ethique.fr/english/avis/a_058.htm.
- No.60. Reexamination of the Law on Bioethics (June 1998), available in English at http://www.ccne-ethique.fr/english/avis/a_060.htm.
- No.61. Opinion on Ethics and Xenotransplantation (June 1999), available in English at http://www.ccne-ethique.fr/english/avis/a_061.htm.
- No.64. Opinion on a Preliminary Draft Law Incorporating Transposition into the Code of Intellectual Property, of the European Parliament and Council Directive 98/44/CE, dated July 6, 1998, on the Legal Protection of Biotechnological Inventions (June 2000), available in English at http://www.ccne-ethique.fr/english/avis/a_064.htm.
- No.67. Opinion on the Preliminary Draft Revision of the Laws on Bioethics (January 2001), available in English at http://www.ccne-ethique.fr/english/avis/a_067.htm.
- No.72. Reflections on an Extension of Preimplantation Genetic Diagnosis (July 2002), available in English at http://www.ccne-ethique.fr/english/avis/a_072.htm.
- No.74. Opinion on Umbilical Cord Blood Banks for Autologous Use or for Research (December 2002), available in English at http://www.ccne-ethique.fr/english/avis/a_074.htm.
- No.75. Ethical Issues Raised by the Development of ICSI (December 2002), available in English at http://www.ccne-ethique.fr/english/avis/a_075.htm.
- No.76. Apropos of the Duty to Circulate Genetic Information within the Family for Urgent Medical Reasons (April 2003), available in English at http://www.ccne-ethique.fr/english/avis/a_076.htm.
- For more, see the website http://www.ccne-ethique.fr/english/start.htm.

3.2 Bioethics Committees in the United Kingdom

3.2.1. Nuffield Council on Bioethics

- Genetic Screening: Ethical Issues (December 1993), available at http://www.nuffieldbioethics.org/filelibrary/pdf/genetic_screening.pdf.
- Human Tissue: Ethical and Legal Issues (April 1995), available at http://www.nuffieldbioethics.org/filelibrary/pdf/human_tissue.pdf.
- Animal-to-Human Transplants: The Ethics of Xenotransplantation (March 1996), available at http://www.nuffieldbioethics.org/filelibrary/pdf/xenotransplantation.pdf.

- Mental Disorders and Genetics: The Ethical Context (September 1998), available at http://www.nuffieldbioethics.org/filelibrary/pdf/mentaldisorders2.pdf.
- Stem Cell Therapy: The Ethical Issues (April 2000), the discussion paper is available at http://www.nuffieldbioethics.org/filelibrary/doc/stem_cell_therapy2.doc.
- The Ethics of Research Related to Healthcare in Developing Countries (April 2002), available at http://www.nuffieldbioethics.org/filelibrary/pdf/errhdc_fullreport.pdf.
- The Ethics of Patenting DNA (July 2002), available at http://www.nuffieldbioethics.org/filelibrary/pdf/theethicsofpatentingdna.pdf.
- Genetics and Human Behaviour (September 2002), available at http://www.nuffieldbioethics.org/filelibrary/pdf/nuffieldgeneticsrep.pdf.
- Pharmacogenetics: Ethical Issues (September 2003), available at http://www.nuffieldbioethics.org/filelibrary/pdf/pharmacogenetics_report.pdf.
- For more, see the website http://www.nuffieldbioethics.org/publications/index.asp.

3.2.2 Advisory Committee on Genetic Testing (ACGT, Department of Health, 1996–1999)

- Code of Practice and Guidance on Human Genetic Testing Services Supplied Direct to the Public (September 1997), available at http://www.doh.gov.uk/pub/docs/doh/hgts.pdf.
- Report on Genetic Testing for Late Onset Disorders (July 1998), available at http://www.doh.gov.uk/pub/docs/doh/lodrep.pdf.
- Advice to Research Ethics Committees: Points to Consider in Ethical Review of Medical Research Involving Genetic Testing (October 1998), available at http://www.doh.gov.uk/pub/docs/doh/recrev3.pdf.
- For more, see the website http://www.doh.gov.uk/genetics/acgt/publications.htm.

3.2.3 Gene Therapy Advisory Committee (GTAC, Department of Health)

- Writing Information Leaflets for Patients Participating in Gene Therapy Research (August 1995), available at http://www.doh.gov.uk/pub/docs/doh/gtacinf.pdf.
- Report on the Potential Use of Gene Therapy *in Utero* (November 1998), available at http://www.doh.gov.uk/pub/docs/doh/inutero.pdf.
- Report of the GTAC Adenovirus Working Party (June 2000), available at http://www.doh.gov.uk/pdfs/gtacwpreport.pdf.

- Summary Table of UK Gene Therapy Research, 1993–2002 (September 2002), available at http://www.doh.gov.uk/genetics/gtac/gtacresearch932002.pdf.
- Recommendations of the GTAC/CSM Working Party on Retroviruses (April 2003), available at http://www.doh.gov.uk/genetics/gtac/gtac-csm-recommendations.pdf.
- For more, see the website http://www.doh.gov.uk/genetics/gtac/publications.htm.

3.2.4 Genetics and Insurance Committee (GAIC, Department of Health)

- Decision of the Genetics and Insurance Committee (GAIC) Concerning the Application for Approval to Use Genetic Test Results for Life Insurance Risk Assessment in Huntington's Disease (October 2000), available at http://www.doh.gov.uk/pdfs/gaichuntington.pdf.
- Genetics and Insurance. Government Response to the Report of the House of Commons Science and Technology Committee (October 2001), available at http://www.doh.gov.uk/genetics/gaicgovrespoct2001.pdf.
- For more, see the website http://www.doh.gov.uk/genetics/gaic/index.htm.

3.2.5 Human Genetics Advisory Commission (HGAC, Department of Health, 1996–1999)

- The Implications of Genetic Testing for Insurance (December 1997, the Government response to this report was published in October 1998), available at http://www.doh.gov.uk/hgac/papers/paperb1.htm.
- Cloning Issues in Reproduction, Science, and Medicine (December 1998), available at http://www.doh.gov.uk/hgac/papers/paperd1.htm.
- Human Genetics: Learning for the Millennium and Beyond (March 1997), available at http://www.doh.gov.uk/hgac/papers/papere1.htm.
- The Implications of Genetic Testing for Employment (July 1999, the Government response to this report was published in July 2000), available at http://www.doh.gov.uk/hgac/papers/paperg1.htm.
- For more, see the website http://www.doh.gov.uk/hgac/papers.htm.

3.2.6 Human Genetics Commission (HGC, Department of Health, from 1999)

- The Human Genetics Commission was established in December 1999 and subsumed the work of three other advisory bodies: the Advisory Committee on Genetic Testing (ACGT), the Advisory Group on Scientific Advances in Genetics (AGSAG), and the Human Genetics Advisory Commission (HGAC). See their reports and recommendations separately.

- Protection of Genetic Information: An International Comparison, A Report to the HGC (September 2000), available at http://www.hgc.gov.uk/business_publications_international_regulations.pdf.
- Public Attitudes to the Use of Human Genetic Information, A Report to the HGC (September 2000), available at http://www.hgc.gov.uk/business_publications_public_attitudes.pdf.
- The Use of Genetic Information in Insurance: Interim Recommendations of the Human Genetics Commission (May 2001), available at http://www.hgc.gov.uk/business_publications_hgcinsurance.pdf.
- Outcome of the Public Consultation on Prenatal Genetic Diagnosis, A Joint Report with the Human Fertilisation and Embryology Authority (HFEA) (November 2001), http://www.hgc.gov.uk/business_publications_pgdoutcome.pdf.
- Inside Information: Balancing Interests in the Use of Personal Genetic Data, (May 2002), available at http://www.hgc.gov.uk/insideinformation/index.htm.
- Memorandum to the House of Commons Science and Technology Committee on the Medical Research Council/Wellcome Trust/Department of Health UK Biobank Study, (November 2002), available at http://www.hgc.gov.uk/business_publications_memorandum_ukbiobank.pdf.
- Genes Direct: Ensuring the Effective Oversight of Genetic Tests Supplied Directly to the Public (April 2003), available at http://www.hgc.gov.uk/genesdirect/index.htm.
- For more, see the website http://www.hgc.gov.uk/

3.2.7 The Genetics, Embryology and Assisted Conception Branch (The Department of Health)

- Laboratory Services for Genetics: Report of an Expert Working Group to the NHS Executive and the Human Genetics Commission (August 2000), available at http://www.doh.gov.uk/pdfs/geneticslab.pdf.
- Stem Cell Research: Medical Progress with Responsibility (August 2000), available at http://www.doh.gov.uk/cegc/stemcellreport.pdf.
- Code of Practice and Guidance on Genetic Paternity Testing Services (March 2001), available at http://www.doh.gov.uk/pdfs/geneticspaternity.pdf.

- Pre-Implantation Genetic Diagnosis (PGD): Guiding Principles for Commissioners of NHS services (September 2002), available at http://www.doh.gov.uk/genetics/pdfs/pgdpaper.pdf.
- Government Response to the HGC's Report "Inside Information: Balancing Interests in the Use of Personal Genetic Data" (June 2003), available at http: //www.doh.gov.uk/genetics/govresp.htm.
- For more, see the website http://www.doh.gov.uk/genetics/index.htm.

3.3 National Ethics Council (Der Nationaler Ethikrat, Germany)

- Opinion on the Import of Human Embryonic Stem Cells (December 2001), available in English at http://www.ethikrat.org/_english/publications/stem_cells/stem_cells_index.html.
- Genetic Diagnosis Before and During Pregnancy. Opinion (January 2003), available in English at http://www.ethikrat.org/_english/publications/Stn_PID_engl.pdf.
- For more, see the website http://www.ethikrat.org/_english/publications/opinions.html.

3.4 National Bioethics Committee (Comitato Nazionale per la Bioetica, Italy)

- Bioethical Guidelines for Genetic Testing (November 1999), available in English at http://www.palazzochigi.it/bioetica/eng/opinions/genetictest.html.
- Opinion of the National Bioethics Committee on the European Protocol on Biomedical Research (November 1999), available in English at http://www.palazzochigi.it/bioetica/eng/opinions/protocol.html.
- Declaration on the Possibility of Patenting Human Embryo Cells (February 2000), available in English at http://www.palazzochigi.it/bioetica/eng/opinions/declaration.html.
- Protection of the Human Embryo and Fetus. The Italian National Bioethics committee statement concerning the preliminary draft Protocol of the Bioethics Committee of the Council of Europe (March 2000), available in English at http://www.palazzochigi.it/bioetica/eng/opinions/embryo.html.
- Opinion of the National Bioethics Committee on the Therapeutic Use of Stem Cells (October 2000), available in English at http://www.palazzochigi.it/bioetica/eng/opinions/cells.html.
- Identity and Status of the Human Embryo (June 1996).

- Opinion of the National Bioethics Committee on the Convention for the Protection of Human Rights and Biomedicine and Preliminary Draft of the Universal Declaration on the Human Genome and Human Rights (February 1997).
- Opinion on Cloning (October 1997).
- For more, see the website http://www.palazzochigi.it/bioetica/eng/opinions.html.

3.5 Medical Research Council (Egészségügyi Tudományos Tanács, Hungary)

- MRC Statement on the Legal and Ethical Norms of Establishing, Operating, and Using Human Cell, Tissue, DNA, and Genetic Databanks (2002), available in Hungarian only.
- MRC Statement on the Medical, Ethical and Legal Aspects of Using Stem Cells Originated from Umbilical Cord Blood (2002), available in Hungarian only.
- Recommendations for the Legal Regulation of Activities in the Field of Human Genetics (2003), available in Hungarian only.

4. International Professional Associations

4.1 European Society of Human Genetics (ESHG)

The Professional and Public Policy Committee (PPPC) of ESHG

- Population Genetic Screening Programmes: Technical, Social, and Ethical Issues. Recommendations of the ESHG (June 2000), available at http://www.eshg.org/ESHGscreeningrec.pdf.
- Genetic Information and Testing in Insurance and Employment: Technical, Social, and Ethical Issues. Recommendations of the ESHG (November 2000), available at http://www.eshg.org/ESHGinsurancerec.pdf.
- Data Storage and DNA Banking for Biomedical Research: Technical, Social, and Ethical Issues. Recommendations of the ESHG (2001), available at http://www.eshg.org/ESHGDNAbankingrec.pdf.
- Provision of Genetic Services in Europe: Current Practices and Issues. Recommendations of the ESHG (2001), available at http://www.eshg.org/ESHGgeneticservicesrec.pdf.

4.2 The Human Genome Organization (HUGO)

- Statement on the Patenting of DNA Sequences (January 1995), available at http://www.hugo-international.org/hugo/patent.htm.
- Statement on the Principled Conduct of Genetics Research (March 1996), available at http://www.hugo-international.org/hugo/conduct.htm.
- Statement on DNA Sampling: Control and Access (February 1998), available at http://www.hugo-international.org/hugo/sampling.html.
- Statement on Cloning (March 1999), available at http://www.biol.tsukuba.ac.jp/~macer/hugoclone.html.
- Statement on the Patenting of DNA Sequences, in Particular Response to the European Biotechnology Directive (April 2000), available at http://www.hugo-international.org/hugo/patent2000.html.
- Statement on Benefit Sharing (April 2000) available at http://www.hugo-international.org/hugo/benefit.html.
- Statement on Human Genomic Databases (December, 2002), available at http://www.hugo-international.org/hugo/HEC_Dec02.html.
- Draft Statement on Gene Therapy Research (2003), available at http://www.hugo-international.org/hugo/genetherapy.htm.
- For more, see the webpages http://www.hugo-international.org/hugo/ethics.html and http://www.hugo-international.org/hugo/statements.html.

4.3 Nordic Committee for Bioethics (Nordic countries)

- Opinion on Human Stem Cell Research (October 2000), available at http://www.ncbio.org/Html/eng_stem_cell.htm.

Universal Declaration on the Human Genome and Human Rights[1] (UNESCO, 1997)

The General Conference,

Recalling that the Preamble of UNESCO's Constitution refers to "the democratic principles of the dignity, equality and mutual respect of men," rejects any "doctrine of the inequality of men and races," stipulates "that the wide diffusion of culture, and the education of humanity for justice and liberty and peace are indispensable to the dignity of men and constitute a sacred duty which all the nations must fulfill in a spirit and mutual assistance and concern," proclaims that "peace must be founded of upon the intellectual and moral solidarity of mankind," and states that the Organization seeks to advance "through the educational and scientific and cultural relations of the peoples of the world, the objectives of international peace and of the common welfare of mankind for which the United Nations Organization was established and which its Charter proclaims";

Solemnly recalling its attachment to the universal principles of human rights, affirmed in particular in the Universal Declaration of Human Rights of 10 December 1948 and in the two International United Nations Covenants on Economic, Social and Cultural Rights and on Civil and Political Rights of 16 December 1966, in the United Nations Convention on the Prevention and Punishment of the Crime of Genocide of 9 December 1948, the International United Nations Convention on the Elimination of All Forms of Racial Discrimination of 21 December 1965, the United Nations Declaration on the Rights of Mentally Retarded Persons of 20 December 1971, the United Nations Declaration on the Rights of Disabled Persons of 9 December 1975, the United Nations Convention on the Elimination of All Forms of Discrimination Against Women of 18 December 1979, the United Nations Declaration of Basic Principles of Justice for Victims of Crime and Abuse of Power of 29 November 1985, the United Nations Convention on the Rights of the Child of 20 November 1989, the United Nations Standard Rules on the Equalization of Opportunities for Persons with Disabilities of 20 December 1993, the Convention on the

1. Adopted unanimously and by acclamation on November 11, 1997 by the 29th session of the General Conference of the UNESCO.

Prohibition of the Development, Production and Stockpiling of Bacteriological (Biological) and Toxin Weapons and on their Destruction of 16 December 1971, the UNESCO Convention against Discrimination in Education of 14 December 1960, the UNESCO Declaration of the Principles of International Cultural Cooperation of 4 November 1966, the UNESCO Recommendation on the Status of Scientific Researchers of 20 November 1974, the UNESCO Declaration on Race and Racial Prejudice of 27 November 1978, the ILO Convention (No. 111) concerning Discrimination in Respect of Employment and Occupation of 25 June 1958 and the ILO Convention (No. 169) concerning Indigenous and Tribal Peoples in Independent Countries of 27 June 1989;

Bearing in mind, and without prejudice to, the international instruments which could have a bearing on the applications of genetics in the field of intellectual property, inter alia the Bern Convention for the Protection of Literary and Artistic Works of 9 September 1886 and the UNESCO Universal Copyright Convention of 6 September 1952, as last revised in Paris on 24 July 1971, the Paris Convention for the Protection of Industrial Property of 20 March 1883, as last revised at Stockholm on 14 July 1967, the Budapest Treaty of the WIPO on International Recognition of the Deposit of Micro-organisms for the Purposes of Patent Procedures of 28 April 1977, and the Trade Related Aspects of Intellectual Property Rights Agreement (TRIPs) annexed to the Agreement establishing the World Trade Organization, which entered into force on 1st January 1995;

Bearing in mind also the United Nations Convention on Biological Diversity of 5 June 1992 and emphasizing in that connection that the recognition of the genetic diversity of humanity must not give rise to any interpretation of a social or political nature which could call into question 'the inherent dignity and . . . the equal and inalienable rights of all members of the human family', in accordance with the Preamble to the Universal Declaration of Human Rights;

Recalling 22 C/Resolution 13.1, 23 C/Resolution 13.1, 24 C/Resolution 13.1, 25 C/Resolutions 5.2 and 7.3, 27 C/Resolution 5.15 and 28 C/Resolutions 0.12, 2.1 and 2.2, urging UNESCO to promote and develop ethical studies, and the actions arising out of them, on the consequences of scientific and technological progress in the fields of biology and genetics, within the framework of respect for human rights and fundamental freedoms;

Recognizing that research on the human genome and the resulting applications open up vast prospects for progress in improving the health of individuals and of humankind as a whole, but emphasizing that such research should fully respect human dignity, freedom and human rights, as well as the prohibition of all forms of discrimination based on genetic characteristics;

Proclaims the principles that follow and *adopts* the present Declaration:

A. Human Dignity and the Human Genome

Article 1

The human genome underlies the fundamental unity of all members of the human family, as well as the recognition of their inherent dignity and diversity. In a symbolic sense, it is the heritage of humanity.

Article 2

(a) Everyone has a right to respect for their dignity and for their rights regardless of their genetic characteristics.

(b) That dignity makes it imperative not to reduce individuals to their genetic characteristics and to respect their uniqueness and diversity.

Article 3

The human genome, which by its nature evolves, is subject to mutations. It contains potentialities that are expressed differently according to each individual's natural and social environment including the individual's state of health, living conditions, nutrition and education.

Article 4

The human genome in its natural state shall not give rise to financial gains.

B. Rights of the Persons Concerned

Article 5

(a) Research, treatment or diagnosis affecting an individual's genome shall be undertaken only after rigorous and prior assessment of the potential risks and benefits pertaining thereto and in accordance with any other requirement of national law.

(b) In all cases, the prior, free and informed consent of the person concerned shall be obtained. If the latter is not in a position to consent, consent or authorization shall be obtained in the manner prescribed by law, guided by the person's best interest.

(c) The right of each individual to decide whether or not to be informed of the results of genetic examination and the resulting consequences should be respected.

(d) In the case of research, protocols shall, in addition, be submitted for prior review in accordance with relevant national and international research standards or guidelines.

(e) If according to the law a person does not have the capacity to consent, research affecting his or her genome may only be carried out for his or her direct health benefit, subject to the authorization and the protective conditions prescribed by law. Research which does not have an expected direct health benefit may only be undertaken by way of exception, with the utmost restraint, exposing the person only to a minimal risk and minimal burden and if the research is intended to contribute to the health benefit of other persons in the same age category or with the same genetic condition, subject to the conditions prescribed by law, and provided such research is compatible with the protection of the individual's human rights.

Article 6

No one shall be subjected to discrimination based on genetic characteristics that is intended to infringe or has the effect of infringing human rights, fundamental freedoms and human dignity.

Article 7

Genetic data associated with an identifiable person and stored or processed for the purposes of research or any other purpose must be held confidential in the conditions set by law.

Article 8

Every individual shall have the right, according to international and national law, to just reparation for any damage sustained as a direct and determining result of an intervention affecting his or her genome.

Article 9

In order to protect human rights and fundamental freedoms, limitations to the principles of consent and confidentiality may only be prescribed by law, for compelling reasons within the bounds of public international law and the international law of human rights.

C. Research on the Human Genome

Article 10

No research or research applications concerning the human genome, in particular in the fields of biology, genetics and medicine, should prevail over respect for the human rights, fundamental freedoms and human dignity of individuals or, where applicable, of groups of people.

Article 11

Practices which are contrary to human dignity, such as reproductive cloning of human beings, shall not be permitted. States and competent international organizations are invited to cooperate in identifying such practices and in taking, at national or international level, the measures necessary to ensure that the principles set out in this Declaration are respected.

Article 12

(a) Benefits from advances in biology, genetics and medicine, concerning the human genome, shall be made available to all, with due regard for the dignity and human rights of each individual.

(b) Freedom of research, which is necessary for the progress of knowledge, is part of freedom of thought. The applications of research, including applications in biology, genetics and medicine, concerning the human genome, shall seek to offer relief from suffering and improve the health of individuals and humankind as a whole.

D. Conditions for the Exercise of Scientific Activity

Article 13

The responsibilities inherent in the activities of researchers, including meticulousness, caution, intellectual honesty and integrity in carrying out their research as well as in the presentation and utilization of their findings, should be the subject of particular attention in the framework of research on the human genome, because of its ethical and social implications. Public and private science policy-makers also have particular responsibilities in this respect.

Article 14

States should take appropriate measures to foster the intellectual and material conditions favorable to freedom in the conduct of research on the human genome and to consider the ethical, legal, social and economic implications of such research, on the basis of the principles set out in this Declaration.

Article 15

States should take appropriate steps to provide the framework for the free exercise of research on the human genome with due regard for the principles set out in this Declaration, in order to safeguard respect for human rights, fundamental freedoms and human dignity and to protect public health. They should seek to ensure that research results are not used for non-peaceful purposes.

Article 16

States should recognize the value of promoting, at various levels, as appropriate, the establishment of independent, multidisciplinary and pluralist ethics committees to assess the ethical, legal and social issues raised by research on the human genome and its applications.

E. Solidarity and International Cooperation

Article 17

States should respect and promote the practice of solidarity towards individuals, families and population groups who are particularly vulnerable to or affected by disease or disability of a genetic character. They should foster, *inter alia*, research on the identification, prevention and treatment of genetically-based and genetically-influenced diseases, in particular rare as well as endemic diseases which affect large numbers of the world's population.

Article 18

States should make every effort, with due and appropriate regard for the principles set out in this Declaration, to continue fostering the international dissemination of scientific knowledge concerning the human genome, human diversity and genetic research and, in that regard, to foster scientific and cultural cooperation, particularly between industrialized and developing countries.

Article 19

(a) In the framework of international cooperation with developing countries, States should seek to encourage measures enabling:
   (i) assessment of the risks and benefits pertaining to research on the human genome to be carried out and abuse to be prevented;

(ii) the capacity of developing countries to carry out research on human biology and genetics, taking into consideration their specific problems, to be developed and strengthened;
(iii) developing countries to benefit from the achievements of scientific and technological research so that their use in favor of economic and social progress can be to the benefit of all;
(iv) the free exchange of scientific knowledge and information in the areas of biology, genetics and medicine to be promoted.

(b) Relevant international organizations should support and promote the initiatives taken by States for the above-mentioned purposes.

F. Promotion of the Principles Set Out in the Declaration

Article 20

States should take appropriate measures to promote the principles set out in the Declaration, through education and relevant means, *inter alia* through the conduct of research and training in interdisciplinary fields and through the promotion of education in bioethics, at all levels, in particular for those responsible for science policies.

Article 21

States should take appropriate measures to encourage other forms of research, training and information dissemination conducive to raising the awareness of society and all of its members of their responsibilities regarding the fundamental issues relating to the defense of human dignity which may be raised by research in biology, in genetics and in medicine, and its applications. They should also undertake to facilitate on this subject an open international discussion, ensuring the free expression of various socio-cultural, religious and philosophical opinions.

G. Implementation of the Declaration

Article 22

States should make every effort to promote the principles set out in this Declaration and should, by means of all appropriate measures, promote their implementation.

Article 23

States should take appropriate measures to promote, through education, training and information dissemination, respect for the above-mentioned principles and to foster their recognition and effective application. States should also encourage exchanges and networks among independent ethics committees, as they are established, to foster full collaboration.

Article 24

The International Bioethics Committee of UNESCO should contribute to the dissemination of the principles set out in this Declaration and to the further examination of issues raised by their applications and by the evolution of the technologies in question. It should organize appropriate consultations with parties concerned, such as vulnerable groups. It should make recommendations, in accordance with UNESCO's statutory procedures, addressed to the General Conference and give advice concerning the follow-up of this Declaration, in particular regarding the identification of practices that could be contrary to human dignity, such as germ-line interventions.

Article 25

Nothing in this Declaration may be interpreted as implying for any State, group or person any claim to engage in any activity or to perform any act contrary to human rights and fundamental freedoms, including the principles set out in this Declaration.

International Declaration on Human Genetic Data[1] (UNESCO, 2003)

The General Conference,

Recalling the Universal Declaration of Human Rights of 10 December 1948, the two United Nations International Covenants on Economic, Social and Cultural Rights and on Civil and Political Rights of 16 December 1966, the United Nations International Convention on the Elimination of All Forms of Racial Discrimination of 21 December 1965, the United Nations Convention on the Elimination of All Forms of Discrimination against Women of 18 December 1979, the United Nations Convention on the Rights of the Child of 20 November 1989, the United Nations Economic and Social Council Resolutions 2001/39 on Genetic Privacy and Non-Discrimination of 26 July 2001 and 2003/232 on Genetic Privacy and Non-Discrimination of 22 July 2003, the ILO Convention (No. 111) concerning Discrimination in Respect of Employment and Occupation of 25 June 1958, the UNESCO Universal Declaration on Cultural Diversity of 2 November 2001, the Trade Related Aspects of Intellectual Property Rights Agreement (TRIPs) annexed to the Agreement establishing the World Trade Organization, which entered into force on 1 January 1995, the Doha Declaration on the TRIPs Agreement and Public Health of 14 November 2001 and the other international human rights instruments adopted by the United Nations and the specialized agencies of the United Nations system;

Recalling more particularly the Universal Declaration on the Human Genome and Human Rights which it adopted, unanimously and by acclamation, on 11 November 1997 and which was endorsed by the United Nations General Assembly on 9 December 1998 and the Guidelines for the implementation of the Universal Declaration on the Human Genome and Human Rights which it endorsed on 16 November 1999 by 30 C/Resolution 23;

1. Adopted unanimously and by acclamation on October 16, 2003 by the 32nd session of the General Conference of UNESCO.

Welcoming the broad public interest worldwide in the Universal Declaration on the Human Genome and Human Rights, the firm support it has received from the international community and its impact in Member States drawing upon it for their legislation, regulations, norms, and standards, and ethical codes of conduct and guidelines;

Bearing in mind the international and regional instruments, national laws, regulations and ethical texts relating to the protection of human rights and fundamental freedoms and to respect for human dignity as regards the collection, processing, use, and storage of scientific data, as well as of medical data and personal data;

Recognizing that genetic information is part of the overall spectrum of medical data and that the information content of any medical data, including genetic data and proteomic data, is highly contextual and dependent on the particular circumstances;

Also recognizing that human genetic data have a special status on account of their sensitive nature since they can be predictive of genetic predispositions concerning individuals and that the power of predictability can be stronger than assessed at the time of deriving the data; they may have a significant impact on the family, including offspring, extending over generations, and in some instances on the whole group; they may contain information the significance of which is not necessarily known at the time of the collection of biological samples; and they may have cultural significance for persons or groups;

Emphasizing that all medical data, including genetic data and proteomic data, regardless of their apparent information content, should be treated with the same high standards of confidentiality;

Noting the increasing importance of human genetic data for economic and commercial purposes;

Having regard to the special needs and vulnerabilities of developing countries and the need to reinforce international cooperation in the field of human genetics;

Considering that the collection, processing, use and storage of human genetic data are of paramount importance for the progress of life sciences and medicine, for their applications and for the use of such data for non-medical purposes;

Also considering that the growing amount of personal data collected makes genuine irretrievability increasingly difficult;

Aware that the collection, processing, use and storage of human genetic data have potential risks for the exercise and observance of human rights and fundamental freedoms and respect for human dignity;

Noting that the interests and welfare of the individual should have priority over the rights and interests of society and research;

Reaffirming the principles established in the Universal Declaration on the Human Genome and Human Rights and the principles of equality, justice, solidarity and responsibility as well as respect for human dignity, human rights and fundamental freedoms, particularly freedom of thought and expression, including freedom of research, and privacy and security of the person, which must underlie the collection, processing, use and storage of human genetic data;

Proclaims the principles that follow and *adopts* the present Declaration:

A. General Provisions

Article 1. Aims and Scope

(a) The aims of this Declaration are: to ensure the respect of human dignity and protection of human rights and fundamental freedoms in the collection, processing, use and storage of human genetic data, human proteomic data and of the biological samples from which they are derived, referred to hereinafter as "biological samples", in keeping with the requirements of equality, justice, and solidarity, while giving due consideration to freedom of thought and expression, including freedom of research; to set out the principles which should guide States in the formulation of their legislation and their policies on these issues; and to form the basis for guidelines of good practices in these areas for the institutions and individuals concerned.

(b) Any collection, processing, use and storage of human genetic data, human proteomic data and biological samples shall be consistent with the international law of human rights.

(c) The provisions of this Declaration apply to the collection, processing, use and storage of human genetic data, human proteomic data and biological samples, except in the investigation, detection and prosecution of criminal offenses and in parentage testing that are subject to domestic law that is consistent with the international law of human rights.

Article 2. Use of Terms

For the purposes of this Declaration, the terms used have the following meanings:

(i) *Human genetic data*: Information about heritable characteristics of individuals obtained by analysis of nucleic acids or by other scientific analysis.

(ii) *Human proteomic data*: Information pertaining to an individual's proteins including their expression, modification and interaction.

(iii) *Consent*: Any freely given specific, informed and express agreement of an individual to his or her genetic data being collected, processed, used and stored.

(iv) *Biological samples*: Any sample of biological material (for example blood, skin and bone cells or blood plasma) in which nucleic acids are present and which contains the characteristic genetic make-up of an individual.

(v) *Population-based genetic study*: A study which aims at understanding the nature and extent of genetic variation among a population or individuals within a group or between individuals across different groups.

(vi) *Behavioral genetic study*: A study that aims at establishing possible connections between genetic characteristics and behavior.

(vii) *Invasive procedure*: Biological sampling using a method involving intrusion into the human body, such as obtaining a blood sample by using a needle and syringe.

(viii) *Non-invasive procedure*: Biological sampling using a method which does not involve intrusion into the human body, such as oral smears.

(ix) *Data linked to an identifiable person*: Data that contain information, such as name, birth date and address, by which the person from whom the data were derived can be identified.

(x) *Data unlinked to an identifiable person*: Data that are not linked to an identifiable person, through the replacement of, or separation from, all identifying information about that person by use of a code.

(xi) *Data irretrievably unlinked to an identifiable person*: Data that cannot be linked to an identifiable person, through destruction of the link to any identifying information about the person who provided the sample.

(xii) *Genetic testing*: A procedure to detect the presence or absence of, or change in, a particular gene or chromosome, including an indirect test for a gene product or other specific metabolite that is primarily indicative of a specific genetic change.

(xiii) *Genetic screening*: Large-scale systematic genetic testing offered in a programme to a population or subsection thereof intended to detect genetic characteristics in asymptomatic people.

(xiv) *Genetic counselling*: A procedure to explain the possible implications of the findings of genetic testing or screening, its advantages and risks and where applicable to assist the individual in the long-term handling of the consequences. It takes place before and after genetic testing and screening.

(xv) *Cross-matching*: Matching of information about an individual or a group contained in various data files set up for different purposes.

Article 3. Person's Identity

Each individual has a characteristic genetic make-up. Nevertheless, a person's identity should not be reduced to genetic characteristics, since it involves complex educational, environmental and personal factors and emotional, social, spiritual and cultural bonds with others and implies a dimension of freedom.

Article 4. Special Status

(a) Human genetic data have a special status because:

(i) they can be predictive of genetic predispositions concerning individuals;

(ii) they may have a significant impact on the family, including offspring, extending over generations, and in some instances on the whole group to which the person concerned belongs;

(iii) they may contain information the significance of which is not necessarily known at the time of the collection of the biological samples;

(iv) they may have cultural significance for persons or groups.

(b) Due consideration should be given to the sensitivity of human genetic data and an appropriate level of protection for these data and biological samples should be established.

Article 5. Purposes

Human genetic data and human proteomic data may be collected, processed, used and stored only for the purposes of:

(i) diagnosis and health care, including screening and predictive testing;

(ii) medical and other scientific research, including epidemiological, especially population-based genetic studies, as well as anthropological or archaeological studies, collectively referred to hereinafter as "medical and scientific research";

(iii) forensic medicine and civil, criminal and other legal proceedings, taking into account the provisions of Article 1(c);

(iv) or any other purpose consistent with the Universal Declaration on the Human Genome and Human Rights and the international law of human rights.

Article 6. Procedures

(a) It is ethically imperative that human genetic data and human proteomic data be collected, processed, used, and stored on the basis of transparent and ethically acceptable procedures. States should endeavor to involve society at large in the decision-making process concerning broad policies for the collection, processing, use and storage of human genetic data and human proteomic data and the evaluation of their management, in particular in the case of population-based genetic studies. This decision-making process, which may benefit from international experience, should ensure the free expression of various viewpoints.

(b) Independent, multidisciplinary, and pluralist ethics committees should be promoted and established at national, regional, local, or institutional levels, in accordance with the provisions of Article 16 of the Universal Declaration on the Human Genome and Human Rights. Where appropriate, ethics committees at national level should be consulted with regard to the establishment of standards, regulations and guidelines for the collection, processing, use and storage of human genetic data, human proteomic data, and biological samples. They should also be consulted concerning matters where there is no domestic law. Ethics committees at institutional or local levels should be consulted with regard to their application to specific research projects.

(c) When the collection, processing, use and storage of human genetic data, human proteomic data or biological samples are carried out in two or more States, the ethics committees in the States concerned, where appropriate, should be consulted and the review of these questions at the appropriate level should be based on the principles set out in this Declaration and on the ethical and legal standards adopted by the States concerned.

(d) It is ethically imperative that clear, balanced, adequate and appropriate information shall be provided to the person whose prior, free, informed, and express consent is sought. Such information shall, alongside with providing other necessary details, specify the purpose for which human genetic data and human proteomic data are being derived from biological samples, and are used and stored. This information should indicate, if necessary, risks and consequences. This information should also indicate that the person concerned can withdraw his or her consent, without coercion, and this should entail neither a disadvantage nor a penalty for the person concerned.

Article 7. Non-Discrimination and Non-Stigmatization

(a) Every effort should be made to ensure that human genetic data and human proteomic data are not used for purposes that discriminate in a way that is intended to infringe, or has the effect of infringing human rights, fundamental freedoms, or human dignity of an individual or for purposes that lead to the stigmatization of an individual, a family, a group, or communities.

(b) In this regard, appropriate attention should be paid to the findings of population-based genetic studies and behavioral genetic studies and their interpretations.

B. Collecting

Article 8. Consent

(a) Prior, free, informed and express consent, without inducement by financial or other personal gain, should be obtained for the collection of human genetic data, human proteomic data, or biological samples, whether through invasive or non-invasive procedures, and for their subsequent processing, use, and storage, whether carried out by public or private institutions. Limitations on this principle of consent should only be prescribed for compelling reasons by domestic law consistent with the international law of human rights.

(b) When, in accordance with domestic law, a person is incapable of giving informed consent, authorization should be obtained from the legal representative, in accordance with domestic law. The legal representative should have regard to the best interest of the person concerned.

(c) An adult not able to consent should as far as possible take part in the authorization procedure. The opinion of a minor should be taken into consideration as an increasingly determining factor in proportion to age and degree of maturity.

(d) In diagnosis and health care, genetic screening and testing of minors and adults not able to consent will normally only be ethically acceptable when it has important implications for the health of the person and has regard to his or her best interest.

Article 9. Withdrawal of Consent

(a) When human genetic data, human proteomic data, or biological samples are collected for medical and scientific research purposes, consent may be withdrawn by the person concerned unless such data are irretrievably unlinked to an identifiable person. In accordance with the provisions of Article 6(d), withdrawal of consent should entail neither a disadvantage nor a penalty for the person concerned.

(b) When a person withdraws consent, the person's genetic data, proteomic data, and biological samples should no longer be used unless they are irretrievably unlinked to the person concerned.

(c) If not irretrievably unlinked, the data and biological samples should be dealt with in accordance with the wishes of the person. If the person's wishes cannot be determined or are not feasible or are unsafe, the data and biological samples should either be irretrievably unlinked or destroyed.

Article 10. The Right to Decide Whether or Not to Be Informed about Research Results

When human genetic data, human proteomic data or biological samples are collected for medical and scientific research purposes, the information provided at the time of consent should indicate that the person concerned has the right to decide whether or not to be informed of the results. This does not apply to research on data irretrievably unlinked to identifiable persons or to data that do not lead to individual findings concerning the persons who have participated in such a research. Where appropriate, the right not to be informed should be extended to identified relatives who may be affected by the results.

Article 11. Genetic Counseling

It is ethically imperative that when genetic testing that may have significant implications for a person's health is being considered, genetic counseling should be made available in an appropriate manner. Genetic counseling should be non-directive, culturally adapted, and consistent with the best interest of the person concerned.

Article 12. Collection of Biological Samples for Forensic Medicine or in Civil, Criminal, and Other Legal Proceedings

When human genetic data or human proteomic data are collected for the purposes of forensic medicine or in civil, criminal and other legal proceedings, including parentage testing, the collection of biological samples, *in vivo* or postmortem, should be made only in accordance with domestic law consistent with the international law of human rights.

C. Processing

Article 13. Access

No one should be denied access to his or her own genetic data or proteomic data unless such data are irretrievably unlinked to that person as the identifiable source or unless domestic law limits such access in the interest of public health, public order, or national security.

Article 14. Privacy and Confidentiality

(a) States should endeavor to protect the privacy of individuals and the confidentiality of human genetic data linked to an identifiable person, a family or, where appropriate, a group, in accordance with domestic law consistent with the international law of human rights.

(b) Human genetic data, human proteomic data, and biological samples linked to an identifiable person should not be disclosed or made accessible to third parties, in particular, employers, insurance companies, educational institutions, and the family, except for an important public interest reason in cases restrictively provided for by domestic law consistent with the international law of human rights or where the prior, free, informed, and express consent of the person concerned has been obtained, provided that such consent is in accordance with domestic law and the international law of human rights. The privacy of an individual participating in a study using human genetic data, human proteomic data, or biological samples should be protected and the data should be treated as confidential.

(c) Human genetic data, human proteomic data, and biological samples collected for the purposes of scientific research should not normally be linked to an identifiable person. Even when such data or biological samples are unlinked to an identifiable person, the necessary precautions should be taken to ensure the security of the data or biological samples.

(d) Human genetic data, human proteomic data, and biological samples collected for medical and scientific research purposes can remain linked to an identifiable person, only if necessary to carry out the research and provided that the privacy of the individual and the confidentiality of the data or biological samples concerned are protected in accordance with domestic law.

(e) Human genetic data and human proteomic data should not be kept in a form which allows the data subject to be identified for any longer than is necessary for achieving the purposes for which they were collected or subsequently processed.

Article 15. Accuracy, Reliability, Quality, and Security

The persons and entities responsible for the processing of human genetic data, human proteomic data and biological samples should take the necessary measures to ensure the accuracy, reliability, quality, and security of these data and the processing of biological samples. They should exercise rigor, caution, honesty, and integrity in the processing and interpretation of human genetic data, human proteomic data, or biological samples, in view of their ethical, legal, and social implications.

D. Use

Article 16. Change of Purpose

(a) Human genetic data, human proteomic data, and the biological samples collected for one of the purposes set out in Article 5 should not be used for a different purpose that is incompatible with the original consent, unless the prior, free, informed, and express consent of the person concerned is obtained according to the provisions of Article 8(a) or unless the proposed use, decided by domestic law, corresponds to an important public interest reason and is consistent with the international law of human rights. If the person concerned lacks the capacity to consent, the provisions of Article 8(b) and (c) should apply *mutatis mutandis*.

(b) When prior, free, informed, and express consent cannot be obtained or in the case of data irretrievably unlinked to an identifiable person, human genetic data may be used in accordance with domestic law or following the consultation procedures set out in Article 6(b).

Article 17. Stored Biological Samples

(a) Stored biological samples collected for purposes other than set out in Article 5 may be used to produce human genetic data or human proteomic data with the prior, free, informed, and express consent of the person concerned. However, domestic law may provide that if such data have significance for medical and scientific research purposes, e.g. epidemiological studies, or public health purposes, they may be used for those purposes, following the consultation procedures set out in Article 6(b).

(b) The provisions of Article 12 should apply *mutatis mutandis* to stored biological samples used to produce human genetic data for forensic medicine.

Article 18. Circulation and International Cooperation

(a) States should regulate, in accordance with their domestic law and international agreements, the cross-border flow of human genetic data, human proteomic data, and biological samples so as to foster international medical and scientific cooperation and ensure fair access to this data. Such a system should seek to ensure that the receiving party provides adequate protection in accordance with the principles set out in this Declaration.

(b) States should make every effort, with due and appropriate regard for the principles set out in this Declaration, to continue fostering the international dissemination of scientific knowledge concerning human genetic data and human proteomic data and, in that regard, to foster scientific and cultural cooperation, particularly between industrialized and developing countries.

(c) Researchers should endeavor to establish cooperative relationships, based on mutual respect with regard to scientific and ethical matters and, subject to the provisions of Article 14, should encourage the free circulation of human genetic data and human proteomic data in order to foster the sharing of scientific knowledge, provided that the principles set out in this Declaration are observed by the parties concerned. To this end, they should also endeavor to publish in due course the results of their research.

Article 19

(a) In accordance with domestic law or policy and international agreements, benefits resulting from the use of human genetic data, human proteomic data, or biological samples collected for medical and scientific research should be shared with the society as a whole and the international community. In giving effect to this principle, benefits may take any of the following forms:

(i) special assistance to the persons and groups that have taken part in the research;

(ii) access to medical care;

(iii) provision of new diagnostics, facilities for new treatments, or drugs stemming from the research;

(iv) support for health services;

(v) capacity-building facilities for research purposes;

(vi) development and strengthening of the capacity of developing countries to collect and process human genetic data, taking into consideration their specific problems;

(vii) any other form consistent with the principles set out in this Declaration.

(b) Limitations in this respect could be provided by domestic law and international agreements.

E. Storage

Article 20. Monitoring and Management Framework

States may consider establishing a framework for the monitoring and management of human genetic data, human proteomic data and biological samples based on the principles of independence, multidisciplinarity, pluralism and transparency as well as the principles set out in this Declaration. This framework could also deal with the nature and purposes of the storage of these data.

Article 21. Destruction

(a) The provisions of Article 9 apply *mutatis mutandis* in the case of stored human genetic data, human proteomic data, and biological samples.

(b) Human genetic data, human proteomic data, and the biological samples collected from a suspect in the course of a criminal investigation should be destroyed when they are no longer necessary, unless otherwise provided for by domestic law consistent with the international law of human rights.

(c) Human genetic data, human proteomic data and biological samples should be available for forensic purposes and civil proceedings only for as long as they are necessary for those proceedings, unless otherwise provided for by domestic law consistent with the international law of human rights.

Article 22. Cross-Matching

Consent should be essential for the cross-matching of human genetic data, human proteomic data, or biological samples stored for diagnostic and health care purposes and for medical and other scientific research purposes, unless otherwise provided for by domestic law for compelling reasons and consistent with the international law of human rights.

F. Promotion and Implementation of the Declaration

Article 23. Implementation

(a) States should take all appropriate measures, whether of a legislative, administrative or other character, to give effect to the principles set out in this Declaration, in accordance with the international law of human rights. Such measures should be supported by action in the sphere of education, training and public information.

(b) In the framework of international cooperation, States should endeavor to enter into bilateral and multilateral agreements enabling developing countries to build up their capacity to participate in generating and sharing scientific knowledge concerning human genetic data and of the related know-how.

Article 24. Ethics Education, Training, and Information

In order to promote the principles set out in this Declaration, States should endeavor to foster all forms of ethics education and training at all levels as well as to encourage information and knowledge dissemination programs about human genetic data. These measures should aim at specific audiences, in particular researchers and members of ethics committees, or be addressed to the public at large. In this regard, States should encourage the participation of international and regional intergovernmental organizations and international, regional and national non-governmental organizations in this endeavor.

Article 25. Roles of the International Bioethics Committee (IBC) and the Intergovernmental Bioethics Committee (IGBC)

The International Bioethics Committee (IBC) and the Intergovernmental Bioethics Committee (IGBC) shall contribute to the implementation of this Declaration and the dissemination of the principles set out therein. On a collaborative basis, the two Committees should be responsible for its monitoring and for the evaluation of its implementation, *inter alia,* on the basis of reports provided by States. The two Committees should be responsible, in particular, for the formulation of any opinion or proposal likely to further the effectiveness of this Declaration. They should make recommendations in accordance with UNESCO's statutory procedures, addressed to the General Conference.

Article 26. Follow-Up Action by UNESCO

UNESCO shall take appropriate action to follow up this Declaration so as to foster progress of the life sciences and their applications through technologies, based on respect for human dignity and the exercise and observance of human rights and fundamental freedoms.

Article 27. Denial of Acts Contrary to Human Rights, Fundamental Freedoms and Human Dignity

Nothing in this Declaration may be interpreted as implying for any State, group or person any claim to engage in any activity or to perform any act contrary to human rights, fundamental freedoms and human dignity, including, in particular, the principles set out in this Declaration.

Convention on Human Rights and Biomedicine[1] (Council of Europe, 1997)

Preamble

The member States of the Council of Europe, the other States and the European Community, signatories hereto,

Bearing in mind the Universal Declaration of Human Rights proclaimed by the General Assembly of the United Nations on 10 December 1948;

Bearing in mind the Convention for the Protection of Human Rights and Fundamental Freedoms of 4 November 1950;

Bearing in mind the European Social Charter of 18 October 1961;

Bearing in mind the International Covenant on Civil and Political Rights and the International Covenant on Economic, Social and Cultural Rights of 16 December 1966;

Bearing in mind the Convention for the Protection of Individuals with regard to Automatic Processing of Personal Data of 28 January 1981;

Bearing also in mind the Convention on the Rights of the Child of 20 November 1989;

Considering that the aim of the Council of Europe is the achievement of a greater unity between its members and that one of the methods by which that aim is to be pursued is the maintenance and further realization of human rights and fundamental freedoms;

1. Convention for the Protection of Human Rights and Dignity of the Human Being with regard to the Application of Biology and Medicine: Convention on Human Rights and Biomedicine (ETS No. 164). Signed in Oviedo, April 4, 1997. See http://conventions.coe.int/treaty/en/treaties/html/164.htm

Conscious of the accelerating developments in biology and medicine;

Convinced of the need to respect the human being both as an individual and as a member of the human species and recognizing the importance of ensuring the dignity of the human being;

Conscious that the misuse of biology and medicine may lead to acts endangering human dignity;

Affirming that progress in biology and medicine should be used for the benefit of present and future generations;

Stressing the need for international cooperation so that all humanity may enjoy the benefits of biology and medicine;

Recognizing the importance of promoting a public debate on the questions posed by the application of biology and medicine and the responses to be given thereto;

Wishing to remind all members of society of their rights and responsibilities;

Taking account of the work of the Parliamentary Assembly in this field, including Recommendation 1160 (1991) on the preparation of a convention on bioethics;

Resolving to take such measures as are necessary to safeguard human dignity and the fundamental rights and freedoms of the individual with regard to the application of biology and medicine;

Have agreed as follows:

Chapter I. General Provisions

Article 1. Purpose and Object

Parties to this Convention shall protect the dignity and identity of all human beings and guarantee everyone, without discrimination, respect for their integrity and other rights and fundamental freedoms with regard to the application of biology and medicine.

Each Party shall take in its internal law the necessary measures to give effect to the provisions of this Convention.

Article 2. Primacy of the Human Being

The interests and welfare of the human being shall prevail over the sole interest of society or science.

Article 3. Equitable Access to Health Care

Parties, taking into account health needs and available resources, shall take appropriate measures with a view to providing, within their jurisdiction, equitable access to health care of appropriate quality.

Article 4. Professional Standards

Any intervention in the health field, including research, must be carried out in accordance with relevant professional obligations and standards.

Chapter II. Consent

Article 5. General Rule

An intervention in the health field may only be carried out after the person concerned has given free and informed consent to it.

This person shall beforehand be given appropriate information as to the purpose and nature of the intervention as well as on its consequences and risks.

The person concerned may freely withdraw consent at any time.

Article 6. Protection of Persons Not Able to Consent

(1) Subject to Articles 17 and 20 below, an intervention may only be carried out on a person who does not have the capacity to consent, for his or her direct benefit.

(2) Where, according to law, a minor does not have the capacity to consent to an intervention, the intervention may only be carried out with the authorization of his or her representative or an authority or a person or body provided for by law.

The opinion of the minor shall be taken into consideration as an increasingly determining factor in proportion to his or her age and degree of maturity.

(3) Where, according to law, an adult does not have the capacity to consent to an intervention because of a mental disability, a disease or for similar reasons, the intervention may only be carried out with the authorization of his or her representative or an authority or a person or body provided for by law.

The individual concerned shall as far as possible take part in the authorization procedure.

(4) The representative, the authority, the person or the body mentioned in paragraphs 2 and 3 above shall be given, under the same conditions, the information referred to in Article 5.

(5) The authorisation referred to in paragraphs 2 and 3 above may be withdrawn at any time in the best interests of the person concerned.

Article 7. Protection of Persons Who Have a Mental Disorder

Subject to protective conditions prescribed by law, including supervisory, control and appeal procedures, a person who has a mental disorder of a serious nature may be subjected, without his or her consent, to an intervention aimed at treating his or her mental disorder only where, without such treatment, serious harm is likely to result to his or her health.

Article 8. Emergency Situation

When because of an emergency situation the appropriate consent cannot be obtained, any medically necessary intervention may be carried out immediately for the benefit of the health of the individual concerned.

Article 9. Previously Expressed Wishes

The previously expressed wishes relating to a medical intervention by a patient who is not, at the time of the intervention, in a state to express his or her wishes shall be taken into account.

Chapter III. Private Life and Right to Information

Article 10. Private Life and Right to Information

(1) Everyone has the right to respect for private life in relation to information about his or her health.

(2) Everyone is entitled to know any information collected about his or her health. However, the wishes of individuals not to be so informed shall be observed.

(3) In exceptional cases, restrictions may be placed by law on the exercise of the rights contained in paragraph 2 in the interests of the patient.

Chapter IV. Human Genome

Article 11. Non-Discrimination

Any form of discrimination against a person on grounds of his or her genetic heritage is prohibited.

Article 12. Predictive Genetic Tests

Tests which are predictive of genetic diseases or which serve either to identify the subject as a carrier of a gene responsible for a disease or to detect a genetic predisposition or susceptibility to a disease may be performed only for health purposes or for scientific research linked to health purposes, and subject to appropriate genetic counseling.

Article 13. Interventions on the Human Genome

An intervention seeking to modify the human genome may only be undertaken for preventive, diagnostic or therapeutic purposes and only if its aim is not to introduce any modification in the genome of any descendants.

Article 14. Non-Selection of Sex

The use of techniques of medically assisted procreation shall not be allowed for the purpose of choosing a future child's sex, except where serious hereditary sex-related disease is to be avoided.

Chapter V. Scientific Research

Article 15. General Rule

Scientific research in the field of biology and medicine shall be carried out freely, subject to the provisions of this Convention and the other legal provisions ensuring the protection of the human being.

Article 16. Protection of Persons Undergoing Research

Research on a person may only be undertaken if all the following conditions are met:

(i) there is no alternative of comparable effectiveness to research on humans;

(ii) the risks which may be incurred by that person are not disproportionate to the potential benefits of the research;

(iii) the research project has been approved by the competent body after independent examination of its scientific merit, including assessment of the importance of the aim of the research, and multidisciplinary review of its ethical acceptability;

(iv) the persons undergoing research have been informed of their rights and the safeguards prescribed by law for their protection;

(v) the necessary consent as provided for under Article 5 has been given expressly, specifically and is documented. Such consent may be freely withdrawn at any time.

Article 17. Protection of Persons Not Able to Consent to Research

(1) Research on a person without the capacity to consent as stipulated in Article 5 may be undertaken only if all the following conditions are met:

(i) the conditions laid down in Article 16(i) to (iv) are fulfilled;

(ii) the results of the research have the potential to produce real and direct benefit to his or her health;

(iii) research of comparable effectiveness cannot be carried out on individuals capable of giving consent;

(iv) the necessary authorization provided for under Article 6 has been given specifically and in writing; and

(v) the person concerned does not object.

(2) Exceptionally and under the protective conditions prescribed by law, where the research has not the potential to produce results of direct benefit to the health of the person concerned, such research may be authorized subject to the conditions laid down in paragraph 1(i), (iii), (iv) and (v) above, and to the following additional conditions:

(i) the research has the aim of contributing, through significant improvement in the scientific understanding of the individual's condition, disease or disorder, to the ultimate attainment of results capable of conferring benefit to the person concerned or to other persons in the same age category or afflicted with the same disease or disorder or having the same condition;

(ii) the research entails only minimal risk and minimal burden for the individual concerned.

Article 18. Research on Embryos *in Vitro*

(1) Where the law allows research on embryos *in vitro*, it shall ensure adequate protection of the embryo.

(2) The creation of human embryos for research purposes is prohibited.

Chapter VI.
Organ and Tissue Removal from Living Donors for Transplantation Purposes

Article 19. General Rule

(1) Removal of organs or tissue from a living person for transplantation purposes may be carried out solely for the therapeutic benefit of the recipient and where there is no suitable organ or tissue available from a deceased person and no other alternative therapeutic method of comparable effectiveness.

(2) The necessary consent as provided for under Article 5 must have been given expressly and specifically either in written form or before an official body.

Article 20. Protection of Persons Not Able to Consent to Organ Removal

(1) No organ or tissue removal may be carried out on a person who does not have the capacity to consent under Article 5.

(2) Exceptionally and under the protective conditions prescribed by law, the removal of regenerative tissue from a person who does not have the capacity to consent may be authorized provided the following conditions are met:
   (i) there is no compatible donor available who has the capacity to consent;
   (ii) the recipient is a brother or sister of the donor;
   (iii) the donation must have the potential to be life-saving for the recipient;
   (iv) the authorization provided for under paragraphs 2 and 3 of Article 6 has been given specifically and in writing, in accordance with the law and with the approval of the competent body;
   (v) the potential donor concerned does not object.

Chapter VII.
Prohibition of Financial Gain and Disposal of a Part of the Human Body

Article 21. Prohibition of Financial Gain

The human body and its parts shall not, as such, give rise to financial gain.

Article 22. Disposal of a Removed Part of the Human Body

When in the course of an intervention any part of a human body is removed, it may be stored and used for a purpose other than that for which it was removed, only if this is done in conformity with appropriate information and consent procedures.

Chapter VIII. Infringements of the Provisions of the Convention

Article 23. Infringement of the Rights or Principles

The Parties shall provide appropriate judicial protection to prevent or to put a stop to an unlawful infringement of the rights and principles set forth in this Convention at short notice.

Article 24. Compensation for Undue Damage

The person who has suffered undue damage resulting from an intervention is entitled to fair compensation according to the conditions and procedures prescribed by law.

Article 25. Sanctions

Parties shall provide for appropriate sanctions to be applied in the event of infringement of the provisions contained in this Convention.

Chapter IX. Relation Between This Convention and Other Provisions

Article 26. Restrictions on the Exercise of the Rights

(1) No restrictions shall be placed on the exercise of the rights and protective provisions contained in this Convention other than such as are prescribed by law and are necessary in a democratic society in the interest of public safety, for the prevention of crime, for the protection of public health or for the protection of the rights and freedoms of others.

(2) The restrictions contemplated in the preceding paragraph may not be placed on Articles 11, 13, 14, 16, 17, 19, 20 and 21.

Article 27. Wider Protection

None of the provisions of this Convention shall be interpreted as limiting or otherwise affecting the possibility for a Party to grant a wider measure of protection with regard to the application of biology and medicine than is stipulated in this Convention.

Chapter X. Public Debate

Article 28. Public Debate

Parties to this Convention shall see to it that the fundamental questions raised by the developments of biology and medicine are the subject of appropriate public discussion in the light, in particular, of relevant medical, social, economic, ethical and legal implications, and that their possible application is made the subject of appropriate consultation.

Chapter XI. Interpretation and Follow-Up of the Convention

Article 29. Interpretation of the Convention

The European Court of Human Rights may give, without direct reference to any specific proceedings pending in a court, advisory opinions on legal questions concerning the interpretation of the present Convention at the request of:

(i) the Government of a Party, after having informed the other Parties;

(ii) the Committee set up by Article 32, with membership restricted to the Representatives of the Parties to this Convention, by a decision adopted by a two-thirds majority of votes cast.

Article 30. Reports on the Application of the Convention

On receipt of a request from the Secretary General of the Council of Europe any Party shall furnish an explanation of the manner in which its internal law ensures the effective implementation of any of the provisions of the Convention.

Chapter XII. Protocols

Article 31. Protocols

Protocols may be concluded in pursuance of Article 32, with a view to developing, in specific fields, the principles contained in this Convention.

The Protocols shall be open for signature by Signatories of the Convention. They shall be subject to ratification, acceptance or approval. A Signatory may not ratify, accept or approve Protocols without previously or simultaneously ratifying accepting or approving the Convention.

Chapter XIII. Amendments to the Convention

Article 32. Amendments to the Convention

(1) The tasks assigned to "the Committee" in the present article and in Article 29 shall be carried out by the Steering Committee on Bioethics (CDBI), or by any other committee designated to do so by the Committee of Ministers.

(2) Without prejudice to the specific provisions of Article 29, each member State of the Council of Europe, as well as each Party to the present Convention which is not a member of the Council of Europe, may be represented and have one vote in the Committee when the Committee carries out the tasks assigned to it by the present Convention.
(3) Any State referred to in Article 33 or invited to accede to the Convention in accordance with the provisions of Article 34 which is not Party to this Convention may be represented on the Committee by an observer. If the European Community is not a Party it may be represented on the Committee by an observer.
(4) In order to monitor scientific developments, the present Convention shall be examined within the Committee no later than five years from its entry into force and thereafter at such intervals as the Committee may determine.
(5) Any proposal for an amendment to this Convention, and any proposal for a Protocol or for an amendment to a Protocol, presented by a Party, the Committee or the Committee of Ministers shall be communicated to the Secretary General of the Council of Europe and forwarded by him to the member States of the Council of Europe, to the European Community, to any Signatory, to any Party, to any State invited to sign this Convention in accordance with the provisions of Article 33 and to any State invited to accede to it in accordance with the provisions of Article 34.
(6) The Committee shall examine the proposal not earlier than two months after it has been forwarded by the Secretary General in accordance with paragraph 5. The Committee shall submit the text adopted by a two-thirds majority of the votes cast to the Committee of Ministers for approval. After its approval, this text shall be forwarded to the Parties for ratification, acceptance or approval.
(7) Any amendment shall enter into force, in respect of those Parties which have accepted it, on the first day of the month following the expiration of a period of one month after the date on which five Parties, including at least four member States of the Council of Europe, have informed the Secretary General that they have accepted it.

In respect of any Party which subsequently accepts it, the amendment shall enter into force on the first day of the month following the expiration of a period of one month after the date on which that Party has informed the Secretary General of its acceptance.

Chapter XIV. Final Clauses

Article 33. Signature, Ratification and Entry into Force

(1) This Convention shall be open for signature by the member States of the Council of Europe, the non-member States which have participated in its elaboration and by the European Community.

(2) This Convention is subject to ratification, acceptance or approval. Instruments of ratification, acceptance or approval shall be deposited with the Secretary General of the Council of Europe.

(3) This Convention shall enter into force on the first day of the month following the expiration of a period of three months after the date on which five States, including at least four member States of the Council of Europe, have expressed their consent to be bound by the Convention in accordance with the provisions of paragraph 2 of the present article.

(4) In respect of any Signatory which subsequently expresses its consent to be bound by it, the Convention shall enter into force on the first day of the month following the expiration of a period of three months after the date of the deposit of its instrument of ratification, acceptance or approval.

Article 34. Non-Member States

(1) After the entry into force of this Convention, the Committee of Ministers of the Council of Europe may, after consultation of the Parties, invite any non-member State of the Council of Europe to accede to this Convention by a decision taken by the majority provided for in Article 20(d) of the Statute of the Council of Europe, and by the unanimous vote of the representatives of the Contracting States entitled to sit on the Committee of Ministers.

(2) In respect of any acceding State, the Convention shall enter into force on the first day of the month following the expiration of a period of three months after the date of deposit of the instrument of accession with the Secretary General of the Council of Europe.

Article 35. Territories

(1) Any Signatory may, at the time of signature or when depositing its instrument of ratification, acceptance or approval, specify the territory or territories to which this Convention shall apply. Any other State may formulate the same declaration when depositing its instrument of accession.

(2) Any Party may, at any later date, by a declaration addressed to the Secretary General of the Council of Europe, extend the application of this Convention to any other territory specified in the declaration and for whose international

relations it is responsible or on whose behalf it is authorized to give undertakings. In respect of such territory the Convention shall enter into force on the first day of the month following the expiration of a period of three months after the date of receipt of such declaration by the Secretary General.

(3) Any declaration made under the two preceding paragraphs may, in respect of any territory specified in such declaration, be withdrawn by a notification addressed to the Secretary General. The withdrawal shall become effective on the first day of the month following the expiration of a period of three months after the date of receipt of such notification by the Secretary General.

Article 36. Reservations

(1) Any State and the European Community may, when signing this Convention or when depositing the instrument of ratification, acceptance, approval or accession, make a reservation in respect of any particular provision of the Convention to the extent that any law then in force in its territory is not in conformity with the provision. Reservations of a general character shall not be permitted under this article.

(2) Any reservation made under this article shall contain a brief statement of the relevant law.

(3) Any Party which extends the application of this Convention to a territory mentioned in the declaration referred to in Article 35(2), may, in respect of the territory concerned, make a reservation in accordance with the provisions of the preceding paragraphs.

(4) Any Party which has made the reservation mentioned in this article may withdraw it by means of a declaration addressed to the Secretary General of the Council of Europe. The withdrawal shall become effective on the first day of the month following the expiration of a period of one month after the date of its receipt by the Secretary General.

Article 37. Denunciation

(1) Any Party may at any time denounce this Convention by means of a notification addressed to the Secretary General of the Council of Europe.

(2) Such denunciation shall become effective on the first day of the month following the expiration of a period of three months after the date of receipt of the notification by the Secretary General.

Article 38. Notifications

The Secretary General of the Council of Europe shall notify the member States of the Council, the European Community, any Signatory, any Party and any other State which has been invited to accede to this Convention of:

(a) any signature;

(b) the deposit of any instrument of ratification, acceptance, approval or accession;

(c) any date of entry into force of this Convention in accordance with Articles 33 or 34;

(d) any amendment or Protocol adopted in accordance with Article 32, and the date on which such an amendment or Protocol enters into force;

(e) any declaration made under the provisions of Article 35;

(f) any reservation and withdrawal of reservation made in pursuance of the provisions of Article 36;

(g) any other act, notification or communication relating to this Convention.

In witness whereof the undersigned, being duly authorized thereto, have signed this Convention. Done at Oviedo (Asturias), this fourth day of April 1997, in English and French, both texts being equally authentic, in a single copy which shall be deposited in the archives of the Council of Europe. The Secretary General of the Council of Europe shall transmit certified copies to each member State of the Council of Europe, to the European Community, to the non-member States which have participated in the elaboration of this Convention, and to any State invited to accede to this Convention.

Additional Protocol on the Prohibition of Cloning Human Beings[1] (Council of Europe, 1998)

The member States of the Council of Europe, the other States and the European Community Signatories to this Additional Protocol to the Convention for the Protection of Human Rights and Dignity of the Human Being with regard to the Application of Biology and Medicine,

Noting scientific developments in the field of mammal cloning, particularly through embryo splitting and nuclear transfer;

Mindful of the progress that some cloning techniques themselves may bring to scientific knowledge and its medical application;

Considering that the cloning of human beings may become a technical possibility;

Having noted that embryo splitting may occur naturally and sometimes result in the birth of genetically identical twins;

Considering however that the instrumentalization of human beings through the deliberate creation of genetically identical human beings is contrary to human dignity and thus constitutes a misuse of biology and medicine;

Considering also the serious difficulties of a medical, psychological and social nature that such a deliberate biomedical practice might imply for all the individuals involved;

Considering the purpose of the Convention on Human Rights and Biomedicine, in particular the principle mentioned in Article 1 aiming to protect the dignity and identity of all human beings;

1. Additional Protocol to the Convention for the Protection of Human Rights and Dignity of the Human Being with regard to the Application of Biology and Medicine, on the Prohibition of Cloning Human Beings (ETS No. 168). Signed in Paris, January 12, 1998. See http://conventions.coe.int/treaty/en/treaties/html/168.htm.

Have agreed as follows:

Article 1.

(1) Any intervention seeking to create a human being genetically identical to another human being, whether living or dead, is prohibited.

(2) For the purpose of this article, the term human being 'genetically identical' to another human being means a human being sharing with another the same nuclear gene set.

Article 2.

No derogation from the provisions of this Protocol shall be made under Article 26(1) of the Convention.

Article 3.

As between the Parties, the provisions of Articles 1 and 2 of this Protocol shall be regarded as additional articles to the Convention and all the provisions of the Convention shall apply accordingly.

Article 4.

This Protocol shall be open for signature by Signatories to the Convention. It is subject to ratification, acceptance or approval. A Signatory may not ratify, accept or approve this Protocol unless it has previously or simultaneously ratified, accepted or approved the Convention. Instruments of ratification, acceptance or approval shall be deposited with the Secretary General of the Council of Europe.

Article 5.

(1) This Protocol shall enter into force on the first day of the month following the expiration of a period of three months after the date on which five States, including at least four member States of the Council of Europe, have expressed their consent to be bound by the Protocol in accordance with the provisions of Article 4.

(2) In respect of any Signatory which subsequently expresses its consent to be bound by it, the Protocol shall enter into force on the first day of the month following the expiration of a period of three months after the date of the deposit of the instrument of ratification, acceptance or approval.

Article 6.

(1) After the entry into force of this Protocol, any State which has acceded to the Convention may also accede to this Protocol.

(2) Accession shall be effected by the deposit with the Secretary General of the Council of Europe of an instrument of accession which shall take effect on the first day of the month following the expiration of a period of three months after the date of its deposit.

Article 7.

(1) Any Party may at any time denounce this Protocol by means of a notification addressed to the Secretary General of the Council of Europe.

(2) Such denunciation shall become effective on the first day of the month following the expiration of a period of three months after the date of receipt of such notification by the Secretary General.

Article 8.

The Secretary General of the Council of Europe shall notify the member States of the Council of Europe, the European Community, any Signatory, any Party and any other State which has been invited to accede to the Convention of:

(a) any signature;

(b) the deposit of any instrument of ratification, acceptance, approval or accession;

(c) any date of entry into force of this Protocol in accordance with Articles 5 and 6;

(d) any other act, notification or communication relating to this Protocol.

In witness whereof the undersigned, being duly authorized thereto, have signed this Protocol.

Done at Paris, this twelfth day of January 1998, in English and in French, both texts being equally authentic, in a single copy which shall be deposited in the archives of the Council of Europe. The Secretary General of the Council of Europe shall transmit certified copies to each member State of the Council of Europe, to the non-member States which have participated in the elaboration of this Protocol, to any State invited to accede to the Convention and to the European Community.

European Biotechnological Directive[1] (European Parliament and Council of Europe, 1998)

Directive 98/44/EC of the European Parliament and of the Council of 6 July 1998 on the Legal Protection of Biotechnological Inventions

Preamble

The European Parliament and the Council of the European Union,

Having regard to the Treaty establishing the European Community, and in particular Article 100a thereof,

Having regard to the proposal from the Commission,[2]

Having regard to the opinion of the Economic and Social Committee,[3]

Acting in accordance with the procedure laid down in Article 189b of the Treaty,[4]

(1) Whereas biotechnology and genetic engineering are playing an increasingly important role in a broad range of industries and the protection of biotechnological inventions will certainly be of fundamental importance for the Community's industrial development;

(2) Whereas, in particular in the field of genetic engineering, research and development require a considerable amount of high-risk investment and therefore only adequate legal protection can make them profitable;

1. Directive 98/44/EC of the European Parliament and of the Council of 6 July 1998 on the Legal Protection of Biotechnological Inventions, signed in Brussels. Published in the *Official Journal of the European Communities* (hereinafter: OJ) on July 30, 1998. For the official document, see http://europa.eu.int/eur-lex/pri/en/oj/dat/1998/l_213/l_21319980730en00130021.pdf.
2. *OJ C 296, 8.10.1996*, p.4 and *OJ C 311, 11.10.1997*, p.12.
3. *OJ C 295, 7.10.1996*, p.11.
4. Opinion of the European Parliament of July 16, 1997 (*OJ C 286, 22.9.1997*, p.87). Council Common Position of February 26, 1998 (*OJ C 110, 8.4.1998*, p.17) and Decision of the European Parliament of May 12, 1998 (*OJ C 167, 1.6.1998*). Council Decision of June 16, 1998.

(3) Whereas effective and harmonized protection throughout the Member States is essential in order to maintain and encourage investment in the field of biotechnology;

(4) Whereas following the European Parliament's rejection of the joint text, approved by the Conciliation Committee, for a European Parliament and Council Directive on the legal protection of biotechnological inventions,[5] the European Parliament and the Council have determined that the legal protection of biotechnological inventions requires clarification;

(5) Whereas differences exist in the legal protection of biotechnological inventions offered by the laws and practices of the different Member States; whereas such differences could create barriers to trade and hence impede the proper functioning of the internal market;

(6) Whereas such differences could well become greater as Member States adopt new and different legislation and administrative practices, or whereas national case-law interpreting such legislation develops differently;

(7) Whereas uncoordinated development of national laws on the legal protection of biotechnological inventions in the Community could lead to further disincentives to trade, to the detriment of the industrial development of such inventions and of the smooth operation of the internal market;

(8) Whereas legal protection of biotechnological inventions does not necessitate the creation of a separate body of law in place of the rules of national patent law; whereas the rules of national patent law remain the essential basis for the legal protection of biotechnological inventions given that they must be adapted or added to in certain specific respects in order to take adequate account of technological developments involving biological material which also fulfill the requirements for patentability;

(9) Whereas in certain cases, such as the exclusion from patentability of plant and animal varieties and of essentially biological processes for the production of plants and animals, certain concepts in national laws based upon international patent and plant variety conventions have created uncertainty regarding the protection of biotechnological and certain microbiological inventions; whereas harmonization is necessary to clarify the said uncertainty;

(10) Whereas regard should be had to the potential of the development of biotechnology for the environment and in particular the utility of this technology for the development of methods of cultivation which are less

5. *OJ C 68, 20.3.1995*, p.26.

polluting and more economical in their use of ground; whereas the patent system should be used to encourage research into, and the application of, such processes;

(11) Whereas the development of biotechnology is important to developing countries, both in the field of health and combating major epidemics and endemic diseases and in that of combating hunger in the world; whereas the patent system should likewise be used to encourage research in these fields; whereas international procedures for the dissemination of such technology in the Third World and to the benefit of the population groups concerned should be promoted;

(12) Whereas the Agreement on Trade-Related Aspects of Intellectual Property Rights (TRIPs)[6] signed by the European Community and the Member States, has entered into force and provides that patent protection must be guaranteed for products and processes in all areas of technology;

(13) Whereas the Community's legal framework for the protection of biotechnological inventions can be limited to laying down certain principles as they apply to the patentability of biological material as such, such principles being intended in particular to determine the difference between inventions and discoveries with regard to the patentability of certain elements of human origin, to the scope of protection conferred by a patent on a biotechnological invention, to the right to use a deposit mechanism in addition to written descriptions and lastly to the option of obtaining non-exclusive compulsory licenses in respect of interdependence between plant varieties and inventions, and conversely;

(14) Whereas a patent for invention does not authorize the holder to implement that invention, but merely entitles him to prohibit third parties from exploiting it for industrial and commercial purposes; whereas, consequently, substantive patent law cannot serve to replace or render superfluous national, European or international law which may impose restrictions or prohibitions or which concerns the monitoring of research and of the use or commercialization of its results, notably from the point of view of the requirements of public health, safety, environmental protection, animal welfare, the preservation of genetic diversity and compliance with certain ethical standards;

(15) Whereas no prohibition or exclusion exists in national or European patent law (Munich Convention) which precludes *a priori* the patentability of biological matter;

6. *OJ L 336, 23.12.1994*, p.213.

(16) Whereas patent law must be applied so as to respect the fundamental principles safeguarding the dignity and integrity of the person; whereas it is important to assert the principle that the human body, at any stage in its formation or development, including germ cells, and the simple discovery of one of its elements or one of its products, including the sequence or partial sequence of a human gene, cannot be patented; whereas these principles are in line with the criteria of patentability proper to patent law, whereby a mere discovery cannot be patented;

(17) Whereas significant progress in the treatment of diseases has already been made thanks to the existence of medicinal products derived from elements isolated from the human body and/or otherwise produced, such medicinal products resulting from technical processes aimed at obtaining elements similar in structure to those existing naturally in the human body and whereas, consequently, research aimed at obtaining and isolating such elements valuable to medicinal production should be encouraged by means of the patent system;

(18) Whereas, since the patent system provides insufficient incentive for encouraging research into and production of biotechnological medicines which are needed to combat rare or 'orphan' diseases, the Community and the Member States have a duty to respond adequately to this problem;

(19) Whereas account has been taken of Opinion No. 8 of the Group of Advisers on the Ethical Implications of Biotechnology to the European Commission;

(20) Whereas, therefore, it should be made clear that an invention based on an element isolated from the human body or otherwise produced by means of a technical process, which is susceptible of industrial application, is not excluded from patentability, even where the structure of that element is identical to that of a natural element, given that the rights conferred by the patent do not extend to the human body and its elements in their natural environment;

(21) Whereas such an element isolated from the human body or otherwise produced is not excluded from patentability since it is, for example, the result of technical processes used to identify, purify and classify it and to reproduce it outside the human body, techniques which human beings alone are capable of putting into practice and which nature is incapable of accomplishing by itself;

(22) Whereas the discussion on the patentability of sequences or partial sequences of genes is controversial; whereas, according to this Directive, the granting of a patent for inventions which concern such sequences or partial

sequences should be subject to the same criteria of patentability as in all other areas of technology: novelty, inventive step and industrial application; whereas the industrial application of a sequence or partial sequence must be disclosed in the patent application as filed;

(23) Whereas a mere DNA sequence without indication of a function does not contain any technical information and is therefore not a patentable invention;

(24) Whereas, in order to comply with the industrial application criterion it is necessary in cases where a sequence or partial sequence of a gene is used to produce a protein or part of a protein, to specify which protein or part of a protein is produced or what function it performs;

(25) Whereas, for the purposes of interpreting rights conferred by a patent, when sequences overlap only in parts which are not essential to the invention, each sequence will be considered as an independent sequence in patent law terms;

(26) Whereas if an invention is based on biological material of human origin or if it uses such material, where a patent application is filed, the person from whose body the material is taken must have had an opportunity of expressing free and informed consent thereto, in accordance with national law;

(27) Whereas if an invention is based on biological material of plant or animal origin or if it uses such material, the patent application should, where appropriate, include information on the geographical origin of such material, if known; whereas this is without prejudice to the processing of patent applications or the validity of rights arising from granted patents;

(28) Whereas this Directive does not in any way affect the basis of current patent law, according to which a patent may be granted for any new application of a patented product;

(29) Whereas this Directive is without prejudice to the exclusion of plant and animal varieties from patentability; whereas on the other hand inventions which concern plants or animals are patentable provided that the application of the invention is not technically confined to a single plant or animal variety;

(30) Whereas the concept 'plant variety' is defined by the legislation protecting new varieties, pursuant to which a variety is defined by its whole genome and therefore possesses individuality and is clearly distinguishable from other varieties;

(31) Whereas a plant grouping which is characterized by a particular gene (and not its whole genome) is not covered by the protection of new varieties and is therefore not excluded from patentability even if it comprises new varieties of plants;

(32) Whereas, however, if an invention consists only in genetically modifying a particular plant variety, and if a new plant variety is bred, it will still be excluded from patentability even if the genetic modification is the result not of an essentially biological process but of a biotechnological process;

(33) Whereas it is necessary to define for the purposes of this Directive when a process for the breeding of plants and animals is essentially biological;

(34) Whereas this Directive shall be without prejudice to concepts of invention and discovery, as developed by national, European or international patent law;

(35) Whereas this Directive shall be without prejudice to the provisions of national patent law whereby processes for treatment of the human or animal body by surgery or therapy and diagnostic methods practiced on the human or animal body are excluded from patentability;

(36) Whereas the TRIPs Agreement provides for the possibility that members of the World Trade Organization may exclude from patentability inventions, the prevention within their territory of the commercial exploitation of which is necessary to protect *ordre public* or morality, including to protect human, animal or plant life or health or to avoid serious prejudice to the environment, provided that such exclusion is not made merely because the exploitation is prohibited by their law;

(37) Whereas the principle whereby inventions must be excluded from patentability where their commercial exploitation offends against *ordre public* or morality must also be stressed in this Directive;

(38) Whereas the operative part of this Directive should also include an illustrative list of inventions excluded from patentability so as to provide national courts and patent offices with a general guide to interpreting the reference to *ordre public* and morality; whereas this list obviously cannot presume to be exhaustive; whereas processes, the use of which offend against human dignity, such as processes to produce chimeras from germ cells or totipotent cells of humans and animals, are obviously also excluded from patentability;

(39) Whereas *ordre public* and morality correspond in particular to ethical or moral principles recognized in a Member State, respect for which is particularly important in the field of biotechnology in view of the potential scope of inventions in this field and their inherent relationship to living matter; whereas such ethical or moral principles supplement the standard legal examinations under patent law regardless of the technical field of the invention;

(40) Whereas there is a consensus within the Community that interventions in the human germ line and the cloning of human beings offends against *ordre*

public and morality; whereas it is therefore important to exclude unequivocally from patentability processes for modifying the germ-line genetic identity of human beings and processes for cloning human beings;

(41) Whereas a process for cloning human beings may be defined as any process, including techniques of embryo splitting, designed to create a human being with the same nuclear genetic information as another living or deceased human being;

(42) Whereas, moreover, uses of human embryos for industrial or commercial purposes must also be excluded from patentability; whereas in any case such exclusion does not affect inventions for therapeutic or diagnostic purposes which are applied to the human embryo and are useful to it;

(43) Whereas pursuant to Article F(2) of the Treaty on European Union, the Union is to respect fundamental rights, as guaranteed by the European Convention for the Protection of Human Rights and Fundamental Freedoms signed in Rome on November 4, 1950 and as they result from the constitutional traditions common to the Member States, as general principles of Community law;

(44) Whereas the Commission's European Group on Ethics in Science and New Technologies evaluates all ethical aspects of biotechnology; whereas it should be pointed out in this connection that that Group may be consulted only where biotechnology is to be evaluated at the level of basic ethical principles, including where it is consulted on patent law;

(45) Whereas processes for modifying the genetic identity of animals which are likely to cause them suffering without any substantial medical benefit in terms of research, prevention, diagnosis or therapy to man or animal, and also animals resulting from such processes, must be excluded from patentability;

(46) Whereas, in view of the fact that the function of a patent is to reward the inventor for his creative efforts by granting an exclusive but time-bound right, and thereby encourage inventive activities, the holder of the patent should be entitled to prohibit the use of patented self-reproducing material in situations analogous to those where it would be permitted to prohibit the use of patented, non-self-reproducing products, that is to say the production of the patented product itself;

(47) Whereas it is necessary to provide for a first derogation from the rights of the holder of the patent when the propagating material incorporating the protected invention is sold to a farmer for farming purposes by the holder of the patent or with his consent; whereas that initial derogation must authorize the farmer to use the product of his harvest for further multiplication or propagation on his own farm; whereas the extent and the condi-

tions of that derogation must be limited in accordance with the extent and conditions set out in Council Regulation (EC) No. 2100/94 of July 27, 1994 on Community Plant Variety Rights[7];

(48) Whereas only the fee envisaged under Community law relating to plant variety rights as a condition for applying the derogation from Community plant variety rights can be required of the farmer;

(49) Whereas, however, the holder of the patent may defend his rights against a farmer abusing the derogation or against a breeder who has developed a plant variety incorporating the protected invention if the latter fails to adhere to his commitments;

(50) Whereas a second derogation from the rights of the holder of the patent must authorize the farmer to use protected livestock for agricultural purposes;

(51) Whereas the extent and the conditions of that second derogation must be determined by national laws, regulations and practices, since there is no Community legislation on animal variety rights;

(52) Whereas, in the field of exploitation of new plant characteristics resulting from genetic engineering, guaranteed access must, on payment of a fee, be granted in the form of a compulsory license where, in relation to the genus or species concerned, the plant variety represents significant technical progress of considerable economic interest compared to the invention claimed in the patent;

(53) Whereas, in the field of the use of new plant characteristics resulting from new plant varieties in genetic engineering, guaranteed access must, on payment of a fee, be granted in the form of a compulsory license where the invention represents significant technical progress of considerable economic interest;

(54) Whereas Article 34 of the TRIPs Agreement contains detailed provisions on the burden of proof which is binding on all Member States; whereas, therefore, a provision in this Directive is not necessary;

(55) Whereas following Decision 93/626/EEC,[8] the Community is party to the Convention on Biological Diversity of June 5, 1992; whereas, in this regard, Member States must give particular weight to Article 3 and Article 8(j), the second sentence of Article 16(2) and Article 16(5) of the

7. *OJ L 227, 1.9.1994*, p.1. Regulation as amended by Regulation (EC) No. 2506/95 (*OJ L 258, 28.10.1995*, p.3).

8. *OJ L 309, 31.12.1993*, p.1.

Convention when bringing into force the laws, regulations and administrative provisions necessary to comply with this Directive;

(56) Whereas the Third Conference of the Parties to the Biodiversity Convention, which took place in November 1996, noted in Decision III/17 that "further work is required to help develop a common appreciation of the relationship between intellectual property rights and the relevant provisions of the TRIPs Agreement and the Convention on Biological Diversity, in particular on issues relating to technology transfer and conservation and sustainable use of biological diversity and the fair and equitable sharing of benefits arising out of the use of genetic resources, including the protection of knowledge, innovations and practices of indigenous and local communities embodying traditional lifestyles relevant for the conservation and sustainable use of biological diversity,"

Have adopted this directive:

Chapter I. Patentability

Article 1

(1) Member States shall protect biotechnological inventions under national patent law. They shall, if necessary, adjust their national patent law to take account of the provisions of this Directive.

(2) This Directive shall be without prejudice to the obligations of the Member States pursuant to international agreements, and in particular the TRIPs Agreement and the Convention on Biological Diversity.

Article 2

(1) For the purposes of this Directive,
   (a) 'biological material' means any material containing genetic information and capable of reproducing itself or being reproduced in a biological system;
   (b) 'microbiological process' means any process involving or performed upon or resulting in microbiological material.

(2) A process for the production of plants or animals is essentially biological if it consists entirely of natural phenomena such as crossing or selection.

(3) The concept of 'plant variety' is defined by Article 5 of Regulation (EC) No. 2100/94.

Article 3

(1) For the purposes of this Directive, inventions which are new, which involve an inventive step and which are susceptible of industrial application shall be patentable even if they concern a product consisting of or containing biological material or a process by means of which biological material is produced, processed or used.

(2) Biological material which is isolated from its natural environment or produced by means of a technical process may be the subject of an invention even if it previously occurred in nature.

Article 4

(1) The following shall not be patentable:
   (a) plant and animal varieties;
   (b) essentially biological processes for the production of plants or animals.

(2) Inventions which concern plants or animals shall be patentable if the technical feasibility of the invention is not confined to a particular plant or animal variety.

(3) Paragraph 1(b) shall be without prejudice to the patentability of inventions which concern a microbiological or other technical process or a product obtained by means of such a process.

Article 5

(1) The human body, at the various stages of its formation and development, and the simple discovery of one of its elements, including the sequence or partial sequence of a gene, cannot constitute patentable inventions.

(2) An element isolated from the human body or otherwise produced by means of a technical process, including the sequence or partial sequence of a gene, may constitute a patentable invention, even if the structure of that element is identical to that of a natural element.

(3) The industrial application of a sequence or a partial sequence of a gene must be disclosed in the patent application.

Article 6

(1) Inventions shall be considered unpatentable where their commercial exploitation would be contrary to *ordre public* or morality; however, exploitation shall not be deemed to be so contrary merely because it is prohibited by law or regulation.

(2) On the basis of paragraph 1, the following, in particular, shall be considered unpatentable:

(a) processes for cloning human beings;
(b) processes for modifying the germ-line genetic identity of human beings;
(c) uses of human embryos for industrial or commercial purposes;
(d) processes for modifying the genetic identity of animals which are likely to cause them suffering without any substantial medical benefit to man or animal, and also animals resulting from such processes.

Article 7

The Commission's European Group on Ethics in Science and New Technologies evaluates all ethical aspects of biotechnology.

Chapter II. Scope of Protection

Article 8

(1) The protection conferred by a patent on a biological material possessing specific characteristics as a result of the invention shall extend to any biological material derived from that biological material through propagation or multiplication in an identical or divergent form and possessing those same characteristics.
(2) The protection conferred by a patent on a process that enables a biological material to be produced possessing specific characteristics as a result of the invention shall extend to biological material directly obtained through that process and to any other biological material derived from the directly obtained biological material through propagation or multiplication in an identical or divergent form and possessing those same characteristics.

Article 9

The protection conferred by a patent on a product containing or consisting of genetic information shall extend to all material, save as provided in Article 5(1), in which the product in incorporated and in which the genetic information is contained and performs its function.

Article 10

The protection referred to in Articles 8 and 9 shall not extend to biological material obtained from the propagation or multiplication of biological material placed on the market in the territory of a Member State by the holder of the patent or with his consent, where the multiplication or propagation necessarily results from the application for which the biological material was marketed, provided that the material obtained is not subsequently used for other propagation or multiplication.

Article 11

(1) By way of derogation from Articles 8 and 9, the sale or other form of commercialization of plant propagating material to a farmer by the holder of the patent or with his consent for agricultural use implies authorization for the farmer to use the product of his harvest for propagation or multiplication by him on his own farm, the extent and conditions of this derogation corresponding to those under Article 14 of Regulation (EC) No. 2100/94.

(2) By way of derogation from Articles 8 and 9, the sale or any other form of commercialization of breeding stock or other animal reproductive material to a farmer by the holder of the patent or with his consent implies authorization for the farmer to use the protected livestock for an agricultural purpose. This includes making the animal or other animal reproductive material available for the purposes of pursuing his agricultural activity but not sale within the framework or for the purpose of a commercial reproduction activity.

(3) The extent and the conditions of the derogation provided for in paragraph 2 shall be determined by national laws, regulations and practices.

Chapter III. Compulsory Cross-Licensing

Article 12

(1) Where a breeder cannot acquire or exploit a plant variety right without infringing a prior patent, he may apply for a compulsory license for non-exclusive use of the invention protected by the patent inasmuch as the license is necessary for the exploitation of the plant variety to be protected, subject to payment of an appropriate royalty. Member States shall provide that, where such a license is granted, the holder of the patent will be entitled to a cross-license on reasonable terms to use the protected variety.

(2) Where the holder of a patent concerning a biotechnological invention cannot exploit it without infringing a prior plant variety right, he may apply for a compulsory license for non-exclusive use of the plant variety protected by that right, subject to payment of an appropriate royalty. Member States shall provide that, where such a license is granted, the holder of the variety right will be entitled to a cross-license on reasonable terms to use the protected invention.

(3) Applicants for the licenses referred to in paragraphs 1 and 2 must demonstrate that:
   (a) they have applied unsuccessfully to the holder of the patent or of the plant variety right to obtain a contractual license;
   (b) the plant variety or the invention constitutes significant technical progress of considerable economic interest compared with the invention claimed in the patent or the protected plant variety.
(4) Each Member State shall designate the authority or authorities responsible for granting the license. Where a license for a plant variety can be granted only by the Community Plant Variety Office, Article 29 of Regulation (EC) No. 2100/94 shall apply.

Chapter IV. Deposit, Access and Re-Deposit of a Biological Material

Article 13

(1) Where an invention involves the use of or concerns biological material which is not available to the public and which cannot be described in a patent application in such a manner as to enable the invention to be reproduced by a person skilled in the art, the description shall be considered inadequate for the purposes of patent law unless:
   (a) the biological material has been deposited no later than the date on which the patent application was filed with a recognized depository institution. At least the international depository authorities which acquired this status by virtue of Article 7 of the Budapest Treaty of April 28, 1977 on the international recognition of the deposit of micro-organisms for the purposes of patent procedure, hereinafter referred to as the 'Budapest Treaty', shall be recognized;
   (b) the application as filed contains such relevant information as is available to the applicant on the characteristics of the biological material deposited;
   (c) the patent application states the name of the depository institution and the accession number.
(2) Access to the deposited biological material shall be provided through the supply of a sample:
   (a) up to the first publication of the patent application, only to those persons who are authorized under national patent law;
   (b) between the first publication of the application and the granting of the patent, to anyone requesting it or, if the applicant so requests, only to an independent expert;

(c) after the patent has been granted, and notwithstanding revocation or cancellation of the patent, to anyone requesting it.

(3) The sample shall be supplied only if the person requesting it undertakes, for the term during which the patent is in force:

(a) not to make it or any material derived from it available to third parties; and

(b) not to use it or any material derived from it except for experimental purposes, unless the applicant for or proprietor of the patent, as applicable, expressly waives such an undertaking.

(4) At the applicant's request, where an application is refused or withdrawn, access to the deposited material shall be limited to an independent expert for 20 years from the date on which the patent application was filed. In that case, paragraph 3 shall apply.

(5) The applicant's requests referred to in point (b) of paragraph 2 and in paragraph 4 may only be made up to the date on which the technical preparations for publishing the patent application are deemed to have been completed.

Article 14

(1) If the biological material deposited in accordance with Article 13 ceases to be available from the recognized depository institution, a new deposit of the material shall be permitted on the same terms as those laid down in the Budapest Treaty.

(2) Any new deposit shall be accompanied by a statement signed by the depositor certifying that the newly deposited biological material is the same as that originally deposited.

Chapter V. Final Provisions

Article 15

(1) Member States shall bring into force the laws, regulations and administrative provisions necessary to comply with this Directive not later than July 30, 2000. They shall forthwith inform the Commission thereof. When Member States adopt these measures, they shall contain a reference to this Directive or shall be accompanied by such reference on the occasion of their official publication. The methods of making such reference shall be laid down by Member States.

(2) Member States shall communicate to the Commission the text of the provisions of national law which they adopt in the field covered by this Directive.

Article 16

The Commission shall send the European Parliament and the Council:

(a) every five years as from the date specified in Article 15(1) a report on any problems encountered with regard to the relationship between this Directive and international agreements on the protection of human rights to which the Member States have acceded;

(b) within two years of entry into force of this Directive, a report assessing the implications for basic genetic engineering research of failure to publish, or late publication of, papers on subjects which could be patentable;

(c) annually as from the date specified in Article 15(1), a report on the development and implications of patent law in the field of biotechnology and genetic engineering.

Article 17

This Directive shall enter into force on the day of its publication in the *Official Journal of the European Communities.*

Article 18

This Directive is addressed to the Member States.

Done at Brussels, July 6, 1998.

For the European Parliament
The President
J. M. Gil-Robles

For the Council
The President
R. Edlinger

Health Sector Database Act[1] (Iceland, 1998)

Section I. General Terms

Article 1. Objectives

The objective of this legislation is to authorize the creation and operation of a centralized database of non-personally identifiable health data with the aim of increasing knowledge in order to improve health and health services.

Article 2. Scope

This legislation extends to the creation and operation of a centralized Health Sector Database. The legislation does not apply to the medical record systems of individual health and research institutions, data collections made in connection with scientific research into individual diseases or groups of diseases, nor to records kept by health and social security authorities on users of the health service and operation of the health service. The legislation does not apply to the storage or handling of, or access to, biological samples.

Article 3. Definitions

In this legislation the following definitions apply:

(1) *Health sector database*: A collection of data containing information on health and other related information, recorded in a standardized systematic fashion on a single centralized database, intended for processing and as a source of information.

(2) *Personal data*: all data on a personally identified or personally identifiable individual. An individual shall be counted as personally identifiable if he can be identified, directly or indirectly, especially by reference to an identity number, or one or more factors specific to his physical, physiological, mental, economic, cultural or social identity.

(3) *Non-personally identifiable data*: data on a person who is not personally identifiable as defined in clause 2.

1. *Lög um gagnagrunn á heilbrigðissviði*. Act No. 139 of 1998 on a Health Sector Database. Passed by the Alþingi (Parliament) on December 17, 1998. See http://government.is/interpro/htr/htr.nsf/pages/gagngr-log-ensk for the official English text.

(4) *Coding*: the transformation of words or numbers into an incomprehensible series of symbols.
(5) *One-way coding*: the transformation of words or series of digits into an incomprehensible series of symbols that cannot be traced by means of a decoding key.
(6) *Health data*: information on the health of individuals, including genetic information.
(7) *Genetic data*: any data; of whatever type, concerning the hereditary characteristics of an individual or concerning the pattern of inheritance of such characteristics within a related group of individuals. It also refers to all data on the carrying of any genetic information (genes) in an individual or genetic line relating to any aspect of health or disease, whether present as identifiable characteristics or not.

Section II. License and Committee on the Creation and Operation of a Health Sector Database

Article 4. Grant of Operating License and Payments by Licensee

The creation and operation of a Health Sector Database are only permitted to those who have an operating license by the terms of this legislation.

When an application has been received, the Minister of Health may grant an operating license to create and operate a Health Sector Database subject to the further terms of this legislation.

The licensee shall pay a fee for the grant of the license in order to meet the costs of preparing and issuing the license. The licensee shall also pay a yearly fee equivalent to the costs of the work of the committee under the terms of Article 6, and other costs pertaining to service and monitoring of the operation, including monitoring by the Data Protection Commission under the terms of legislation on the recording and handling of personal data, and costs of publication and publicity cp. Article 8.

The licensee shall pay all costs of processing information for entry onto the database, cp. Article 5(8).

The minister and licensee may agree on further payments to the Treasury, which shall be devoted to promoting the health service, research and development.

Article 5. Conditions of License etc.

An operating license for the creation and operation of a Health Sector Database is contingent upon the following conditions:

(1) The database must be located exclusively here in Iceland.
(2) Technical, security and organizational standards meet the requirements of the Data Protection Commission.
(3) The recording and processing of health data shall be carried out by, or under the supervision of, people who are professionally qualified in the health sector.
(4) Detailed information shall be available on the area of activity and projects of the applicant for a license.
(5) A detailed work plan from the applicant shall be available, which shall fulfill the conditions and objectives of this Act regarding working arrangements and progress.
(6) The operation of the database shall be financially separate from the licensee's other business.
(7) The Ministry of Health and Social Security and the Director General of Public Health shall at all times have access to statistical data from the database in accessible form, so that they will be of use in statistical processing for compiling health reports and planning, policy-making and other projects of the parties specified.
(8) The licensee shall pay all costs of processing data from health institutions and self-employed health workers for entry onto the database. The data shall be processed in a manner that fulfills the needs of the relevant institution or self-employed health worker for a standardized information system, the needs of medical specialist fields and the requirements of health authorities, cp. clause 7, and so that it can be used in scientific research.
(9) The license shall be temporary, and it shall not be granted for more than 12 years at a time.
(10) The licensee shall hand over to the committee cp. Article 6 a copy of the database, which shall be updated regularly, to be further specified in the license. A copy of the database shall always be stored in a bank safety deposit box, or in some other secure manner, to be further specified in the license.
(11) The licensee shall ensure that after the expiration of the period of the license, the Minister of Health and Social Security, or the party assigned by the Minister to operate the database, shall receive indefinite use of all software and right required for the maintenance and operation of the database.

The Minister may make the license subject to further conditions than those specified above.

At the end of the period of the license by the terms of the license, the Minister shall make a decision on the operation of the database, after receiving the opinion of the committee cp. Article 6 and the Data Protection Commission. The same applies if the license is revoked or if the license is withdrawn from the licensee by the terms of this legislation.

The license and database under the terms of this legislation cannot be transferred, nor can they be subjected to attachment for debt. Neither the license nor the database may be used as collateral for financial liabilities.

Article 6. Committee on the Creation and Operation of a Health Service Database

The Minister shall appoint a committee on the creation and operation of a database under the terms of this legislation. The committee shall comprise three people and three substitutes, appointed for four years at a time. One shall be a health sector worker with a knowledge of epidemiology, another shall have knowledge of information technology and/or computer science, and the third shall be a lawyer, and shall chair the committee. Their substitutes shall fulfill the same conditions.

The role of the committee is to ensure that the creation and operation of the database are in keeping with the terms of this legislation, regulations made on the basis of the legislation, and conditions laid down in the operating license, in so far as this does not fall within the ambit of the Data Protection Commission. The committee shall supervise the negotiation of contracts between the licensee, on the one hand, and health institutions and self-employed health workers, on the other. It shall protect the interests of health authorities, health institutions, self-employed health workers and scientists in the drawing up of agreements. The sum to be paid by the licensee under the terms of Article 4(3) shall be negotiated by the committee, as shall recompense in the form of access to data from the database for health institutions, self-employed health workers and their staff for purposes of scientific research.

The committee shall advise the Ministry of Health and the Director General of Public Health on the utilization of data from the database. Should the operating license be revoked or the license withdrawn from the licensee, the database shall be operated by the committee until the Minister has reached a decision on its long-term operation, cp. Article 5(3).

The committee shall be provided with staff and working facilities. The committee shall seek specialist assistance as deemed necessary.

The committee shall inform the Minister and the Data Protection Commission without delay if it believes that there is some defect in the operation of the database.

The committee shall, no later than March 1 each year, submit a report to the Minister on the operations of the past year.

Section III. Collection of Information

Article 7. Access to Data from Health Records

With the consent of health institutions or self-employed health workers, the licensee may be provided with data derived from medical records for entry onto a Health Sector Database. The health institutions shall confer with the physicians' council and specialist management of the relevant institution before contracts are concluded with the licensee.

In the handling of records, other data and information, the conditions deemed necessary by the Data Protection Commission at any time shall be complied with. Personal identification shall be coded before entry on the database, so that it is ensured that the licensee's staff work only with non-personally identifiable data. The staff of the relevant health institution or self-employed health workers shall prepare the data for entry on the health-sector database. Health data shall be transferred in coded form in order to ensure their security. Personal identification shall be coded one-way, i.e. by coding that cannot be traced using a decoding key. The Data Protection Commission shall carry out further coding of personal identification, using those methods that the commission deems to ensure confidentiality best.

With regard to access to data from medical records, this shall otherwise be subject to the Acts on the rights of patients, on physicians, on the health service and on the recording and handling of personal data.

Article 8. Rights of Patients

A patient may request at any time that information on him/her not be entered onto the health-sector database. The patient's request may apply to all existing information on him/her or that which may be recorded in the future, or to some specific information. Such a request must be complied with. The patient shall inform the Director General of Public Health of his/her wish. The Director General of Public Health shall produce forms for giving such notice, and shall ensure that these are available at health institutions and at the premises of self-employed health workers. The Director General of Public Health shall ensure that a coded register of the relevant patients is always accessible for those who carry out the entry of data onto the health-sector database.

The Director General of Public Health shall ensure that information on the Health Sector Database and on the rights of patients cp. paragraph 1 shall be accessible to the public. Health institutions and self-employed health workers shall have this information available to patients on their premises.

Section IV. Access to the Database and Utilization of Data, etc.

Article 9. Access by Health Authorities to Data on the Health Sector Database

The Ministry of Health and Director General of Public Health shall always be entitled to statistical data from the Health Sector Database so that it may be used in statistical processing for the making of health reports and planning, policy-making and other projects of these bodies. This information to the specified parties shall be provided free of charge.

Article 10. Utilization of the Health Sector Database

Data recorded or acquired by processing on the Health Sector Database may be used to develop new or improved methods of achieving better health, prediction, diagnosis and treatment of disease, to seek the most economic ways of operating health services, and for making reports in the health sector.

The licensee shall be authorized to process data on the Health Sector Database from the health data recorded there, provided that data are processed and connected in such a way that they cannot be linked to identifiable individuals. The licensee shall develop methods and protocols that meet the requirements of the Data Protection Commission in order to ensure confidentiality in connecting data from the Health Sector Database, from a database of genealogical data, and from a database of genetic data. With regard to linking the data on the Health Sector Database with other databases than those specified here, the Act on recording and handling of personal data shall apply. It is not permissible to give information on individuals, and this shall be ensured e.g. by limitation of access.

The licensee may not grant direct access to data on the database.

The licensee is authorized during the period of the license to use the data on the database for purposes of financial profit, under the conditions laid down in this legislation and the license.

The Health Sector Database may not be transported out of Iceland, and processing of it may only be carried out in Iceland.

Article 11. Confidentiality

Employees of the licensee, including contractors, are bound by an obligation of confidentiality on matters that they become aware of in their work, which should remain confidential, by law or by their nature. They shall sign an oath of confidentiality before they begin work. The obligation of confidentiality remains in force, even if employment ceases.

Section V. Monitoring

Article 12. Monitoring of the Creation and Operation of a Health Sector Database

The Data Protection Commission shall monitor the creation and operation of the Health Sector Database with regard to recording and handling of personal data and the security of data on the database, and is responsible for monitoring compliance with conditions laid down by the commission.

The committee on the operation of the database, cp. Article 6, shall be responsible for monitoring the compliance in every way of the activities of the Health Sector Database with the terms of this legislation, regulations issued under the terms of this legislation, and the conditions of the license. The committee shall monitor all questions to and processing from the database. It shall regularly send to the Science Ethics Committee a record of all questions processed on the database, together with information on the inquirers.

The minister shall issue regulations on an interdisciplinary ethics committee, which shall assess studies, carried out within the licensee's company and questions that are received. The committee's evaluation must reveal that there is no scientific or ethical reason to prevent the study in question being carried out, or the questions processed from the database.

Section VI. Penalties

Article 13. Revocation of License

The Minister may revoke the license under the terms of this legislation if the licensee or the licensee's employees violate the terms of legislation, if the conditions of the license are not fulfilled, or if the licensee becomes unable to operate the database. Should the licensee violate the terms of this legislation or not comply with the conditions of the license, the Minister shall give the licensee a written warning, allowing a reasonable period of grace to rectify matters. Should the licensee not comply with such a warning, the license shall be revoked. In the case of deliberate violation or gross negligence, the Minister may revoke the license without notice and without allowing time for rectification.

Article 14. Penalties

Violation of the terms of this legislation entails fines or imprisonment for up to three years, unless a more severe penalty is prescribed in other legislation.

The same penalties apply to failure to comply with the conditions for granting of an operating license under the terms of this legislation, or government regulations under the terms of the legislation, or failure to comply with a command or prohibition under the terms of the legislation, or government regulations under the terms of the legislation.

A legal entity may be sentenced to pay fines due to violation of this Act or regulations based on it. A legal entity may be fined regardless of the guilt of its employees. The legal entity shall be responsible for payment of a fine imposed upon an employee of the legal entity, provided that the offense is connected to the employee's work for the legal entity.

Article 15. Withdrawal of License etc.

The licensee may, in addition to the penalties specified in Article 14, be subject to revocation of the license by legal verdict, in the case of deliberate violation or gross negligence.

Equipment which has been used for serious violation of this legislation may be confiscated, together with the profits of the violation, cp. Article 69 of the Penal Code No. 19 of 1940.

Article 16

Attempted violation, and participation in violation, of this legislation are subject to penalties as stated in Section III of the Penal Code, No. 19 of 1940.

Article 17. Compensation

Should the licensee, an employee of the licensee or a person assigned to process data violate the provisions of this Act with regard to confidentiality, regulations issued on the basis of them, or the conditions laid down by the Data Protection Commission, the licensee shall compensate the person to whom the data relate for financial loss which this has caused.

The licensee, however, is not obliged to compensate for loss which the licensee proves not to be attributable to a mistake or negligence on the licensee's part, or that of an employee or processor.

Section VII. Various Provisions

Article 18. Regulations

The Minister may prescribe further terms on the practice of this Act by issuing regulations.

The Minister shall issue regulations on the activity of the committee on operation of a Health Sector Database under Article 6, and on limitation of access under Article 10(2).

Article 18. Enactment

This Act shall take force immediately.

This Act shall be reviewed no later than 10 years after its enactment.

Provisional Clauses

(1) The licensee's license fee under Article 4(3) shall for the first year be based upon estimated costs pertaining to the preparation and monitoring of the operations of the Health Sector Database.

(2) The entry of data onto the Health Sector Database shall not commence until six months after the enactment of this Act.

(3) Before processing begins on the Health Sector Database, the committee on the operation of the database cp. Article 6 shall ensure that the assessment of an independent expert on the security of information systems has been sought.

Biobanks Act[1] (Iceland, 2000)

Section I. General Provisions

Article 1. Objectives

The objective of the Act is to authorize the collection, keeping, handling and utilization of biological samples from human beings, in such a way that confidentiality is ensured, the interests of donors of biological samples is safeguarded and that the utilization of the biological samples serves the purposes of science and medicine, and is conducive to the public good.

The interests of science and of the community shall never be given priority over the interests of the donor of a biological sample. It is prohibited to discriminate against a donor of a biological sample on the grounds of data derived from a biological sample.

Article 2. Scope

This Act applies to the collection of biological samples, and their keeping, handling, utilization and storage in biobanks.

The Act does not apply to temporary keeping of biological samples taken for purposes of clinical testing, treatment, or for specific scientific study, provided such samples are destroyed when the tests, treatment or research are completed. Temporary keeping means storage for up to five years, unless the National Bioethics Committee authorizes a longer period of storage. Should the long-term preservation of such samples be desired, they shall be stored in a biobank.

The Act does not apply to the storage of gametes and embryos under the provisions of the Act on Artificial Procreation, to organs under the provisions of the Act on Organ Removal, or to bodily remains under the terms of the National Heritage Act.

1 *Lög um lífsýnasöfn*. Act No. 110 of 2000 on Biobanks. Passed by the Alþingi (Parliament) on May 13, 2000. See http://government.is/interpro/htr/htr.nsf/pages/Act-biobanks for the official English text.

Article 3. Definitions

In this Act the following terms have the following meanings:

(1) *Biological sample*: organic material from a human being, alive or deceased, which may provide biological information about him/her.

(2) *Biobank*: a collection of biological samples that are permanently preserved.

(3) *Scientific study*: a study whose primary aim is to add to knowledge, with the purpose among other things of improving health and curing disease.

(4) *Clinical test*: test carried out in order to provide health service to an individual.

(5) *Free, informed consent*: consent granted in writing of the person's own free will, after the donor of a biological sample has been informed of the purpose of taking the biological sample, its usefulness, risks attendant upon the process, and that the biological sample will be permanently preserved in a biobank for use under the terms of Article 9.

(6) *Assumed consent*: Consent that consists in the donor of a biological sample not expressing any unwillingness for a biological sample taken from him/her for a clinical test to be permanently preserved in a biobank for use by the terms of Article 9, information in writing on this possibility having been available to him/her.

(7) *Donor of a biological sample*: A person from whom a biological sample is taken.

(8) *Licensee*: Individual or legal entity granted a license by the Minister to operate a biobank under the terms of Article 4 of this Act.

Section II. Establishment and Operation of Biobanks

Article 4. Authority to Found and Operate a Biobank

The establishment and operation of a biobank, i.e. collection, keeping, handling, utilization and storage of biological samples, is permissible only for those who have been granted a licensee from the Minister under the provisions of this Act, following the receipt of recommendations from the Director General of Health and the National Bioethics Committee.

Article 5. Conditions of License

A license for the establishment and operation of a biobank is contingent upon he following conditions:

(1) The terms of this Act, and government directives on the basis of the Act, shall be complied with.

(2) The biobank shall be located in Iceland.
(3) The objectives of the operation of the biobank, and the operational basis of the bank, shall be clearly defined.
(4) Conditions of storage for biological samples shall be described.
(5) Protocols of the biobank shall have been drawn up, including regulations of the biobank on arrangements for collaboration with foreign parties.
(6) A governing board shall be nominated, as provided in Article 6, and one individual shall be nominated to be answerable for the biobank.
(7) The answerable party for the biobank shall be a physician and shall have practiced independent research and development work within the health sector. In the case of the biobank comprising exclusively biological samples gathered for purposes of scientific study, the answerable party is not required to be a physician.
(8) That evaluation of security, and security measures in gathering of biological samples, shall be consistent with the rules laid down by the Data Protection Authority on security of personal data in biobanks.

The Minister may lay down further conditions.

Article 6. Board of a Biobank

The licensee shall appoint a board of at least three people for each biobank, which shall monitor its operations. The board shall be under an obligation to keep the Director General of Health, the Data Protection Authority and the National Bioethics Committee informed regarding the biological samples and operations of the biobank.

Section III. Collection, Handling and Access to Biological Samples

Article 7. Consent of Donor of a Biological Sample and Withdrawal of Consent

In connection with collection of a biological sample for preservation in a biobank, the free, informed consent of the person giving the biological sample shall be sought. This consent shall be given freely and in writing after the donor of a biological sample has been informed of the objective of the sample collection, the benefits, risks associated with its collection, and that the biological sample will be permanently stored at a biobank for use as provided in Article 9. In addition the provisions of Article 20 of the Act on personal privacy and handling of personal data shall be observed where applicable.

A donor of a biological sample can at any time withdraw his/her consent under the terms of paragraph 1, and the biological sample shall then be destroyed.

Material that has been produced from a biological sample by performance of a study or the results of studies already carried out shall, however, not be destroyed.

If biological samples have been collected for the purpose of clinical tests or treatment, the consent of the patient may be assumed for the storage of the biological sample in a biobank for use as provided in Article 9, provided that general information on this is provided by a health care professional or health institution.

A donor of a biological sample may at any time withdraw his/her assumed consent for his/her biological sample to be stored in a biobank for use as provided in Article 9, in which case it shall thereafter only be used in the interests of the donor of a biological sample or by his/her specific permission, but see also Article 9(4). The request of a donor of a biological sample may apply to all biological samples that have been taken or may be taken from him/her. Such a request must be complied with. The donor of a biological sample shall inform the Director General of Health of his/her request. The Director General of Health shall be responsible for preparation of forms for giving such notice, and shall ensure that these are available at health institutions, and at the premises of self-employed health care professionals. The Director General of Health shall ensure that a coded register of those who have opted out in this way shall always be available to the boards of biobanks. Staff of the Director General of Health who carry out this work are subject to an obligation of confidentiality regarding information they may become aware of in the course of their work, which should remain confidential by law or by its nature. Such staff shall sign an oath of confidentiality before their employment begins. The obligation of confidentiality remains in force after employment ceases.

Article 8. Preservation of Biological Samples

Biological samples shall be kept securely and labeled, but stored without personal identification. The linking of biological samples with personal identification shall be in keeping with standards laid down by the Data Protection Authority.

Biological samples shall be stored in such a way that they are not lost or damaged, and that they are not accessible to those who are not entitled to use them.

Should the licensee decide to cease operation of the biobank, the license having been revoked as provided in Article 14, the Minister shall, after receiving recommendations from the Director General of Health, the Data Protection Authority and the National Bioethics Committee, decide on the future of the biobank, taking into account the wishes and proposals of the licensee.

Article 9. Access to Biobank and Use of Biological Samples

Biological samples shall be acquired for clearly defined and lawful purposes, and not used for other purposes, but see paragraphs 2, 3 and 4 below.

The answerable party for the biobank grants access to biological samples for further diagnosis of diseases. He/she may also grant access to biological samples for purposes of quality control, development of methods and tuition, provided that they are not personally identified.

The board of the biobank shall negotiate with scientists on access to biological samples. Access to biological samples for scientific studies may not, however, be granted until the permission of the Data Protection Authority has been granted on the basis of the Act on personal privacy and handling of personal data, and a research protocol has been approved by the National Bioethics Committee or the ethics committee of the relevant health institution, as provided in the Act on the Rights of Patients and of regulations issued on the basis of the Act.

The board of the biobank may, if approved by the Data Protection Authority and the National Bioethics Committee, authorize the use of biological samples for other purposes than those for which the samples were originally collected, provided that important interests are at stake, and that the potential benefit outweighs any potential inconvenience to the donor of a biological sample or other parties.

The Minister shall, having received proposals from the Director General of Health, the National Bioethics Committee and the Data Protection Authority, issue regulations defining more precisely the use of biological samples.

Article 10. Rights and Fees

The licensee shall not be counted as the owner of the biological samples, but has rights over them, with the limitations laid down by law, and is responsible for their handling being consistent with the provisions of this Act, and of government directives based on it. The licensee may thus not pass the biological samples on to another party, nor use them as collateral for financial liabilities, and they are not subject to attachment for debt.

The licensee may take a fee for a biological sample, or access to a biological sample, equivalent to the cost of gathering, storage and access to the sample. Any further fee is prohibited.

A biological sample may be sent out of the country in the interests of the donor of a biological sample, for diagnosis or quality control. Other transportation out of the country of biological samples is subject to the approval of the National Bioethics Committee and the Data Protection Authority and on the conditions they lay down.

Article 11. Confidentiality

All staff of biobanks and those who have access to them shall preserve confidentiality regarding matters relating to their work, which should be kept confidential, by law or by their nature. The obligation of confidentiality remains in force after employment, research or tuition ceases.

Section IV. Monitoring and Obligation to Supply Information

Art.12. Monitoring

The answerable party for the biobank shall be responsible for the implementation of internal monitoring and security assessments shall be carried out regularly, in accord with the provisions of Articles 11 and 12 of the Act on personal privacy and handling of personal data.

The Data Protection Authority shall monitor the security of personal data in biobanks. The Data Protection Authority's monitoring of biobanks is subject to the terms of Article 35(4), Article 37(2) and (4), and Articles 38 to 43 of the Act on Personal Privacy and Handling of Personal Data.

The Director General of Health shall monitor biobanks in so far as this monitoring does not fall within the ambit of the Data Protection Authority or the National Bioethics Committee.

Article 13. Obligation to Supply Information: Government and Biobank Boards

The Director General of Health is under an obligation to promulgate in detail to the general public the terms of this Act on biobanks, especially the provisions on assumed consent of a donor of a biological sample regarding a clinical test, and also the rights of the individual by the terms of Article 7 and of Article 7(3).

The Director General of Health shall annually issue a register of biobanks, their purposes, activities and protocols. The register shall contain information on the membership of the board of each bank, and the identity of the answerable party. This register shall be made public and shall be accessible to the general public.

The board of the biobank or the Director General of Health is obliged to provide individuals with information on whether biological samples from him or her are stored in a biobank, and on the nature of such biological samples.

Section V. Penalties

Article 14.

The Minister may revoke the license under the terms of this Act, if the licensee or its employees violate the terms of the Act or government directives on the basis of the Act, if the conditions of the license are not fulfilled, or if the licensee proves unable to operate the biobank. Should the licensee violate the terms of this legislation or not comply with the conditions of the license, the Minister shall give the licensee a written warning, allowing a reasonable period of grace to rectify matters. Should the licensee not comply with such a warning, the license shall be revoked. In the case of deliberate violation or gross negligence, the Minister may revoke the license without notice and without allowing time for rectification.

Article 15.

Violation of the terms of this Act or government directives based on it entails fines or imprisonment for up to three years, unless a more severe penalty is prescribed in other legislation.

A legal entity or an individual may be sentenced to pay fines due to violation of this Act. A legal entity may be fined whether or not the guilt of an employee of the legal entity has been proved. Should a member of the staff of the legal entity violate the terms of this Act or of government directives based on it, the legal entity may also be fined. The legal entity shall be responsible for payment of a fine imposed upon a member of its staff, for violation of the terms of this Act, provided that the offense is connected to the employee's work for the legal entity.

Section VI. Various Provisions

Article 16. Government Directives

The Minister may issue regulations on the further implementation of this Act.

The Minister shall issue regulation on how information on assumed consent by the terms of Article 7(3) shall be provided, on how to ensure that withdrawal of assumed consent by a donor of a biological sample by the terms of Article 7(4) is complied with, on the register of those opting out and its form cp. Article 7(4), and how to ensure equal treatment of those who request access to biobanks for purposes of scientific studies, cp. Article 9(3).

Article 17. Entry into Force

This Act shall take force on January 1, 2001.

Provisional Clauses

(1) Before the Act comes into force, the Minister of Health and Social Security shall assign the Directorate General of Health to carry out detailed publicity among the general public on biobanks and regulations applying to collection and utilization of biological samples.

(2) Biological samples gathered before this Act came into force may be stored in a biobank, unless the donor of a biological sample declares his/her opposition to this. Otherwise the provisions of the Act shall apply to the storage, handling and utilization of such biological samples.

Human Genes Research Act[1]
(Estonia, 2001)

Chapter 1. General Provisions

Article 1. Purpose and Scope of Application of Act

(1) The objectives of this Act are to regulate the establishment and maintenance of a Gene Bank, to organize the genetic research necessary therefor, to ensure the voluntary nature of gene donation and the confidentiality of the identity of gene donors, and to protect persons from misuse of genetic data and from discrimination based on interpretation of the structure of their DNA and the genetic risks arising therefrom.

(2) The following are provided for in this Act:

1) the conditions for processing tissue samples, descriptions of DNA, descriptions of state of health, and genealogies in the Gene Bank;
2) the rights and obligations of gene donors, the chief processor and authorized processors of the Gene Bank and genetic researchers relating to tissue samples, descriptions of DNA, descriptions of state of health, and genealogies;
3) the conditions for the establishment and maintenance of the Gene Bank;
4) the restrictions on the use of tissue samples, descriptions of DNA, descriptions of state of health and genealogies collected in the Gene Bank;
5) the conditions for genetic research relating to the Gene Bank and the organization of supervision thereof.

Article 2. Definitions

In this Act, the following definitions are used:

(1) *DNA* means a molecule of deoxyribonucleic acid which contains the genetic information of a human being;

1. *Inimgeeniuuringute seadus* (Human Genes Research Act), passed on December 13, 2000 (as RT I 2000, 104, 685), entered into force on January 8, 2001. RT is *Riigi Teataja* (State Gazette). See http: //www.legaltext.ee/text/en/X50010.htm for the official English translation.

(2) *tissue sample* means cells, intercellular substance and body fluids taken from a human being for the purposes of genetic research;

(3) *description of DNA* means a digital model of DNA created as a result of genetic research;

(4) *gene donor* means a person who provides a tissue sample in accordance with this Act and with regard to whom a description of state of health and genealogy are prepared;

(5) *description of state of health* means data collected for use in genetic research which reflects the state of health of a gene donor, illnesses he or she has suffered from and treatment thereof, and his or her lifestyle, physical and social environment and hereditary characteristics;

(6) *genealogy* means information in the Gene Bank concerning the names, dates of birth and blood relationships of the ascendants and descendants of a gene donor;

(7) *genetic research* means the scientific study and description of DNA, other constituents of tissue samples, descriptions of state of health, and genealogies and determination of the links between them with the purpose of obtaining data on the genes, gene products and hereditary characteristics of human beings;

(8) *genetic researcher* means a natural or legal person or a state or local government agency who performs genetic research;

(9) *genetic data* means data concerning the genes, gene products or hereditary characteristics of a human being, including a description of DNA or a part thereof, which is obtained as a result of genetic research;

(10) *Gene Bank* means a database established and maintained by the chief processor consisting of tissue samples, descriptions of DNA, descriptions of state of health, genealogies, genetic data, and data enabling the identification of gene donors;

(11) *coding* means the replacement with a unique code of such data of a tissue sample, description of DNA, description of state of health, or genealogy which enables the identification of the person;

(12) *decoding* means the reverse identification of a gene donor through the unique code given to a tissue sample, description of DNA, description of state of health or genealogy or the identification of the tissue sample, description of DNA, or description of state of health of a gene donor based on the identity of the gene donor.

Article 3. Chief Processor of the Gene Bank

(1) The chief processor of the Gene Bank is a non-profit foundation founded by the Republic of Estonia on the basis of the Foundation of and Participation in Legal Persons in Private Law by the State Act (RT I 1996, 48, 942; 73, correction notice; 1998, 59, 941) within the area of government of the Ministry of Social Affairs which has the right to organize the taking of tissue samples, to prepare descriptions of state of health and genealogies, to code, decode, store, destroy and issue descriptions of state of health and genealogies, to perform genetic research and to collect, store, destroy and issue genetic data.

(2) The objectives of the chief processor are to:

1) promote the development of genetic research;
2) collect information on the health of the Estonian population and genetic information concerning the Estonian population;
3) use the results of genetic research to improve public health.

(3) In order to achieve its objectives, the competence of the chief processor includes the establishment and maintenance of the Gene Bank. The chief processor has the right to delegate the rights of processing, except for coding and decoding, to an authorized processor on the basis of a contract in the cases and under the conditions prescribed in this Act.

Article 4. Supervisory Board of the Chief Processor of the Gene Bank

(1) The supervisory board of the chief processor of the Gene Bank shall consist of nine members. The *Riigikogu*[2] shall, on the proposal of the Social Affairs Committee of the *Riigikogu*, appoint three members of the supervisory board. The Government of the Republic shall appoint three members of the supervisory board. The Board of the Estonian Academy of Sciences shall appoint three members of the supervisory board.

(2) The term of authority of a member of the supervisory board is from one year to five years. The due date for termination of the authority of a member of the supervisory board shall be determined upon appointment of the member of the supervisory board. If a member of the *Riigikogu* or the Government of the Republic is appointed to the supervisory board, the authority of such member of the supervisory board shall terminate upon the termination of his or her authority as a member of the *Riigikogu* or the Government of the Republic.

2. *Riigikogu* is the Parliament of Estonia.

(3) The body which appointed a member of the supervisory board may remove the member of the supervisory board with good reason, which is, above all, failure to perform his or her duties to a material extent, a constant conflict of interests or inability to participate in the work of the supervisory board or the causing of significant damage to the interests of the chief processor of the Gene Bank in any other manner, and also the commencement of bankruptcy proceedings against the member of the supervisory board.

Article 5. Authorized Processor of the Gene Bank

An authorized processor of the Gene Bank is a natural or legal person or an Estonian state agency or local government agency who may receive all processing rights, except the right to code and decode, from the chief processor on the basis of a contract. The requirements for authorized processors shall be established by the Government of the Republic.

Article 6. Genetic Research and Genetic Testing

(1) Genetic research relating to the Gene Bank is permitted in order to study and describe the links between genes, the physical and social environment and the lifestyles of people, to find medicinal products or methods of treatment on the basis thereof, to assess individual health hazards and to prevent illnesses.

(2) Genetic testing to which the provisions of Chapters 2, 3 and 4 of this Act do not apply may be performed pursuant to the procedure and for the purposes provided by law. Tissue samples taken from people in the course of genetic testing and the results of research performed on such samples shall not be added to the Gene Bank.

Article 7. Application of Provisions that Regulate the Processing of Personal Data

(1) The provisions of the Personal Data Protection Act (RT I 1996, 48, 944; 1998, 59, 941; 111, 1833; 2000, 50, 317; 92, 597) that regulate the processing of personal data, together with the specifications provided for in this Act, apply to the taking of tissue samples, preparation of descriptions of state of health, coding, decoding, and the maintenance of databases by the chief processor.

(2) The provisions regulating the processing of personal data do not apply to the processing of coded tissue samples, coded descriptions of DNA, and coded descriptions of state of health if such tissue samples, descriptions of DNA, and descriptions of state of health are processed as a set of data and

on the condition that the set of data to be processed contains DNA samples, descriptions of DNA, or descriptions of state of health of at least five gene donors at a time.

Chapter 2. Rights of Gene Donors

Article 8. Confidentiality of Identity of Gene Donors

(1) The identity of a gene donor shall remain classified after coding.

(2) A gene donor has the right to permit disclosure of his or her identity.

(3) The fact of a person being or not being a gene donor shall remain classified. Only a person himself or herself has the right to disclose the fact of being or not being a gene donor and the circumstances thereof, unless otherwise prescribed by law.

(4) Coding and decoding shall be performed in accordance with the requirements of this Act.

Article 9. Voluntary Nature of Gene Donation

(1) It is prohibited to take a tissue sample and prepare a description of state of health or genealogy without the specific knowledge and voluntary consent of the person.

(2) It is prohibited to influence a person's decision to become a gene donor, including by threatening the person with negative consequences, promising material benefits or providing subjective information.

(3) The chief processor and authorized processors are permitted to disseminate only general information concerning the Gene Bank and genetic research.

Article 10. Right to Apply for the Destruction of Material

(1) Pursuant to the provisions of Article 21 of this Act, gene donors have the right to apply, at any time, to the chief processor for the destruction of data that enables decoding.

(2) If the identity of a gene donor is unlawfully disclosed, the gene donor has the right to apply to the chief processor for the destruction of the tissue sample, description of DNA and description of the state of health pursuant to the provisions of Article 21 of this Act.

Article 11. Other Rights of Gene Donors

(1) Gene donors have the right not to know their genetic data.

(2) Gene donors have the right to access personally their data stored in the Gene Bank. Gene donors do not have the right to access their genealogies.
(3) Gene donors shall not be charged for accessing their data stored in the Gene Bank.
(4) Gene donors have the right to genetic counseling upon accessing their data stored in the Gene Bank.
(5) Gene donors have the right to submit additional information on themselves to the chief processor.
(6) Gene donors have the right to prohibit the supplementation, renewal and verification of descriptions of their state of health stored in the Gene Bank.

Article 12. Consent to Become a Gene Donor

(1) A person's consent to provide a tissue sample, to have a description of the state of health or the genealogy of the person prepared, to entry of the description of the state of health or the genealogy in the Gene Bank in coded form and to use thereof for genetic research, public health research and statistical purposes shall be prepared in writing and shall be signed by the person who becomes a gene donor.
(2) Partial or conditional consent is not valid.
(3) A document specified in paragraph (1) of this Article shall, in addition to an expression of consent and the signature of the person who becomes a gene donor, set out the following information:
    1) the name and personal identification code or, in the absence thereof, sex, date of birth and residence of the person;
    2) information concerning the medical institution or family physician who is to take the tissue sample;
    3) the part of the body from which and the manner in which the tissue sample is to be taken;
    4) the date on which consent is granted.
(4) The chief processor or authorized processor shall inform a person, prior to the person granting consent, of the following facts:
    1) grant of the consent specified in paragraph (1) of this Article is voluntary;
    2) the gene donor is not entitled to request a fee for providing a tissue sample, preparation and study of a description of his or her state of health or genealogy, or use of the research results;
    3) data on hereditary characteristics and genetic risks obtained as a result of genetic research may be unpleasant for the gene donor;

4) the gene donor has the right not to know his or her genetic data;
5) the gene donor has the right to know his or her genetic data, except genealogy;
6) the gene donor has the right to apply to the chief processor for the destruction of data which enables decoding or, upon unlawful disclosure of his or her identity, for the destruction of the tissue sample, description of DNA, and description of his or her state of health;
7) the gene donor has the right to withdraw his or her consent until his or her tissue sample or the description of his or her state of health is coded.

(5) The chief processor or authorized processor to whom a person who is about to become a gene donor submits his or her consent is required to provide the gene donor promptly with a copy of his or her written consent.

(6) The chief processor shall preserve the written consent of a gene donor for as long as the tissue sample, description of state of health, or description of DNA are stored, except in the case provided for in Article 10(1) of this Act.

(7) The format of written consent to become a gene donor and the procedure for the completion and preservation thereof shall be established by the Minister of Social Affairs.

Article 13. Persons with Restricted Active Legal Capacity and Persons without Capacity to Exercise Will

(1) The consent of a person with restricted active legal capacity to become a gene donor shall be deemed valid if the following conditions are met:
1) the person with restricted active legal capacity and his or her legal representative or guardian have been given the information provided for in Article 12(4) of this Act;
2) the legal representative or guardian has expressed the consent provided for in Article 12(1) of this Act;
3) the person with restricted active legal capacity is not opposed to providing a tissue sample or to the collection of descriptions of his or her state of health.

(2) Persons who are unable to understand the content and meaning of consent shall not be gene donors. In the case of doubt, it is presumed that the person is temporarily unable to understand the content and meaning of consent.

Chapter 3. Processing of Gene Bank

Article 14. Taking of Tissue Samples and Preparation of Descriptions of State of Health

(1) The chief processor shall organize the taking of tissue samples and the preparation of descriptions of state of health itself or through an authorized processor.

(2) The medical procedure of taking a tissue sample shall be performed by a health care professional of the Republic of Estonia.

(3) A description of state of health shall be prepared and supplemented by the chief processor or authorized processor on the basis of statements by the gene donor and data on the gene donor stored in medical institutions. Only data on illnesses suffered from by the gene donor himself or herself may be collected from a gene donor. A gene donor may be asked questions about habits or illnesses that have occurred in his or her family. Such questions shall not enable the identification of a certain person or generation.

(4) The chief processor or authorized processor may prepare a genealogy on the basis of the results of questioning a gene donor and on the basis of other databases and genetic research.

(5) A person taking a tissue sample and a person preparing a description of state of health shall maintain the confidentiality of the identity of the gene donor, his or her tissue sample, the description of his or her state of health and his or her genealogy and shall deliver the personal data of the gene donor, the tissue sample, the description of his or her state of health and his or her genealogy to the chief processor or an authorized processor designated by the chief processor as soon as possible.

Article 15. Right of Ownership of Tissue Samples, Descriptions of State of Health and Genealogies

(1) The chief processor's right of ownership of a tissue sample, description of state of health, other personal data and genealogy is created from the moment the tissue sample or personal data is provided or the moment the state of health or genealogy is prepared.

(2) Tissue samples and uncoded information in the ownership of the chief processor and written consent of gene donors are not transferable.

(3) A gene donor is not entitled to request a fee for providing a tissue sample, preparation and study of a description of his or her state of health or genealogy, or use of the research results.

Article 16. Permission to Use Gene Bank

(1) The Gene Bank may be used only for scientific research, research into and treatment of illnesses of gene donors, public health research and statistical purposes. Use of the Gene Bank for other purposes, especially to collect evidence in civil or criminal proceedings or for surveillance, is prohibited.

(2) The doctor of a gene donor has the right, pursuant to Article 24(2)(7) of this Act, to obtain the decoded description of the state of health of the gene donor from the Gene Bank in order to treat the gene donor.

Article 17. Permission to Use Genealogy

(1) Genealogies may be used only within the Gene Bank for structuring tissue samples, descriptions of DNA, and descriptions of state of health on the basis of blood relationships.

(2) The chief processor may issue, to a gene researcher, the link between coded tissue samples, coded descriptions of DNA, and coded descriptions of state of health based on blood relationships in adherence with the requirements provided for in Article 22(4) of this Act.

Article 18. Storage of Tissue Samples, Descriptions of DNA and Descriptions of State of Health

(1) Tissue samples, descriptions of DNA, and descriptions of state of health shall be stored by the chief processor or an authorized processor.

(2) Coded DNA samples, coded descriptions of DNA, and coded descriptions of state of health may also be stored by a gene researcher who has the resources and conditions necessary for storage. The conditions for storage shall be established by the Minister of Social Affairs.

(3) In order to organize storage, the chief processor shall enter into a contract with an authorized processor or gene researcher which shall set out the following:

1) the tissue samples, descriptions of DNA and descriptions of state of health handed over for preservation;
2) the time of delivery of the tissue samples, descriptions of DNA and descriptions of state of health to the authorized processor or gene researcher and the manner of supplementing the tissue samples, descriptions of DNA and descriptions of state of health;
3) the place and term of storage;
4) the method of storage;
5) the security measures to be applied by the storer;

6) the procedure for the copying, distribution, destruction and issue of tissue samples, descriptions of DNA or descriptions of state of health and for the return thereof to the chief processor or authorized processor;
7) the liability of the authorized processor or gene researcher.

(4) All tissue samples shall be stored in the territory of the Republic of Estonia. The Government of the Republic may, at the request of the chief processor and if good reasons therefor become evident, grant permission for tissue samples to be stored outside the territory of the Republic of Estonia if it is ensured that the chief processor has effective control over the tissue samples and that the tissue samples cannot be used in a manner prohibited by Estonian law.

Article 19. Rights Related to Results of Genetic Research

(1) An authorized processor or gene researcher shall unconditionally deliver descriptions of DNA or parts thereof to the chief processor. The chief processor may grant an authorized processor or gene researcher the right to use a description of DNA or a part thereof for a charge or without charge. The chief processor shall grant gene researchers who are legal persons in public law or state agencies of the Republic of Estonia the right to use descriptions of DNA or parts thereof without charge.

(2) Results of genetic research not specified in paragraph (1) of this Article and intellectual property rights related thereto shall be provided by law.

Article 20. Issue of Tissue Samples, Descriptions of DNA and Descriptions of State of Health, and Records of Issue

(1) The chief processor shall issue tissue samples, descriptions of DNA and descriptions of state of health from the Gene Bank only in coded form, except upon the issue of data on a gene donor to the gene donor or to the doctor of the gene donor in the cases provided for in Articles 11(2) and 16(2) of this Act.

(2) Upon the issue of a tissue sample, description of DNA or description of state of health, the issuer shall prepare a record of issue in which the tissue samples, descriptions of DNA and descriptions of state of health issued and received and the manner and date of delivery are indicated. The record of issue shall be signed by the representatives of the issuer and receiver.

(3) If the issuer is an authorized processor, the issuer shall promptly send a copy of the record of issue to the chief processor.

(4) The procedure for issuing tissue samples, descriptions of DNA, and descriptions of state of health, including the format of the record of issue and the

procedure for the completion and preservation thereof shall be established by the Minister of Social Affairs.

Article 21. Destruction of Data Which Enables Decoding, Tissue Samples, Descriptions of DNA and Descriptions of State of Health, and Records of Destruction

(1) If a gene donor applies to the chief processor for the destruction of data which enables decoding, the chief processor is required to destroy the data that enables decoding within two weeks as of the receipt of the corresponding written application from the gene donor.

(2) If a gene donor applies to the chief processor for the destruction of a tissue sample, description of DNA or description of his or her state of health due to unlawful disclosure of his or her identity, the chief processor is required to organize the destruction of the tissue sample, description of DNA and description of state of health within one month as of the receipt of the corresponding written application from the gene donor unless the chief processor proves that the identity of the gene donor was disclosed as a result of the gene donor's own behavior.

(3) The chief processor may organize the destruction of tissue samples, descriptions of DNA and descriptions of state of health on the basis of records of issue prepared pursuant to Article 20 of this Act, in which case the chief processor shall issue a destruction order to the receiver specified in the record of issue.

(4) All persons who have a specific tissue sample, description of DNA ,or description of state of health in their possession are required, on the order of the chief processor, to destroy the specific tissue sample, description of DNA, or description of state of health promptly and at their own expense.

(5) Upon the destruction of data which enables decoding, tissue samples, descriptions of DNA, or descriptions of state of health, the destroyer shall prepare a record of destruction which sets out the destroyed object and the manner, place and date of destruction. If the destroyer is not the chief processor, the destroyer shall promptly send a copy of the record of destruction to the chief processor.

(6) The procedure for destroying data which enables decoding, tissue samples, descriptions of DNA, and descriptions of state of health, including the format of the record of destruction and the procedure for the completion and preservation thereof shall be established by the Minister of Social Affairs.

Chapter 4. Data Protection

Article 22. General Requirements for Data Protection

(1) Data in the Gene Bank shall be processed in compliance with the highest standard of data protection.

(2) Data enabling identification of gene donors shall not be available through the external computer network of the Gene Bank.

(3) The chief processor shall appoint the specific persons who perform coding or decoding, who have access to the consent of gene donors, to the collection of tissue samples and descriptions of DNA, to descriptions of state of health and to genealogies, and who issue coded or uncoded tissue samples, descriptions of DNA or descriptions of state of health.

(4) The chief processor may issue tissue samples, descriptions of DNA, or descriptions of state of health from the Gene Bank only in coded form, as a set of data and on the condition that samples or data concerning at least five gene donors are issued at a time. The cases provided for in Articles 11(2) and 16(2) of this Act constitute an exception.

(5) In order to verify the authenticity of data to be entered in the Gene Bank before coding, the chief processor and authorized processor are permitted to compare such data with data stored in other databases.

Article 23. Coding

(1) The chief processor shall give each tissue sample, description of DNA, description of state of health, and genealogy a unique code consisting of at least sixteen random characters immediately after receipt of the tissue sample, description of DNA, description of state of health, or genealogy in the Gene Bank. The chief processor shall obtain the approval of the Data Protection Supervision Authority for the method of generating the codes.

(2) The chief processor shall replace with a code all data concerning a tissue sample, description of DNA, and description of state of health which enables the reverse identification of the gene donor, including the name, personal identification code, date of birth and residence.

(3) The chief processor shall indicate the code given to a tissue sample, description of DNA, description of state of health, or genealogy on the written consent of the gene donor. The chief processor shall store the written consent together with the code indicated thereon in the database of the Gene Bank and it shall be the only possible key for decoding.

(4) In the interests of greater security, the chief processor may give an additional code to a coded tissue sample, coded description of DNA, coded description

of state of health, or coded genealogy and this code shall replace the initial code given to the tissue sample, description of DNA, description of state of health or genealogy. The chief processor shall make a record of the link between an initial code and an additional code in a database used solely for that purpose.

(5) The chief processor, authorized processors and genetic researchers are required to encode tissue samples, descriptions of DNA, descriptions of state of health, and genealogies only with the code given to the tissue samples, descriptions of DNA, descriptions of state of health, or genealogies by the chief processor.

Article 24. Decoding

(1) Only persons appointed pursuant to Article 22(3) of this Act may gain access to written consent with a code indicated thereon and have the technical possibility to decode data.

(2) The chief processor is permitted to decode data only in the following cases and for the following purposes:

1) in order to destroy a tissue sample, a description of DNA, or a description of state of health or data which enables decoding;
2) in order to enable access to data on a gene donor stored in the Gene Bank, except genealogies, at the written request of the gene donor;
3) in order to renew, supplement or verify a description of the state of health of a gene donor without contacting the gene donor unless the gene donor has prohibited the supplementation, renewal or verification of the description of his or her state of health;
4) in order to identify a gene donor on the proposal of the chief processor and with the consent of the Ethics Committee, to contact the gene donor and to renew, supplement or verify a description of his or her state of health with his or her written consent;
5) in order to identify a gene donor and, with the gene donor's written consent, to take a new DNA sample if a tissue sample has been destroyed or does not contain sufficient DNA;
6) in order to identify a gene donor within the Gene Bank and amend his or her genealogy if the results of DNA research contradict the previously known genealogy or provide new information concerning the genealogy;
7) in order to issue a description of the state of health of a gene donor to the doctor of the gene donor at the request of the doctor of the gene donor and with the consent of the gene donor or, in urgent cases where

a gene donor is unable to grant consent and has no legal representative or guardian or if the legal representative or guardian of the gene donor is unavailable, without the consent of the gene donor but in his or her interests and according to his or her presumed intentions.

Chapter 5. Prohibition on Discrimination

Article 25. Prohibition on Discrimination

(1) It is prohibited to restrict the rights and opportunities of a person or to confer advantages on a person on the basis of the structure of the person's DNA and the genetic risks resulting therefrom.

(2) It is prohibited to discriminate against a person on the basis of the person being or not being a gene donor.

Article 26. Discrimination in Employment Relationships

(1) Employers are prohibited from collecting genetic data on employees or job applicants and from requiring employees or job applicants to provide tissue samples or descriptions of DNA.

(2) Employers are prohibited from imposing discriminatory working and wages conditions for people with different genetic risks.

Article 27. Discrimination in Insurance Relationships

(1) Insurers are prohibited from collecting genetic data on insured persons or persons applying for insurance cover and from requiring insured persons or persons applying for insurance cover to provide tissue samples or descriptions of DNA.

(2) Insurers are prohibited from establishing different insurance conditions for people with different genetic risks and from establishing preferential tariff rates and determining insured events restrictively.

Chapter 6. Supervision and Resolution of Appeals

Article 28. Data Protection Supervision Authority

Supervision over the collection of descriptions of state of health and genealogical data, the coding and decoding of tissue samples, descriptions of DNA, descriptions of state of health and genealogical data, and the processing of tissue samples, descriptions of DNA, descriptions of state of health and genealogical data shall be exercised by the Data Protection Supervision Authority.

Article 29. Ethics Committee

(1) The processing procedures of the Gene Bank. The assessment of the Ethics Committee is not binding, except in the case provided for in Article 24(2)(4) shall be assessed by the Ethics Committee.

(2) The Ethics Committee shall act pursuant to generally recognized ethical rules and international conventions.

(3) Each member of the Ethics Committee shall be an Estonian citizen with active legal capacity whose permanent residence is in the Republic of Estonia. Each member of the Ethics Committee shall be a recognized specialist in his or her field with the necessary expertise to perform the duties of a member of the Ethics Committee and shall have an impeccable reputation.

(4) The members of the Ethics Committee shall be appointed and the number of members shall be determined by the supervisory board of the chief processor of the Gene Bank. The term of authority of a member of the Ethics Committee is five years.

(5) The supervisory board of the chief processor of the Gene Bank may remove a member of the Ethics Committee prior to expiry of the term of authority of the member with good reason, which is, above all, failure to perform his or her duties to a material extent, inability to participate in the work of the Ethics Committee or the causing of significant damage to the interests of the chief processor in any other manner, and also the commencement of bankruptcy proceedings against the member of the Ethics Committee.

Article 30. Resolution of Appeals Arising from Employment and Insurance Relationships

(1) Complaints concerning discrimination occurring in employment relationships due to genetic risks shall be adjudicated by the Labor Inspectorate, if necessary with the assistance of the chief processor or experts from the Data Protection Supervision Authority in investigating and deciding the matter.

(2) Complaints concerning discrimination occurring in insurance relationships due to genetic risks shall be adjudicated by the Insurance Supervisory Authority, if necessary with the assistance of the chief processor or experts from the Data Protection Supervision Authority in investigating and deciding the matter.

Chapter 7. Implementing Provisions

Article 31. Amendment of Criminal Code

The Criminal Code (RT 1992, 20, 287 and 288; RT I 1999, 38, 485; 57, 595, 597 and 598; 60, 616; 97, 859; 102, 907; 2000, 10, 55; 28, 167; 29, 173; 33; 193; 40, 247; 49, 301 and 305; 54, 351; 57, 373; 58, 376; 84, 533; 92, 597) is amended as follows:

1) Article 1244 is added to the Code worded as follows:
   "Article 1244. Inducing persons to become gene donors
   Inducing a person to consent to providing a tissue sample for the purposes of genetic research by offering material remuneration or by causing damage or threatening to cause damage to the person is punishable by a fine, detention or up to one year imprisonment.";
2) Article 1245 is added to the Code worded as follows:
   "Article 1245. Illegal conduct of human research
   Conduct of medical or scientific research on a person who has not granted his or her valid consent thereto is punishable by a fine, detention or up to one year imprisonment.";
3) Article 1281 is amended and worded as follows:
   "Article 1281. Disclosure of confidential data which became known due to professional activity
   Disclosure by a doctor, medical assistant, nurse, midwife, psychologist, advocate, notary or other person of confidential data relating to the descent, genetic data, artificial insemination, family or health of a person which became known to the offender due to his or her professional activity, if such disclosure violates legislation regulating the professional activity or other legislation, is punishable by a fine or deprivation of the right of employment in a particular position or operation in a particular area of activity or by detention or up to one year imprisonment.";
4) Article 1342 is added to the Code worded as follows:
   "Article 1342. Discrimination based on genetic risks
   Unlawful restriction of the rights of a person or conferral of unlawful preferences on a person based on the genetic risks of the person is punishable by a fine, detention or up to one year imprisonment."

Article 32. Amendment of Personal Data Protection Act

The Personal Data Protection Act (RT I 1996, 48, 944; 1998, 59, 941; 111, 1833; 2000, 50, 317; 92, 597) is amended as follows:

1) Article 4(3)(3) is amended and worded as follows:
   "3) data relating to state of health, genetic information or sexual life;";
2) Article 9(2) is amended and worded as follows:
   "(2) Processing of sensitive personal data revealing ethnic or racial origin, state of health, genetic information or sexual life is permitted without the consent of the person if the processing is carried out:
   1) for performance of obligations prescribed by law;
   2) for protection of the person's life, health or freedom;
   3) for performance of a task in the public or general interest which is assigned by law to a chief processor or a third person to whom the data are disclosed."

Article 33. Amendment of Estonian Academy of Sciences Act

Clause 51 is added to Article 9(3) of the Estonian Academy of Sciences Act (RT I 1997, 35, 539) worded as follows:

"(51) appoint three members of the supervisory board of the chief processor of the Gene Bank;".

Article 34. Establishment of Chief Processor

The Government of the Republic shall decide on the establishment of the non-profit foundation specified in Article 3(1) of this Act within two months as of the entry into force of this Act.

Human Genome Research Act[1] (Latvia, 2003)

Chapter 1. General Provisions

Article 1. Definitions

In this Act the following definitions apply:

1) *Decoding:* reverse identification of a gene donor through a unique code assigned to a tissue sample, a description of DNA, a description of the state of health or genealogy; or identification of a tissue sample, description of DNA, the description of state of health or genealogy based on the identity of a gene donor.
2) *Tissue sample:* cells, intercellular substances, and bodily fluids taken from a human being for the purposes of genetic research.
3) *DNA:* a molecule of deoxyribonucleic acid which contains genetic information.
4) *Description of DNA:* a digital model of DNA created as a result of genetic research.
5) *Genome:* a set of genes in an organism which contains genetic information about the organism's structure, about processes of life therein, and about the organism's reaction toward the environment.[2]
6) *Gene:* a part of a DNA molecule which contains genetic information about the synthesis of an organism's proteins or performs regulatory functions in specifying the structure or life process stage of a specific component of an organism.[3]

1. *Cilvēka genoma izpētes likumā* (Human Genome Research Act), adopted by *Saeima* (the Latvian Parliament) on June 13, 2002 and amended by *Saeima* on June 12, 2003. The 2003 Parliamentary Act shall take effect on January 1, 2004. This is the first English language translation of these documents in a unified form; the paragraphs amended in 2003 were translated by Kārlis Streips.
2. The original text amended in 2003 reads as follows: "a set of genes or genetic information which determines the constitution of an organism, life processes therein and reactions of the organism to environment."
3. The original text amended in 2003 reads as follows: "genetic information on the construction of some individual component of the organism or the stage of a life process."

7) *Gene donor:* a natural person who provides a tissue sample for genetic research.[4]
8) *Genome database:* a set of data including coded DNA descriptions, coded health status descriptions, coded genealogical and genetic data, coded DNA samples, and coded tissue samples to be used in genetic research.[5]
9) *Gene donor database:* a genome database consisting of data that allow the identification of a gene donor and the gene donor's genealogy.[6]
10) *Genetic researcher:* a natural or legal person performing genetic research.
11) *Genealogy:* information on ascendants and descendants of a gene donor: name, surname, date of birth, and blood relationship.
12) *Genetic research:* scientific research of DNA and other tissue components with a purpose to determine the connection between genes, gene products, and hereditary characteristics.
13) *Genetic testing:* gene analysis of a person performed for the purpose of the identification of the person or the diagnosis of diseases, as well as for the selection of prophylactic or treatment measures.
14) *Genetic data:* data concerning genes, gene products, and hereditary characteristics, including a description of DNA or a part thereof, which is obtained as the result of genetic research.
15) *Coding:* replacement of personal data of a gene donor with a unique code enabling to identify the gene donor according to the description of a tissue sample, the description of DNA, the description of the state of health or genealogy.
16) *Description of state of health:* data collected for use in genetic research which provides information on the state of health of a gene donor, the diseases he or she contracted, his or her lifestyle, physical and social environment, and hereditary characteristics.

Article 2. Purpose and Scope of the Act

(1) The purpose of the Act is to regulate genetic research[7] and the establishment and maintenance of a single genome database of the population (hereinafter: genome database); to ensure the voluntary nature of gene donation and

4. The original text amended in 2003 reads as follows: "a natural person who provides a tissue sample for the inclusion of the sample in the genome database and genetic research."
5. A new paragraph added in the 2003 amendment.
6. The original text amended in 2003 reads as follows: "data enabling the identification of a gene donor and genealogy." Paragraphs 9, 10, 11, 12, 13, 14, and 15 in the original 2002 text are now considered, respectively, Paragraphs 10, 11, 12, 13, 14, 15, and 16.
7. The phrase "related thereto" in the original text was excluded in the 2003 amendment.

confidentiality regarding the identity of gene donors; as well as to protect persons from misuse of genetic data and discrimination related to genetic data.

(2) The following provided for in this Act:

1) the conditions for tissue sample processing, preparing descriptions of DNA, descriptions of state of health and genealogy in relation to the genome database;
2) the conditions for performing genetic research[8] and the organization of supervising such research;
3) the rights and obligations of gene donors, of the Main Processor and Authorized Processors of the genome database, as well as of gene researchers;
4) the restrictions on the use of tissue samples, descriptions of DNA, descriptions of state of health and genealogy, collected in the genome database; and
5) the rules of coding and decoding.

Article 3. General Principles of Human Genome Research

(1) In genetic research, the interests, rights, and protection of gene donors shall be set higher than the interests of society and science.

(2) The gene donor shall participate voluntarily in genetic research.

(3) It is prohibited to discriminate a person based on his or her genetic origin and any other data acquired as the result of genetic research, as well as on the basis of the fact that the person is or is not a gene donor.

(4) Human genome research is permitted only for the purpose of acquiring scientifically relevant information that may be utilized for improving the health of a person or the whole society.

(5) Human genome research shall be scientifically grounded, and shall be performed taking into account the professional criteria and guarantees of the relevant field.

(6) No study shall be done on the genome of a deceased individual against the expressed will of the individual during his or her life. In absence of such an expression of will, research on the deceased individual's genome is prohibited.[9]

8. The phrase "in relation to the genome database" in the original text was excluded in the 2003 amendment.

9. Article 3 was supplemented by a new paragraph 6 in the amendment.

Article 4. Main Processor of the Genome Database

(1) The main processor of the genome database (hereinafter: Main Processor) is a scientific authority appointed by the government to establish and maintain the genome database.[10] The government shall determine the terms of operation of the Main Processor.

(2) The tasks of the Main Processor shall be the following:

1) to organize the obtaining and storage of tissue samples; to organize the preparation, storage, and destruction of descriptions of state of health and genealogies;
2) to perform the coding;
3) to conduct genetic research and to collect, store, issue, or destroy genetic data;
4) to provide the State Population Genome Register with genetic data related to a particular gene donor, which is obtained in the course of genetic research;
5) to promote the development of genetic research in Latvia; and
6) to promote the utilization of results of genetic research for the health improvement of individuals and the society as a whole.

(3) The Main Processor shall submit a report each year to the Ministry of Health and the Ministry of Education and Science, in which the Main Processor shall account for the use of government budget funds in the implementation of genetic research programs.[11]

Article 5. The Genome Research Council[12]

(1) The Genome Research Council[13] (hereinafter: the Council) is an authority established by the Cabinet of Ministers (hereinafter: the Government). The main tasks of the Council shall be the following:

1) to evaluate projects and concepts in the field of genetic research, to provide opinions on them, and to coordinate the implementation of these projects;

10. The end of this sentence in the original text, "in which coded descriptions of DNA, coded descriptions of state of health, coded genealogical and genetic data, as well as samples of DNA and tissues to be utilized in genetic research shall be included," was excluded in the 2003 amendment.

11. The original wording of Article 4(3), "the Main Processor may authorize a natural or legal person to perform activities related to the genome database, except coding and decoding," was replaced in the 2003 amendment.

12. The article title "Board of the Main Processor of the Genome Database" in the original text was replaced with the title "Genome Research Council" in the 2003 amendment.

13. The phrase "Board of the Main Processor of the genome database" in the original text was replaced with the words "Genome Research Council" in the 2003 amendment.

2) to promote the provision of information to the public on the purposes and procedures of genetic research; and
3) to represent the interests of the public in the field of genetic research.

(2) The Members of the Council shall include[14]
1) three members of the *Saeima*;
2) three representatives appointed by the Government (including one representative of the Ministry of Health and one representative of the Ministry of Education and Science);
3) one member nominated by the National Human Rights Office;
4) one member nominated by the Latvian Academy of Sciences;
5) one representative of the Latvian Medical Association;
6) one representative of the Latvian Association of Medical Genetics;
7) one representative of the Central Medical Ethics Committee; and
8) one representative of the Office of Patient's Rights Protection.

(3) The Government shall approve of the Members and by-laws of the Council.

Article 6. Authorized Processor of the Genome Database

The authorized processor of the genome database (hereinafter: Authorized Processor) is a natural or legal person authorized by the Main Processor of the genome database to perform particular activities related to the genome database, except coding and decoding. The Government shall determine the terms of operation of the Authorized Processor.

Article 7. State Population Genome Register

(1) The Government shall establish the State Population Genome Register, a government administration organization supervised by the Ministry of Health.[15]

14. This paragraph was reformulated in the 2003 amendment. Article 5(2) in the original 2001 Act read as follows:

"(2) The Board shall include:
1) three members of the *Saeima*;
2) three representatives nominated by the Cabinet;
3) one representative nominated by the Latvian Academy of Science;
4) one representative nominated by the Latvian Medical Association;
5) one representative nominated by the Latvian Bar Association; and
6) one representative of the Central Medical Ethics Committee."

15. The words "Ministry of Welfare" in the original text were replaced with the words "Ministry of Health" in the 2003 amendment.

(2) The tasks of the State Population Genome Register, in relation to the genome database, are the following:[16]
   1) to store genetic data related to a particular gene donor, which is obtained in the course of genetic research;
   2) to establish a database of gene donors; and
   3) to perform decoding.

Article 8. Genetic Research and Genetic Testing

(1) Genetic research is permitted for the purpose of studying and describing the relationship among genes, the health and lifestyle of individuals, their physical and social environment, so as to use this research to develop diagnostic and treatment methods for diseases which help in assessing the health risks of individuals and in preventing the incidence of diseases.[17]

(2) Genetic research shall be performed in accordance with the procedures determined by the Government.

(3) If genetic research is done independently of the genome database, the gene donor's tissue samples, DNA description, and information on his or her health status may be included in the genome database only if the gene donor provides a written authorization. Where such authorization is given, the researcher shall have the right to work with the tissue samples that have been collected and that are stored in the genome database, insofar as the law permits it.[18]

(4) Tissue samples taken from human subjects for the purpose of genetic testing and the results of the related research shall not be included in the genome database.

(5) Genetic testing shall be performed in accordance with a procedure to be approved by the Government.[19]

Article 9. Provisions on Processing Personal Data

Provisions for the processing of a person's data used in genetic research[20], the genome database, and the gene donor database shall be regulated by the Personal Data Protection Act, insofar as not prescribed otherwise by this Act.

16. The phrase "in relation to the genome database" was added in the 2003 amendment.

17. Article 8(1) was rephrased in the 2003 amendment without changing its content.

18. Article 8(2) in the original 2002 Act read as follows: "If genetic research is performed irrespective of the genome database, the information on the state of health of a gene donor, tissue samples and descriptions of DNA shall be included in the genome database."

19. This paragraph was added in the 2003 amendment.

20. The phrase "genetic research" was added in the 2003 amendment.

Chapter 2. Consent to Become a Gene Donor and Rights of Gene Donors

Article 10. Consent to Become a Gene Donor

(1) Before a person participates in the genetic research, a medical doctor shall inform that person on:

1) the purpose, content, and duration of the genome research project;
2) the potential risks involved in such research;
3) rights to express freely his or her consent and revoke it at any time; and
4) opportunities to conduct genetic research outside of Latvia.[21]

(2) A person's consent to provide a tissue sample, to have a description of the state of health or the genealogy of that person prepared and supplemented, to store these in the genome database, to use these for genetic research and statistical purposes, to take these genetic data out of the territory of Latvia and to use them outside Latvia,[22] shall be expressed in writing. The gene donor's consent document shall be prepared in two copies, both of which shall be signed and dated by the gene donor or his or her guardian or trustee and by the Main Processor or the Authorized Processor. One copy of the consent document shall be kept by the State Population Genome Register, the other copy shall be issued to the gene donor or his or her guardian or trustee.

(3) The State Population Genome Register shall retain the gene donor's consent document for 75 years after the last entry concerning the tissue samples, descriptions of state of health, and descriptions of DNA.

(4) The Government shall approve of the template of the consent document, the procedures for filling it out, and the rules of storing it.

(5) If genetic research is done independently of the genome database, the consent document shall be signed and dated by the gene donor, his or her guardian or trustee and by the genetic researcher. One copy of the consent document shall be given to the gene donor or his or her guardian or trustee, the other to the genetic researcher.[23]

21. Article 10, paragraph (1)(4) was added in the 2003 amendment.
22. The reference to the possibility of exporting and using tissue samples and genetic data outside the territory of Latvia was added in the 2003 amendment.
23. This paragraph was added in the 2003 amendment.

Article 11. Rights of the Gene Donors

(1) A gene donor has the following rights:

1) to know about his or her data stored in the genome database or to refuse to know about such data;
2) to prohibit any acts of supplementation, renewal, or verification concerning the description of his or her state of health already stored in the genome database;
3) to withdraw his or her consent to become a gene donor at any time. In such a case the tissue samples, the description of the state of health of the gene donor, and any information that could identify the gene donor shall be destroyed; and
4) to restrict the scope of research performed of his or her genome.

(2) The gene donor does not have the right to request a payment for the transfer of tissue samples, for the preparation of or research on the description of his or her state of health and genealogy, as well as for any utilization of the research results.

Article 12. Genome Research on Persons Recognized as Not Having the Capacity to Act

(1) Genome research on a person recognized as not having the capacity to act shall be permitted only in exceptional circumstances,[24] and only if such research is directly beneficial for the health of the individual and the risks stemming from the research are commensurate with the benefits that are achieved.

(2) Genome research on an adult person recognized as not having the capacity to act may be permitted only after obtaining written agreements from the individual's guardian and from the Central Medical Ethics Committee.

(3) Research on a child's genome shall be permitted only after obtaining the written agreement of the parents (guardians) and the orphanage court (parish court). In taking a decision on this matter with respect to a child who is capable of formulating his or her own views, the orphanage court shall learn the views of the child and evaluate the said views in accordance with the age and maturity of the child.[25]

24. The phrase "only in exceptional circumstances" was added in the 2003 amendment.

25. This paragraph was added in the 2003 amendment.

Chapter 3. Processing the Gene Donor's Tissue Samples, Health Status Descriptions, Genealogies, DNA Descriptions, and Personal Data[26]

Article 13. Collecting Tissue Samples and Preparing Descriptions of State of Health

(1) The Main Processor or the Authorized Processor shall collect the tissue samples and prepare the descriptions of state of health.

(2) The Main Processor or the Authorized Processor shall prepare and supplement the description of state of health based on information provided either by the gene donor or received from a medical institution which has treated the gene donor.

(3) The medical professional who takes the tissue sample of the gene donor shall provide the Main Processor with a completed document (the template of which has been approved by the Minister of Health[27]) and which shall include the following information:

1) the gene donor's name, surname, personal identity number, sex, and place of residence;
2) the name of the medical treatment institution in which the tissue samples were taken; and
3) information on a tissue sample and type by which it was taken.

(4) The Main Processor or the Authorized Processor may prepare the genealogy of a gene donor based on the data provided by the gene donor or the results of genetic research.

(5) During the process of collecting tissue samples and preparing descriptions of state of health, the identity of a gene donor, his or her tissue samples, the description of state of health, and genealogy shall remain confidential.

Article 14. Utilization of Genealogies

(1) Within the scope of the genome database it is permitted to utilize genealogies based on blood relationships only for the purpose of preparing of tissue samples, descriptions of DNA, and descriptions of state of health.

(2) Genetic researchers may receive information from the Main Processor on the correspondence between coded tissue samples, coded DNA samples, and coded descriptions of state of health on the basis of blood relationships in accordance with Article 17(2) of this Act.

26. The original title in the 2002 Act read as follows: "Establishment and Operation of the Genome Database."

27. The words "Minister of Welfare" in the original text were replaced with the words "Minister of Health" in the 2003 amendment.

Article 15. Storage of Tissue Samples, Descriptions of DNA, and Descriptions of State of Health and Their Removal from the Territory of Latvia[28]

(1) Coded tissue samples, coded descriptions of DNA, coded descriptions of state of health, as well as coded genealogies shall be stored in the genome database by the Main Processor[29] in accordance with procedures determined by the Government.

(2) In response to a request from the Main Processor, the Authorized Processor or the genetic researcher, the Council may permit the removal from the territory of Latvia of a part of the coded tissue samples (aliquots) and of copies of coded descriptions of state of health after obtaining a written consent from the gene donor and only for the following purposes:

1) genetic research under the auspices of international medical, genetic, or biopharmaceutical scientific projects;
2) the establishment of an internationally recognized database for medical and ethnogenetic research, if a treaty is signed with Latvia on the establishment of such genetic databases.[30]

(3) The Government shall determine the procedures for issuing authorizations to remove from the territory of Latvia the coded tissue samples (aliquots) and the copies of coded descriptions of state of health, as well as the procedures for controlling the use of tissue samples.[31]

Article 16. Other Rights Related to Genetic Research

(1) The genome database may be used only for scientific research, for research on and treatment of the gene donor's diseases, for public health research, and for statistical purposes. The use of the genome database for any other purpose is prohibited.[32]

28. The phrase "Their Removal from the Territory of Latvia" was added in the 2003 amendment.
29. The reference "and the Authorized Processor" was excluded in the 2003 amendment.
30. In the original 2002 text, Articles 15(2) and 15(3) read as follows:
"(2) All tissue samples and descriptions of state of health shall be stored in Latvia.
(3) If the Main Processor determines that it is not possible to ensure the commenced genetic research to the full extent in Latvia, the Board may, on the basis of a submission of the Main Processor, issue a permit for taking the parts of coded tissue samples (aliquots) and copies of coded descriptions of state of health out of Latvia."
31. Article 15(4) in the original 2002 Act became Article 15(3) in the 2003 amendment.
32. This new Article 16(1) was added in the 2003 amendment.

(2) The Main Processor, the Authorized Processor, and the genetic researcher shall handle the coded tissue samples, coded descriptions of state of health and coded genealogies only within the scope of authorization prescribed in the law.[33]

(3) The gene donor's physician is entitled to receive a decoded description of the gene donor's state of health stored in the genome database in case it is necessary for the treatment of the gene donor and after obtaining the gene donor's written consent.[34]

Article 17. Issuance of Tissue Samples, Descriptions of DNA, Descriptions of State of Health, and Genealogies

(1) The Government shall determine the procedures for the issuance of coded tissue samples, coded descriptions of DNA, coded descriptions of state of health, and coded genealogies that are stored in the genome database. The Government shall also approve the content of the issuance documents as well as the rules of completing and storing these documents.

(2) The Main Processor, after obtaining authorization from the Council, may issue coded tissue samples, coded descriptions of DNA, coded descriptions of state of health, and coded genealogies that are stored in the genome database or parts thereof for genetic research in Latvia that is not related to the genome database, provided that this cohort of data includes tissue samples, descriptions of DNA, coded descriptions of state of health from at least five different gene donors and that the provisions of Article 8 of this Act are observed. A fee may be demanded for issuing such coded tissue samples, coded descriptions of DNA, coded descriptions of state of health, and coded genealogies, but the said fee shall not exceed expenditures that are related to the obtaining, processing, and storing in the genome database of the relevant samples and descriptions. Furthermore, the Main Processor shall gain no

33. Article 16(1) in the original 2002 Act became Article 16(2) in the 2003 amendment.

34. Article 16(2) in the original 2002 Act became Article 16(3) in the 2003 amendment, without the original second sentence which read as follows: "The physician may receive such information without consent of the gene donor for the provision of emergency medical assistance." Article 16(3) in the original 2002 Act, which read as "[i]n cases genetic research is performed independently and irrespective of the genome database, the genetic researcher is entitled to work with tissue samples collected by him or her and stored in the genome database", was excluded in the 2003 amendment.

profit from such use. These provisions shall not apply to the issuance of data to the gene donor or to the gene donor's physician.[35]

(3) [36]

Article 18. Destruction of Personal Data, Tissue Samples, Descriptions of DNA, and Descriptions of State of Health of a Gene Donor

The Government shall determine the procedures for destroying the gene donor's personal data, tissue samples, descriptions of DNA, and descriptions of state of health. The Government shall also approve the content of the official documents for the destruction procedure as well as the rules of completing and storing these documents.

Chapter 4. Coding and Decoding

Article 19. Coding

(1) Immediately after receipt of tissue samples, descriptions of DNA, descriptions of state of health, and genealogies in the genome database, the Main Processor shall assign a unique code consisting of at least sixteen different symbols to each tissue sample, description of DNA, description of state of health, and genealogy. The Main Processor shall receive a permit from the State Data Inspection Agency for the use of selected code assigning method.

(2) The Main Processor shall replace with a code all data concerning tissue samples, descriptions of DNA, and descriptions of state of health providing the possibility of a reverse identification of the gene donor. The Main Processor shall also replace with a code the gene donor's name, surname, personal identity number, and place of residence.

35. Article 17(2) in the original 2002 Act read as follows: "The Main Processor may issue the coded tissue samples, coded descriptions of state of health, and coded genealogies included in the genome database or parts thereof as a set of data for genetic research that is not related to the genome database on condition that such a set of data includes tissue samples, descriptions of DNA and descriptions of state of health of at least five gene donors and that the provisions of Article 8 of this Act are complied with. In such cases the payment may be requested for the utilization of coded tissue samples, coded descriptions of state of health, and coded genealogies. This request shall not apply to the issuance of data to the gene donor or the gene donor's physician."

36. Article 17(3), which read "[t]he Main Processor may issue the coded descriptions of DNA or parts thereof to the Authorized Processor or the genetic researcher for payment or free of charge," was excluded in the text of the 2003 amendment.

(3) The Main Processor shall indicate the code assigned to a tissue sample, description of DNA, description of state of health, or genealogy on the consent document of a gene donor. Within a time period of three days after the coding took place, the Main Processor shall transfer the consent document with the code indicated thereon to the State Population Genome Register and this code shall be the only possible decoding key.

(4) The Main Processor, the Authorized Processor and the genetic researchers are permitted to mark tissue samples, descriptions of DNA, descriptions of state of health, and genealogies only with those codes that were assigned to these tissue samples, descriptions of DNA, descriptions of state of health, and genealogies by the Main Processor.

Article 20. Decoding

The State Population Genome Register is permitted to decode data only for the following purposes:

1) to destroy tissue samples, descriptions of DNA, and descriptions of state of health;
2) to provide access to data on a gene donor stored in the genome database, except genealogy data, upon his or her request;
3) to renew, supplement, or verify the description of gene donor's state of health, except in the case the gene donor has expressed his or her written prohibition to renew, supplement, or verify the description of his or her state of health;
4) to collect a new DNA sample if a tissue sample has been destroyed or does not contain sufficient genetic information, and only after the gene donor's written consent is obtained;
5) to make changes in the gene donor's genealogy if the results of genetic research are in contradiction with the genealogy previously prepared or to provide new information on such genealogy; and
6) to issue the description of a gene donor' state of health to the gene donor's physician upon the physician's written request in which the necessity to acquire information is justified, and only after the gene donor's written consent is obtained.[37]

37. The words "or, for the provision of emergency medical assistance, without a consent of the gene donor" at the end of the paragraph in the original 2002 Act were excluded in the 2003 amendment.

Chapter 5. Supervision and Procedures for Examination of Complaints

Article 21. Data Protection Supervision

The State Data Inspection Agency shall supervise the collection of descriptions of state of health and genealogical data, the coding and decoding of tissue samples, descriptions of DNA, descriptions of state of health, and genealogies, as well as the processing of tissue samples, descriptions of DNA, descriptions of state of health, and genealogical data.

Article 22. Compliance with Ethical Principles

The Central Medical Ethics Committee shall evaluate if genetic research, including the establishment and operation of the genome database is in compliance with the principles of ethics.

Article 23. Examination of Complaints

(1) Individual persons may submit a complaint for any violations of this Act to the State Data Inspection Agency.
(2) Decisions of the State Data Inspection Agency may be appealed to the court.

This Act comes into effect on January 1, 2004.

Transitional provision

Upon the taking effect of this law, Government Regulation No. 561 on "Amendments to the Act on Human Genome Research" (*Latvijas Republikas Saeimas un Ministry Kabineta Ziņotājs*, No. 6, 2003), issued in accordance with the provisions of Article 81 of the Constitution, shall no longer be in effect.

This law was approved by the *Saeima* on June 12, 2003.

V. Vīķe-Freiberga
President of Latvia

Bibliography

Advisory Committee on Genetic Testing (1998) *Report on Genetic Testing for Late Onset Disorders*. London: ACGT.

Agius, Emmanuel (1998) "Patenting Life: Our Responsibilities to Present and Future Generations." In Emmanuel Agius and Salvino Busuttil (eds.) *Germ-Line Intervention and our Responsibilities to Future Generations*. Dordrecht: Kluwer.

Alper, Joseph S., Lisa N. Geller, Carol I. Barash, Paul R. Billings, Vicki Laden, and Marvin R. Natowicz (1994) "Genetic Discrimination and Screening for Hemochromatosis." *Journal of Public Health Policy*, 15(3): 345–359.

American Civil Liberties Union (2002) "ACLU Renews Calls for Congress to Ban Genetic Discrimination." *Workplace Rights, Press Release*, February 13, 2002. New York: ACLU, available at http://www.aclu.org/WorkplaceRights/WorkplaceRights.cfm?ID=9688&c=180.

Andorno, Roberto (2001) "La dignidad humana como noción clave en la Declaración de la UNESCO sobre el genoma humano." *Revista de Derecho y Genoma Humano (Law and the Human Genome Review)*, 14: 41–53.

Annas, George J. (1998) "Why We Should Ban Human Cloning." *The New England Journal of Medicine*, 339(2): 122–125.

Armstrong, David (1982) "The Doctor–Patient Relationship: 1930–80." In Peter Wright and Andrew Treacher (eds.) *The Problem of Medical Knowledge: Examining the Social Construction of Medicine*. Edinburgh: University of Edinburgh Press.

Árnason, Einar (2002) "Personal Identifiability in the Icelandic Health Sector Database." *The Journal of Information, Law and Technology (JILT)*, 7(2). An electronic publication, available at http://elj.warwick.ac.uk/jilt/02-2/arnason.html.

Asch, Adrienne (1989) "Reproductive Technology and Disability." In Sherrill Cohen and Nadine Taub (eds.) *Reproductive Laws for the 1990s*. Clifton, NJ: Humana Press, pp.59–101.

Asch, Adrienne (1999) "Prenatal Diagnosis and Selective Abortion: A Challenge to Practice and Policy." *American Journal of Public Health*, 89(11): 1649–1657.

Asch, Adrienne and Gail Geller (1996) "Feminism, Bioethics, and Genetics." In Susan M. Wolf (ed.) *Feminism and Bioethics: Beyond Reproduction*. Oxford and New York: Oxford University Press, pp.318–350.

Atlan, Henri (1999) "Possibilités biologiques, impossibilités sociales." In Henri Atlan *et al.* (eds.) *Le clonage humain*. Paris: Seuil, p.29.

Bachelard-Jobard, Catherine (2001) *L'eugénisme, la science et le droit.* Paris: Presses Universitaires de France.

Barritt, Jason A., Steen Willadsen, Carol Brenner, and Jacques Cohen (2001) "Epigenetic and Experimental Modifications in Early Mammalian Development, Part II: Cytoplasmic Transfer in Assisted Reproduction." *Human Reproduction Update,* 7(4): 428–435. Available at http://www3.oup.co.uk/humupd/hdb/Volume_07/Issue_04/pdf/070428.pdf.

Baslar, Kemal (1998) *The Concept of the Common Heritage of Mankind in International Law.* The Hague: Nijhoff.

Bayertz, Kurt, Johann S. Ach, and Rainer Paslack (1999) *Genetische Diagnostik: Zukunftsperspektiven und Regelungsbedarf in den Bereichen innerhalb und außerhalb Humangenetik, Allgemeinmedizin, Arbeitsmedizin und Versicherungen.* Münster: Argos und Berlin: TAB.

Bekker, Hilary *et al.* (1994) "The Impact of Population Based Screening for Carriers of Cystic Fibrosis." *Journal of Medical Genetics,* 31(5): 364–368.

Beljin, Sasha and Hermann Fenger (2001) "Nationale Humangenomprojekte." In Stefan F. Winter, Hermann Fenger, Hans-Ludwig Schreiber (eds.) *Genmedizin und Recht. Rahmenbedingungen und Regelungen für Forschung, Entwicklung, Klinik, Verwaltung.* München: C.H. Beck Verlag, pp.269–275.

Benatar, Solomon R. (1998) "A Perspective from Africa on Human Rights and Genetic Engineering." In Justine Burley (ed.) *The Genetic Revolution and Human Rights.* The Oxford Amnesty Lectures. Oxford and New York: Oxford University Press.

Bentham, Martin and Lorraine Fraser (2002) "Move to Outlaw Secret DNA Testing by Fathers." *The Daily Telegraph,* May 19, 2002, available at http://www.telegraph.co.uk/news/main.jhtml?xml=/news/2002/05/19/ndna19.xml.

Bently, Lionel and Brad Sherman (2001) *Intellectual Property Law.* Oxford: Oxford University Press.

Berberich, Kerstin (1998) *Zur Zulässigkeit genetischer Tests in der Lebens- und privaten Krankenversicherung.* Karlsruhe: Verlag Versicherungswirtschaft.

Berg, Kåre Ingmar (1994) "Confidentiality Issues in Medical Genetics" and "The Need for Laws, Rules, and Good Practices to Secure Optimal Disease Control." In *Ethics and Human Genetics: Proceedings of the Second Symposium of the Council of Europe on Bioethics, Strasbourg, November 30–December 2, 1993.* Strasbourg: Council of Europe, CDBI.

Beyleveld, Deryck and Roger Brownsword (2001) *Human Dignity in Bioethics and Biolaw.* Oxford and New York: Oxford University Press.

Billings, Paul R. (1992) "The Scientific Basis of the 'Genetic Revolution': A Selective Review." In *The Genome, Ethics and the Law: Issues in Genetic Testing.* Washington, DC: American Association for the Advancement of Science, pp.23–45.

Binet, Jean-René (2002) *Droit et progrés scientifique. Science du droit, valeurs et biomédecine*. Paris: Presses Universitaires de France.

Bodmer, Sir Walter (1995) "Where Will Genome Analysis Lead Us Forty Years on?" In Donald A. Chambers (ed.) *DNA, The Double Helix: Perspective and Prospective at Forty Years*. New York: New York Academy of Sciences, pp.414–426.

Bostrom, Nick (2002) "A Transhumanist Perspective on Human Genetic Enhancements." Available at http://www.nickbostrom.com/ethics/enhancements.html.

Botkin, Jeffrey R. (1988) "The Legal Concept of Wrongful Life." *JAMA (The Journal of the American Medical Association)*, 259(10): 1541–1545.

Botkin, Jeffrey R. (1995) "Fetal Privacy and Confidentiality." *Hastings Center Reports*, 25(3): 32–39.

British Medical Association (1992) *Our Genetic Future: The Science and Ethics of Genetic Technology*. Oxford: Oxford University Press.

British Medical Association (1998) *Human Genetics. Choice and Responsibility*. Oxford: Oxford University Press.

Brownsword, Roger, William Cornish, and Margaret Llewelyn (eds.) (1998) *Law and Human Genetics: Regulating a Revolution*. Oxford: Hart Publishing.

Byk, Christian (1998) "A Map to a New Treasure Island: The Human Genome and the Concept of Common Heritage." *The Journal of Medicine and Philosophy*, 23(3): 234–246.

Campbell, Fiona A.K. (2001) "Inciting Legal Fictions: 'Disability's' Date with Ontology and the Ableist Body of the Law." *Griffith Law Review*, 10(1): 42.

Caplan, Arthur L. (1992) "If Gene Therapy Is The Cure, What Is the Disease?" In George J. Annas and Sherman Elias (eds.) *Gene Mapping: Using Law and Ethics as Guides*. Oxford and New York: Oxford University Press, pp.128–141.

Capron, Alexander M. (1999) "Genetics and Insurance: Accessing and Using Private Information." In Ellen Frankel Paul, Fred D. Miller, Jr, and Jeffrey Paul (eds.) *The Right to Privacy*, Vol. 17, Part 2. Cambridge: Cambridge University Press, pp.235–275.

Chambers, Donald A., Kenneth B.M. Reid, and Rhonna L. Cohen (1995) "The Biomedical Revolution at Forty Years." In Donald A. Chambers (ed.) *DNA, The Double Helix: Perspective and Prospective at Forty Years*. New York: New York Academy of Sciences, pp.441–472.

Cohen, Jacques, Richard Scott, Tim Schimmel, Jacob Levron, and Steen Willadsen (1997) "Birth of Infant after Transfer of Anucleate Donor Oocyte Cytoplasm into Recipient Eggs." *The Lancet*, 350: 186–187.

Cohen, Jean (2001)"Wrongful Life Verdict in Paris." *Reproductive BioMedicine Online*, 2(1): 2, available at http://www.rbmonline.com/4DCGI/Article/2001/021/wp_2001_021.pdf.

Cole-Turner, Ronald (2000) "Testimony." in *Ethical Issues in Human Stem Cell Research, vol. III. Religious Perspectives*, Rockville, MD: Natioanl Bioethics Advisory Commission, pp.A1–A4.

Coleman, Diane (1996) *Testimony Before the Constitution Subcommittee of the Judiciary Committee of The House of Representatives*, April 29, 1996. Available at http://www.mcil.org/mcil/mcil/testimon.htm.

Comaroff, Jean (1982) "Medicine: Symbol and Ideology." In Peter Wright and Andrew Treacher (eds.) *The Problem of Medical Knowledge: Examining the Social Construction of Medicine*. Edinburgh: University of Edinburgh Press, pp.49–68.

Council for International Organizations of Medical Sciences (CIOMS) (1990) *Declaration of Inuyama on Human Genome Mapping, Genetic Screening and Gene Therapy*. Available at http://www.cioms.ch/1990_texts_of_guidelines.htm.

Council for Responsible Genetics (2000 [1992]) *Paper on Human Germline Manipulation*. Available at http://www.gene-watch.org/programs/cloning/germline-position.html.

Council for Responsible Genetics (2001a) *A Comparison of Enacted State Genetic Discrimination Legislation*. Cambridge, MA: CRG. Available at http://www.gene-watch.org/programs/privacy/GDcomparisonchart.pdf.

Council for Responsible Genetics (2001b) "Genetic Discrimination Legislation in the United States." In *Genetic Testing, Discrimination, and Privacy Program*. Cambridge, MA: CRG, available at http://www.gene-watch.org/programs/privacy/summary2001.html.

Cushman, Laura A. and Marcel P. Dijkers (1990) "Depressed Mood in Spinal Cord Injured Patients: Staff Perceptions and Patient Realities." *Archives of Physical Medicine and Rehabilitation*, 71: 191–196.

Danish Council of Ethics (1993) *Ethics and Mapping the Human Genome*. Copenhagen: The Danish Council of Ethics.

Degener, Theresia (1998) "Chronologie der Bioethik-Konvention und ihrer Streitpunkte." *Kritische Vierteljahresschrift für Gesetzgebung und Rechtswissenschaft*, 1: 7–33.

Degener, Theresia and Gerard Quinn (2000) "A Survey of International, Comparative and Regional Disability Law Reform." Available at http://www.dredf.org/symposium/degener1.html.

Demopulos, Demetrios (2000) "An Eastern Orthodox View of Embryonic Stem Cell Research." In *Ethical Issues in Human Stem Cell Research, vol. III. Religious Perspectives*. Rockville, MD: National Bioethics Advisory Commission, pp.B1–B4.

Deņisova, Raisa (2001) "The Most Ancient Population of Latvia." Available at http://vip.latnet.lv/hss/denisova.htm.

Dietschi, Irène (1999) "Wie versicherbar sind 'schlechte Gene'?" *Neue Zürcher Zeitung*, December 7, 1999, 285: 15. Available at http://www.shv.ch/download/NZZ_Artikel_Dietschi.doc.

Disability Rights Commission (DRC, UK) (2001) "Responses to the 'Whose Hands on Your Genes?' Consultation." Available at http://www.hgc.gov.uk/whoygconsultation_responses/drc.htm.

Doherty, Neil A. and Lisa L. Posey (1998) "On the Value of a Checkup: Adverse Selection, Moral Hazard, and the Value of Information." *Journal of Risk and Insurance*, 65(2): 189–211.

Dorff, Elliot N., Rabbi (2000) "Stem Cell Research." In *Ethical Issues in Human Stem Cell Research, vol. III. Religious Perspectives*. Rockville, MD: National Bioethics Advisory Commission, pp.C1–C5.

Eisenberg, Myron G. and Constance C. Saltz (1991) "Quality of Life Among Aging Spinal-Cord Injured Persons: Long-Term Rehabilitation Outcomes." *Paraplegia*, 29(8): 514–520.

Elliston, Sarah (1996) "If You Know What's Good for You . . . : A Consideration of Refusal of Consent to Medical Treatment by Children." In Sheila A.M. McLean (ed.) *Contemporary Issues in Law, Medicine, and Ethics*. Brookfield, VT and Aldershot: Dartmouth and Ashgate.

European Commission (2003) *Commission Staff Working Paper: Report on Human Embryonic Stem Cell Research*, SEC(2003)441, April 3, 2003. Brussels: Commission of the European Communities, available at http://europa.eu.int/comm/research/conferences/2003/bioethics/pdf/sec2003-441report_en.pdf.

Farley, Margaret (2000) "Roman Catholic Views on Research Involving Human Embryonic Stem Cells." In *Ethical Issues in Human Stem Cell Research, vol. III. Religious Perspectives*. Rockville, MD: National Bioethics Advisory Commission, pp.D1–D5.

Field, Martha A. (1993) "Killing 'the Handicapped'—Before and After Birth." *Harvard Women's Law Journal*, 16: 79–138.

Fine, Michelle and Adrienne Asch (1982) "The Question of Disability: No Easy Answers for the Women's Movement." *Reproductive Rights Newsletter*, 4(3): 19–20.

Finger, Anne (1990) *Past Due: A Story of Disability, Pregnancy, and Birth*. Seattle, WA: Seal Press.

Fletcher, John C. and Dorothy C. Wertz (1990) "Proposed: An International Code of Ethics in Medical Genetics before the Human Genome is Mapped." In Zbigniew Bankowski and Alexander M. Capron (eds.) *Genetics, Ethics, and Human Values: Human Genome Mapping, Genetic Screening, and Therapy*. Geneva: CIOMS.

Fletcher, Joseph (1979) *Humanhood: Essays in Biomedical Ethics.* Buffalo, NY: Prometheus Books.

Fukuyama, Francis (2002) *Our Posthuman Future: Consequences of the Biotechnology Revolution.* New York: Farrar, Straus and Giroux.

Gannon, Pippa and Charlotte Villiers (2000) "Regulation of Genetic Testing." In Laura McGregor, Tony Posser, and Charlotte Villiers (eds.) *Regulation and Markets Beyond 2000.* Aldershot: Ashgate.

Gaus, Gerald F. (1994) "Property Rights and Freedom." In Ellen Frankel Paul, Fred D. Miller Jr., and Jeffrey Paul (eds.) *Property Rights.* Cambridge: Cambridge University Press, pp.209–240.

Gearty, Conor A. (1999) "The Internal and External 'Other' in the Union Legal Order: Racism, Religious Intolerance and Xenophobia in Europe." In Philip Alston, Mara Bustelo, and James Heenan (eds.) *The EU and Human Rights.* Oxford: Oxford University Press, pp.327–356.

Genetic Interest Group (1995) "GIG Response to the Clinical Genetics Society Report: 'The Genetic Testing of Children'." Available at http://www.gig.org.uk/docs/gig_testingchildren.pdf.

Gerhart, Kenneth A., Jane Koziol-McLain, Steven R. Lowenstein, and Gale G. Whiteneck (1994) "Quality of Life Following Spinal Cord Injury: Knowledge and Attitudes of Emergency Care Providers." *Annals of Emergency Medicine,* 23(4): 807–812.

Gewirth, Alan (1978) *Reason and Morality.* Chicago: Chicago University Press.

Glass, Kathleen C., Charles Weijer, Denis Cournoyer, Trudo Lemmens, Roberta M. Palmour, and Stanley H. Shapiro (1997) "Structuring the Review of Human Genetics Protocols, Part III: Gene Therapy Studies." *IRB: A Review of Human Subjects Research,* 21(2): 1–9.

Gostin, Larry (1991) "Genetic Discrimination: The Use of Genetically Based Diagnostic and Prognostic Tests by Employers and Insurers." *American Journal of Law and Medicine,* 17(1–2): 109–144.

Greely, Henry T. (2000) "Iceland's Plan for Genomics Research: Facts and Implications." *Jurimetrics: The Journal of Law, Science, and Technology,* 40(2): 153–191.

Grēns, Elmārs, Valdis Pīrāgs, and Juris Šteinbergs (2002) *Project Draft of the Unified Genome Database of the Latvian Population.* Rīga: University of Latvia.

Gros Espiell, Héctor (1999) "Introduction." In *Genèse de la Déclaration universelle sur le génome humain et les droits de l'homme.* Paris: UNESCO.

Group of Advisers on the Ethical Implications of Biotechnology (GAEIB) (1996) *The Ethical Aspects of Prenatal Diagnosis.* Opinion of the Group of Advisers on the Ethical Implications of Biotechnology to the European Commission, No.6,

February 20, 1996. Brussels: The European Commission, available at http://europa.eu.int/comm/european_group_ethics/gaieb/en/opinion6.pdf.

Habermas, Jürgen (2001) *Die Zukunft der menschlichen Natur. Auf dem Weg zu einer liberalen Eugenik?* Frankfurt am Main: Suhrkamp.

Habermas, Jürgen (2003) *The Future of Human Nature.* Cambridge: Polity Press.

Hall, Mark A. (1999) "Legal Rules and Industry Norms: The Impact of Laws Restricting Health Insurers' Use of Genetic Information." *Jurimetrics: The Journal of Law, Science, and Technology*, 40(1): 93–122.

Hardacre, Helen (1997) *Marketing the Menacing Fetus in Japan.* Berkeley: University of California Press.

Harrer, Herbert (1994) "Aspects of Failed Family Planning in the United States of America and Germany." *The Journal of Legal Medicine*, 15(1): 89–127.

Harris, John (2000) "Is There A Coherent Social Conception of Disability?" *Journal of Medical Ethics*, 26(2): 95–100.

Harris, John and Kirsty Keywood (2001) "Ignorance, Information, and Autonomy." *Theoretical Medicine and Bioethics*, 22(5): 415–436.

Henn, Wolfram (1999) "Genetic Screening with the DNA Chip: A New Pandora's Box?" *Journal of Medical Ethics*, 25(2): 200–203.

Hennen, Leonhard, Thomas Petermann, and Arnold Sauter (2000) *Stand und Perspektiven der genetischen Diagnostik. Sachstandsbericht* (TAB-Arbeitsbericht Nr. 66). Berlin: Büro für Technikfolgen-Abschätzung beim Deutschen Bundestag (TAB).

Hepburn, Elizabeth R. (1996) "Genetic Testing and Early Diagnosis and Intervention: Boon or Burden?" *Journal of Medical Ethics*, 22(2): 105–113.

Hettinger, Edwin C. (1989) "Justifying Intellectual Property." *Philosophy and Public Affairs*, 18(1): 31–52.

Heyd, David (1992) *Genethics: Moral Issues in the Creation of People.* Berkeley: University of California Press.

Hoffman, Diane E. and Eric A. Wulfsberg (1995) "Testing Children for Genetic Predispositions: Is It in Their Best Interest?" *Journal of Law, Medicine, and Ethics*, 23(4): 331–344.

Holland, Suzanne, Karen Lebacqz, and Laurie Zoloth (eds.) (2002) *The Human Embryonic Stem Cell Debate: Science, Ethics, and Public Policy.* Cambridge, MA: MIT Press.

Honnefelder, Ludger (1997) "Das Menschenrechtsübereinkommen zur Biomedizin des Europarats: zur zweiten und endgültigen Fassung des Dokuments." In *Jahrbuch für Wissenschaft und Ethik. Band 2: Beiträge, Berichte, Dokumentation.* Berlin and New York: Walter de Gruyter, pp.305–330.

Honnefelder, Ludger (1998) "Stellungnahme aus ethischer Perspektive zur 'Allgemeinen Erklärung über das menschliche Genom und die Menschenrechte' der UNESCO." In *Jahrbuch für Wissenschaft und Ethik. Band 3*. Berlin: de Gruyter, pp.226.

Honoré, A. M. (Tony) (1961) "Ownership." In A. G. Guest (ed.) *Oxford Essays in Jurisprudence*. Oxford: Clarendon Press, pp.107–147.

House of Commons Select Committee on Science and Technology (1995) *Human Genetics: The Science and Its Consequences*, Volume 41–1. London: Her Majesty's Stationery Office.

House of Commons Select Committee on Science and Technology (2001) *Genetics and Insurance*. London: HMSO.

Hubbard, Ruth (1990) *The Politics of Women's Biology*. New Brunswick, NJ: Rutgers University Press.

Hubbard, Ruth and Elijah Wald (1993) *Exploding the Gene Myth*. Boston: Beacon Press.

Hughes, James J. (1996) "Embracing Change with All Four Arms: A Post-Humanist Defense of Genetic Engineering." *Eubios: Journal of Asian and International Bioethics*, 6(4): 94–101. Available at http://www.changesurfer.com/hlth/genetech.html.

Human Fertilisation and Embryology Authority (1993) *Sex Selection*. A Public Consultation Document. London: HFEA.

Human Genetics Commission (2002) *Inside Information: Balancing Interests in the Use of Personal Genetic Data*. London: HGC, Also available at http://www.hgc.gov.uk/insideinformation/index.htm.

International Human Genome Sequencing Consortium (2001) "Initial Sequencing and Analysis of the Human Genome." *Nature*, 409 (February 15, 2001): 860–921.

Jonas, Hans (1985) *Technik, Medizin und Ethik*. Frankfurt am Main: Insel.

Kalaça, Çağri and Ayşe Akin (1995) "The Issue of Sex Selection in Turkey." *Human Reproduction*, 10(7): 1631–1632.

Kant, Immanuel (1911) *Grundlegung zur Metaphysik der Sitten*, volume IV. Berlin: Akademie-Ausgabe.

Kaplan, Deborah (1994) "Prenatal Screening and Diagnosis: The Impact on Persons with Disabilities." In Karen H. Rothenberg and Elizabeth J. Thompson (eds.) *Women and Prenatal Testing: Facing the Challenges of Genetic Technology*. Columbus: Ohio State University Press, pp.49–61.

Kass, Leon R. (1998) "The Wisdom of Repugnance." In Leon R. Kass and James Q. Wilson, *The Ethics of Human Cloning*. Washington: The AEI Press, pp.3–59.

Kass, Leon R. and James Q. Wilson (1998) *The Ethics of Human Cloning*. Washington, DC: The AEI Press.

Keefer, Christopher M. (1999) "Bridging the Gap Between Life Insurer and Consumer in the Genetic Testing Era: The RF Proposal." *Indiana Law Journal*, 74(4):1375–1397.

Kevles, Daniel J. (1995[1985]) *In the Name of Eugenics: Genetics and the Uses of Human Heredity*. Cambridge, MA: Harvard University Press.

Kinderlerer, Julian and Diane Longley (1998) "Human Genetics: The New Panacea." In Roger Brownsword, William Cornish, and Margaret Llewelyn (eds.) *Law and Human Genetics: Regulating a Revolution*. Oxford: Hart Publishing, pp.11–28.

Knoppers, Bartha Maria (1999) *Le génome humain: patrimoine commun de l'Humanité?* Montreal: Fides.

Kolehmainen, Sophia, Peter Shorett, Sara Gambin, and Paul Billings (2002) "Comments to the National Conference of Insurance Legislators on the Proposed Genetic Discrimination Act." In *Genetic Testing, Discrimination, and Privacy Program*. Cambridge, MA: Council for Responsible Genetics. Available at http://www.gene-watch.org/programs/privacy/insurance.html.

Koppernock, Martin (1997) *Das Grundrecht auf bioethische Selbstbestimmung. Zur Rekonstruktion des allgemeinen Persönlichkeitsrechts*. Baden-Baden: Nomos.

Köhler, Michael (2000) "Europäische Bioethikkonvention: Beitritt unter Vorbehalt." *Zeitschrift für Rechtspolitik*, 1: 8–10.

Kruks, Sergejs and Ilze Šulmane (2001) "Etniskie stereotipi Latvijas presē (Ethnic Stereotypes in Latvian media)." In *Mēdiju analīze II*. Rīga: LU SZF (Faculty of Social Sciences, University of Latvia).

Lammersmann, Birgit (1999) "Medizinische Eingriffe an einwilligungsunfähigen Personen: Die Position der Biomedizin-Konvention des Europarats im Spannungsfeld zwischen Forschungsbedarf der Medizin und Selbstbestimmungsrecht des Patienten." *Recht und Psychiatrie*, 17: 157–164.

Latour, Bruno (1988) *The Pasteurization of France*. Cambridge, MA: Harvard University Press.

Laurie, Graeme T. (2002) *Genetic Privacy: A Challenge to Medico-Legal Norms*. Cambridge: Cambridge University Press.

Laurie, Graeme T. (2002b) " Better to Hesitate at the Threshold of Compulsion: PKU Testing and the Concept of Family Autonomy in Eire." *Journal of Medical Ethics*, 28(3): 136–138.

Law Commission, The (1998) *Legislating the Criminal Code: Misuse of Trade Secrets*. The Law Commission Consultation Paper No. 150. London: The Law Commission, available at http://www.lawcom.gov.uk/351.htm.

Le Bris, Sonia (2001) "Give Me Your DNA, and I'll Tell You Who You Are . . . Or Who You'll Become: Questions Surrounding the Use of Genetic Information in Europe." *Isuma: The Canadian Journal of Policy Research*, 2(3): 90–101. Available at http://www.isuma.net/v02n03/lebris/lebris_e.pdf.

Lee, Carol (1993) "Creating a Genetic Underclass: The Potential for Genetic Discrimination by the Health Insurance Industry." *Pace Law Review*, 13(1): 189–228.

Lemmens, Trudo (2000) "Selective Justice, Genetic Discrimination, and Insurance: Should We Single Out Genes in Our Laws?" *McGill Law Journal*, 45(2): 347–412.

Lenoir, Noëlle (1998) "La Déclaration sur le génome humain, avancée futuriste." *Le Monde,* December 9, 1998.

Levi, Joel (1987) "Wrongful Life Decision in Israel." *Medicine and Law,* 6: 373–374.

Lippman, Abby (1991) "Prenatal Genetic Testing and Screening: Constructing Needs and Reinforcing Inequities." *American Journal of Law and Medicine*, 17(1–2): 15–50.

Liu, Athena Nga-chee (1987) "Wrongful Life: Some of the Problems." *Journal of Medical Ethics*, 13(2): 69–73.

Lock, Margaret (2001) "The Alienation of Body Tissue and the Biopolitics of Immortalized Cell Lines." *Body and Society*, 7(2–3): 63–91.

Lockhart, Lisa K., Peter H. Ditto, Joseph H. Danks, Kristen M. Coppola, William D. Smucker (2001) "The Stability of Older Adults' Judgments of Fates Better And Worse Than Death." *Death Studies*, 25(4): 299–317.

Lulle, Aija (2001) "Iebilst pret genoma izpētes likumprojektu (The Central Medical Ethics Committee Objects the Human Genome Act Project)." *Neatkarīgā Rīta Avīze*, August 2, 2001.

Macpherson, Crawford B. (1975) "Capitalism and the Changing Concept of Property." In Eugene Kamenka and R.S. Neale (eds.) *Feudalism, Capitalism and Beyond.* London: Edward Arnold, pp.104–124.

Macran, Susan and Paul Kind (2001) "'Death' and the Valuation of Health-Related Quality of Life." *Medical Care*, 39(3): 217–227.

Malauzat, Marie-Isabelle (2000) *Le droit face aux pouvoirs de données génétiques.* Aix-en-Provence: Presses Universitaires d'Aix-Marseille.

Mao, Xin and Dorothy C. Wertz (1997) "China's Genetics Services Providers' Attitudes Toward Several Ethical Issues: A Cross-Cultural Survey." *Clinical Genetics*, 52: 100–109.

Mason, John K., R. Alexander McCall Smith, and Graeme T. Laurie (1999) *Law and Medical Ethics,* Fifth edition. London: Butterworths.

Mathieu, Bertrand (2000) *Génome humain et droits fondamentaux*. Paris: Economica and Aix-en-Provence: Presses Universitaires d'Aix-Marseille.

McGleenan, Tony, Urban Wiesing, and Francois Ewald (1999) *Genetics and Insurance*. Oxford: Bios.

McLean, Sheila A.M. (1995) "Genetic Screening of Children: The UK Position." *Journal of Contemporary Health Law and Policy*, 12: 113–130.

McLean, Sheila A.M. (2000) "Intergenerational Justice: A Brave New World or a Leap in the Dark?" *Genetics Law Monitor*, 1(1): 1–3.

Meilaender, Gilbert C. (2000) "Testimony." In *Ethical Issues in Human Stem Cell Research, vol. III. Religious Perspectives*. Rockville, MD: National Bioethics Advisory Commission, pp.E1–E6.

Meilaender, Gilbert C. (2001) "The Point of a Ban: Or How to Think about Stem Cell Research." *The Hastings Center Report*, 31(1): 9–16.

Meyer, Michael J. and Lawrence J. Nelson (2001) "Respecting What We Destroy: Reflections on Human Embryo Research." *The Hastings Center Report*, 31(1): 16–23.

Miringoff, Marque-Luisa (1991) *The Social Costs of Genetic Welfare*. New Brunswick, NJ: Rutgers University Press.

Montgomery, Cal (2002) "Mom Can Refuse Sons' Cochlear Implants, Says Court." *Ragged Edge Magazine*, October 4, 2002. Available at http://www.raggededgemagazine.com/drn/10_02.shtml#446.

Mulholland, William F. and Amy S. Jaeger (1999) "Genetic Privacy and Discri-mination: A Survey of State Legislation." *Jurimetrics Journal*, 39(3): 317–326.

Munson, Ronald and Lawrence H. Davis (1992) "Germ-Line Gene Therapy and the Medical Imperative." *Kennedy Institute of Ethics Journal*, 2(2): 137–158.

Murray, Thomas (1992) "Genetics and the Moral Mission of Health Insurance." *Hastings Center Report*, 22(6): 12–17.

Nagle, Gunita (2000) "The Gene Unifying the Baltic Nations Discovered." *Diena*, December 28, 2000.

National Bioethics Advisory Commission (2000) *Ethical Issues in Human Stem Cell Research*. Volume III. Religious Perspectives. Rockville, MD: NBAC.

National Conference of State Legislatures (2003) "Genetics and Life, Disability, and Long-Term Care Insurance." In *Genetic Technologies Project*. Denver and Washington: NCSL, available at http://www.ncsl.org/programs/health/genetics/ndislife.htm.

National Consultative Ethics Committee for the Health and Life Sciences (CCNE) (1995) *Opinion and Recommendations on "Genetics and Medicine: From Prediction to Prevention."* Report, No.46, October 30, 1995 (Paris: CCNE). Also available at http://www.ccne-ethique.fr/english/avis/a_046.htm.

National Council on Disability (2002) *Position Paper on Genetic Discrimination Legislation*, March 4, 2002. Washington, DC: The National Council on Disability, available at http://www.ncd.gov/newsroom/publications/geneticdiscrimination_positionpaper.html.

Nelson, James R. (1994) *On the New Frontiers of Genetics and Religion*. Grand Rapids: William Eerdmans Co.

Newman, Stuart and Carolyn Raffensperger (2000) "Genetic Manipulation and the Precautionary principle." In Pamela A. Norton and Laura F. Steel (eds.) *Gene Transfer Methods*. Natick, MA: Eaton Publishing.

Norwegian Ministry of Health (2002) *The Norwegian Biobank Report* (NOU 2001: 19). Oslo: Ministry of Health.

Nuffield Council on Bioethics (1993) *Genetic Screening: Ethical Issues*. London: Nuffield Council on Bioethics.

Palsson, Gisli and Kristin E. Hardardottir (2002) "For Whom the Cell Tolls. Debates about Biomedicine." *Current Anthropology*, 43(2): 271–301.

Park, James (2002) "Personhood Bibliography." Available at http://www.tc.umn.edu/~parkx032/b-person.html.

Pelias, Mary (1986) "Tort of Wrongful Birth and Wrongful Life." *American Journal of Medical Genetics*, 25: 71–80.

Pellegrino, Edmund D. (2000) "Testimony." In *Ethical Issues in Human Stem Cell Research, vol. III. Religious Perspectives*. Rockville, MD: National Bioethics Advisory Commission, pp.F1–F5.

Pembrey, Marcus E. (1998) "In the Light of Preimplantation Genetic Diagnosis: Some Ethical Issues in Medical Genetics Revisited." *European Journal of Human Genetics*, 6(1): 4–11.

Penney, Jonathan (2002) "A Constitution for the Disabled or a Disabled Constitution? Toward a New Approach to Disability for the Purposes of Section 15(1)." *Journal of Law and Equality*, 1(1): 83–115, available at http://www.jle.ca/files/v1n1/JLEv1n1art3.pdf

Poissonnet, Claude Michele (2003) *Le dépistage génétique en milieu de travail*. Paris: L'Harmattan.

Pokorski, Robert (1998) "Genetische Tests: Herausforderung für die Versicherungsw irtschaft." *Die Kölnische Rück, Themen 7*. Köln: General Cologne Re.

Pontifical Academy for Life (2000) *Declaration of the Production and the Scientific and Therapeutic Use of Human Embryonic Stem Cells*. Vatican City: Pontifical Academy for Life, available at http://www.vatican.va/roman_curia/pontifical_academies/acdlife/documents/rc_pa_acdlife_doc_20000824_cellule-staminali_en.html.

Putniņa, Aivita (2001) "Kam vajadzīgi mūsu gēni? (Who Needs Our Genes?)" *Neatkarīgā Rīta Avīze*, July 24, 2001.

Putniņa, Aivita (2002a) "Jaunajā pasaulē vecajās čībās (Wearing Old Slippers in the Bright New World)." *Neatkarīgā Rīta Avīze*, June 13, 2002.
Putniņa, Aivita (2002b) "Population Genome Project in Latvia." *Soziale Technik*, 3: 9–11.
Ray, C. and J. West (1984) "Social, Sexual, and Personal Implications of Paraplegia." *Paraplegia*, 22(2): 75–86.
Regenauer, Achim (1997) "Medizinische Risikoprüfung vor der Jahrtausendwende. Internationale Entwicklungen und Trends: dargestellt an einigen typischen Veränderungen." *Zeitschrift für Versicherungswesen*, 201(1): 629–635.
Reich, Charles A. (1964) "The New Property." *Yale Law Journal*, 73(5): 733–787.
Reindal, Solveig Magnus (2002) "Disability, Gene Therapy, and Eugenics: A Challenge to John Harris." *Journal of Medical Ethics*, 26(2): 89–94.
Rivera y Carlo, Roberto (2002) "Targeting the Disabled." *Boundless Webzine*, http: //www.boundless.org/2002_2003/features/a0000685.html.
Robertson, John A. and Julian Savulescu (2001) "Is There a Case in Favour of Predictive Genetic Testing in Young Children?" *Bioethics*, 15(1): 26–49.
Rodríguez, José Antonio Seoane (2002) "De la intimdad genética al derecho a la protección de datos genéticos. La protección iusfundamental de los datos genéticos en el derecho espanol (A propósito de las SSTC 290/2000, de 30 de noviembre)." *Revista de derecho y genoma humano* (Bilbao), 16: 71–105.
Rothman, David J. (1982) "Research Ethics and Social Deprivation: Were Tuskegee and Willowbrook 'Studies in Nature'?" *Hastings Center Report*, 12(2): 5–7.
Rothstein, Mark A. (1995) "Genetic Discrimination in Employment: Ethics, Policy and Comparative Law." In Swiss Institute of Comparative Law (ed.) *Human Genetic Analysis and the Protection of Personality and Privacy*. Proceedings of an International Colloquium. Zürich: Schultheiss Polygraphischer Verlag.
Rothstein, Mark A. (1997) "Genetic Secrets: A Policy Framework." In Mark A. Rothstein (ed.) *Genetic Secrets: Protecting Privacy and Confidentiality in the Genetic Era*. New Haven and London: Yale University Press, pp.451–495.
Royal College of Physicians (1989) *Prenatal Diagnosis and Genetic Screening*. London: RCP.
Sachedina, Abdulasis (2000) "Islamic Perspectives on Research with Human Embryonic Stem Cells." In *Ethical Issues in Human Stem Cell Research, vol. III. Religious Perspectives*. Rockville, MD: National Bioethics Advisory Commission, pp.G1–G6.
Saigal, Saroj, David Feeny, Peter L. Rosenbaum, William Furlong, Elizabeth Burrows, Barbara L. Stoskopf, and Lorraine Hoult (1996) "Self-Perceived Health Status and Health-Related Quality of Life of Extremely Low-Birth-Weight Infants at

Adolescence." *JAMA, The Journal of the American Medical Association,* 276(6): 453–459.

Savalescu, Julian and Mary Spriggs (2002) "The Perruche Judgment and the 'Right Not To Be Born'." *Journal of Medical Ethics,* 28: 63–64.

Saxton, Marsha (1984) "Born and Unborn: The Implications of Reproductive Technologies for People with Disabilities." In Rita Arditti, Renate Duelli-Klein, and Shelley Minden (eds.) *Test-Tube Women: What Future for Motherhood?* Boston, MA: Pandora, pp.298–312.

Schmidtke, Jörg (1998) "Gentests in der Lebensversicherung." *Versicherungsmedizin,* 50(3): 110–111.

Schöffski, Oliver (2000) *Gendiagnostik: Versicherung und Gesundheitswesen. Eine Analyse aus ökonomischer Sicht.* Karlsruhe: Verlag Versicherungswirtschaft.

Schreiber, Hans-Ludwig (2000) *Das europäische Menschenrechtsübereinkommen zur Bioethik: Ein Schritt auf dem Wege der Einheit des Biorechts.* Aulavortrag, September 14, 1999. Tartu: Tartu Ülikool (The University of Tartu Press).

Schroedel, Jean Reith (2000) *Is the Fetus a Person? A Comparison of Policies Across the Fifty States.* Ithaca, NY: Cornell University Press.

Schulz-Weidner, Wolfgang (1993) *Der versicherungsrechtliche Rahmen für eine Verwertung von Genomanalysen: die Bedeutung präklinischer kausalitäts-, Risiko- und Eigenverschuldungsbewertungen.* Baden-Baden: Nomos.

Shannon, Thomas A. (1993) "Method in Ethics, A Scottish Contribution." *Journal of Theological Studies,* 44(1): 272–293.

Shannon, Thomas A. (1998) "Remaking Ourselves? The Ethics of Stem Cell Research." *Commonweal* (December 4, 1998): 9–10.

Sheldon, Tony (2003) "Court Awards Damages to Disabled Child for Having Been Born." *British Medical Journal,* 326(7393): 784, available at http://bmj.bmjjournals.com/cgi/reprint/326/7393/784/b.pdf.

Sherwin, Susan (1992) *No Longer Patient: Feminist Ethics and Health Care.* Phila-delphia: Temple University Press.

Simon, Jürgen (1993a) "Rechtliche und rechtspolitische Aspekte der gegenwärtigen und zukünftig erwartbaren Nutzung genanalytischer Methoden am Menschen. Gutachten im Auftrag des Büros für Technikfolgenabschätzung beim Deutschen Bundestag." In *Bericht des Ausschusses für Forschung, Technologie und Technikfolgenabschätzung zur Technikfolgenabschätzung Genomanalyse,* BT-Drucks, 12/7094, pp.13–83.

Simon, Jürgen (1993b) "Risikoregulierung und Rechtspolitik im Bereich Genomanalyse." In Reinhard Damm and Dieter Hart (eds.) *Rechtliche Regulierung von Gesundheitsrisiken.* Baden-Baden: Nomos, pp.19–37.

Simon, Jürgen (2001) *Gendiagnostik und Versicherung. Die internationale Lage im Vergleich*. Baden-Baden: Nomos.

Spranger, Tade Matthias (2000) "Prädiktive genetische Tests und genetische Diskriminierung im Versicherungswesen." *Versicherungsrecht*, 51: 815–821.

Staley, Kristina (2001) *Giving Your Genes to Biobank UK: Questions to Ask*. A Report for GeneWatch UK, December 2001. Buxton: GeneWatch UK. Available at http://www.genewatch.org/humangen/Publications/Reports/BioRport.pdf.

Stein, Edward (1998) "Choosing the Sexual Orientation of Children?" *Bioethics*, 12(1): 1–24.

Stone, D. and S. Stewart (eds.) (1994) *Towards a Screening Strategy for Scotland*. Glasgow: Scottish Forum for Public Health Medicine.

Stone, Deborah A. (1996) "The Implications of the Human Genome Project for Access to Health Insurance." In Thomas H. Murray, Mark A. Rothstein, and Robert F. Murray, Jr. (eds.) *The Human Genome Project and the Future of Health Care*. Bloomington: Indiana University Press, pp.133–157.

Sullivan, Daniel J. (1977) "Gene Splicing: The Eighth Day of Creation." *America* (December 17, 1977): 440–443.

Szabo, Joan (2001) "Regulating Genetic Testing." *Physicians Financial News*, 19(2): 5, 8.

Szebik, Imre and Kathleen Cranley Glass (2001) "Ethical Issues of Human Germ-Cell Therapy: A Preparation for Public Discussion." *Academic Medicine*, 76: 32–38.

Taupitz, Jochen (2000a) *Genetische Diagnostik und Versicherungsrecht* (Frankfurter Vorträge zum Versicherungswesen, Nr. 32). Karlsruhe: Verlag Versicherungswirtschaft.

Taupitz, Jochen (2000b) "Genetische Tests. Rechtliche Möglichkeiten einer Steuerung ihrer Gefahren." In Claus R. Bartram, Jan P. Beckmann, Friedrich Breyer, Georg H. Fey, Christa Fonatsch, Bernhard Irrgang, Jochen Taupitz, Felex Thiele, and Klaus-M. Seel (eds.) *Humangenetische Diagnostik: Wissenschaftliche Grundlagen und gesellschaftliche Konsequenzen*. Berlin–Heidelberg: Springer, pp.82–125.

Taylor, Allyn L. (1999) "Globalization and Biotechnology: UNESCO and an International Strategy to Advance Human Rights and Public Health." *American Journal of Law and Medicine*, 25(4): 479–541.

Tendler, Moshe Dovid, Rabbi (2000) "Stem Cell Research and Therapy: A Judeo-Biblical Perspective." In *Ethical Issues in Human Stem Cell Research, vol. III. Religious Perspectives*. Rockville, MD: National Bioethics Advisory Commission, pp.H1–H5.

Tinnefeld, Marie-Therese (2000) "Menschenwürde, Biomedizin und Datenschutz." *Zeitschrift für Rechtspolitik*, 1: 10–13

Tintemann, Sven (2000) "Thema: 'Wrongful Life'. Vergleich der Argumentation US-amerikanischer und deutscher höchstrichterlicher Entscheidungen." Avaliable at http://userpage.fu-berlin.de/~stsurfer/ha/methosem-ho.doc.

Townend, David (2003) "Social Welfare Challenges to Intellectual Property Traditions." Manuscript.

UNESCO International Bioethics Committee (2003) *Report of the IBC on Pre-Implantation Genetic Diagnosis and Germ-Line Intervention*, SHS-EST/02/CIB-9/2 (Rev. 3), April 24, 2003. Paris: United Nations Educational, Scientific, and Cultural Organization, available at http://unesdoc.unesco.org/images/0013/001302/130248e.pdf.

US National Academy of Sciences (2002) *Scientific and Medical Aspects of Human Reproductive Cloning*. Washington: National Academies Press.

Van der Laan, Nanette (2001) "France Debates Right *Not* To Be Born." *The Christian Science Monitor*, December 7, 2001. Also available at http://www.csmonitor.com/2001/1207/p1s3-woeu.htm.

Venter, J. Craig *et al.* (2001) "The Sequence of the Human Genome." *Science*, 291(5507, February 16, 2001): 1304–1351.

Wadham, John and Helen Mountfield (1999) *Blackstone's Guide to the Human Rights Act 1998*. London: Blackstone Press Limited.

Walters, LeRoy (1986) "The Ethics of Human Gene Therapy." *Nature*, 320: 225–227.

Wambach, Achim (1999) "Die ökonomischen Auswirkungen von Gentests auf Versicherungsmärkte." Manuscript.

Warnock, Mary (2002) *Making Babies: Is There a Right to Have Children?* Oxford: Oxford University Press.

Wasserman, David and Robert Wachbroit (eds.) (2001) *Genetics and Criminal Behavior*. Cambridge: Cambridge University Press.

Watson, Michael S. and Carol L. Greene (2001) "Points to Consider in Preventing Unfair Discrimination Based on Genetic Disease Risk: A Position Statement of the American College of Medical Genetics." *Genetics in Medicine*, 3(6): 436–437.

Wertz, Dorothy C. (1998) "Eugenics Is Alive and Well. A Survey of Genetic Professionals Around the World." *Science in Context*, 11(3–4): 493–501.

Wertz, Dorothy C. and John C. Fletcher (1993) "A Critique of Some Feminist Challenges to Prenatal Diagnosis." *Journal of Women's Health*, 2(2): 173–188.

Wertz, Dorothy C. and John C. Fletcher (1998) "Ethical and Social Issues in Prenatal Sex Selection: A Survey of Geneticists in 37 Nations." *Social Science and Medicine*, 46(2): 255–273.

Whiteneck, Gale G. *et al.* (1985) *The Rocky Mountain Spinal Cord Injury System*, Report to the National Institute of Handicapped Research. Washington, DC: National Institute of Handicapped Research.

Wildes, Kevin W. (2000) "Testimony." In *Ethical Issues in Human Stem Cell Research, vol. III. Religious Perspectives.* Rockville, MD: National Bioethics Advisory Commission, pp.I1–I4.

Wolbring, Gregor (2000) *Science and the Disadvantaged.* Occasional Papers of the Edmonds Institute. Washington, DC: The Edmonds Institute.

Wolbring, Gregor (2003a) "Disability Rights Approach to Bioethics." *Journal of Disability Policy Studies*, 14(4): 154–189.

Wolbring, Gregor (2003b) "The Social Construct of Health and Medicine." *Health Ethics Today*, 13(1): 5–6.

Wolf, Naomi (1998) "Our Bodies, Our Souls." In Louis P. Pojman and Francis J. Beckwith (eds.) *The Abortion Controversy: 25 Years after Roe v. Wade.* Belmont: Wadsworth, pp.400–413.

Wolff, Gerhard, Dorothy C. Wertz, and Irmgard Nippert (1999) "Ethik und Genetik: Ergebnisse der Umfrage zu Problemaspekten angewandter Humangenetik." *Medizinische Genetik*, 11(3): 308–318.

Working Party of the Clinical Genetics Society (CGS, UK) (1994) "The Genetic Testing of Children." *Journal of Medical Genetics*, 31: 785–797.

Yesley, Michael S. (1998) "Protecting Genetic Difference." *Berkeley Technology Law Journal*, 13(2): 653–665.

Young, Tillie (2002) "The New Jersey Genetic Privacy Act (1996): Some Key Points." In *Genes at Work.* Newark: University of Medicine and Dentistry of New Jersey, available at http://www.umdnj.edu/genesatwork/clinicalserv/08_nj.htm.

Zimmern, Ron L. (1999) "Genetic Testing: A Conceptual Exploration." *Journal of Medical Ethics*, 25(2): 151–156.

Zoloth, Laurie (2000) " The Ethics of the Eighth Day: Jewish Bioethics and Genetic Medicine, A Jewish Contribution to the Discourse." In *Ethical Issues in Human Stem Cell Research, vol. III. Religious Perspectives.* Rockville, MD: National Bioethics Advisory Commission, pp.J1–J26.

Index